AMBER
Diceless
Role-Playing

Erick Wujcik

Roger Zelazny

Stephen Hickman

Michael Kucharski

Amber Titles by Roger Zelazny

The Amber Series

Nine Princes in Amber
The Guns of Avalon
Sign of the Unicorn
The Hand of Oberon
The Courts of Chaos

The Merlin Series

Trumps of Doom
Blood of Amber
Sign of Chaos
Knight of Shadows
Prince of Chaos

Role-Playing Titles by Erick Wujcik

Weapons & Assassins™

Teenage Mutant Ninja Turtles & Other
    Strangeness®
After the Bomb®
Teenage Mutant Ninja Turtles Adventures®
Road Hogs™
Teenage Mutant Ninja Turtles Guide to the
    Universe®
Mutants Down Under™
Transdimensional Teenage Mutant Ninja
    Turtles®
Mutants of the Yucatan™

Revised RECON®
Advanced RECON®

Ninjas & Superspies™

THANKS TO:
   When a game simmers for over six years, there are a lot of people responsible for seeing to print. I'm grateful to all of the following folks.
   Kevin Siembieda, best of friends, I owe him most of all.
   Thanks to the early Amber players. That day, way back in the Fall of 1985, when the first auction for Amber characters was held, these guys were there; Don Anderson (AURELIA), Dan Clemens (DAMIEN), Paul Deckert (ADRIAN), Chuck Knakal (HERDAN), Michael Kucharski (MORGAN), Kevin Lowry (AARON), Alan Moen (SAUREN), Julius Rosenstein (GWYNTBRAWD), Peter Schermerhorn (SOLEM), Eric Snider (HARLAN), John Speck (GODFREY), Tim Treloar (DEREK), Don Woodward (CAROLAN) and Jim Webster (JEREMI). Plus Joshua Darlington, Bruce Roberts, Michael Robinson, Joe Samuels, and Gary Sibley. They made the game come alive.
   Thanks to the first Amber Game Masters, Randy McCall and Don Woodward.
   Thanks to the New Jersey Cabal; Carol Dodd, Felicia Baker, Jim Kenny and Ron Miller.
   Also, thanks to the early Trump artists; Todd Bake, Michael Kucharski, Jim Webster, and Don Woodward; and to the expert chroniclers of the Amber experience; Don Anderson, Dan Clemens, Carol Dodd, Brian Garwood, and Cathy Klessig.
   To Daniel F. Iyama-Kurtycz, M.D., for those handy, grisly details.
   And a tip of the black leather cap also goes to Kirby McCauley, Bill Fawcett, Ken Rolston, Martin Wixted, Eric Goldberg, Greg Costikyan, and Darwin Bromley. And Mike Pondsmith, Lisa Pondsmith, Michael MacDonald and Derek Quintanar.
   First, last, and always, thanks to Roger Zelazny. He created *Amber*. Then he found something in my sketchy outline to believe in. His vote of faith put the game in your hand.

Roger Zelazny's *The Chronicles of Amber* is available through the Science Fiction Book Club. Membership information can be obtained by calling 1-317-542-6354, or writing to Member Service Center, 6550 East 30th Street, P.O. Box 6375, Indianapolis, IN 46206-6375.

The Chronicles of Amber are also available in audio book form, read by Roger Zelazny, complete with sound effects and a musical score.
  For a catalog of tapes write to Sunset Productions, 369 Montezuma, #172, Santa Fe, NM 87501, or call 1-800-829-5723.

AMBER Diceless Role-Playing

Diceless Role-Playing™
System Designed & Written by
Erick Wujcik

based on
The Chronicles of Amber by **Roger Zelazny**

Cover Painting by
Stephen Hickman

Trump Portraits
& Interior Illustrations by
Michael Kucharski

Editing Help from
**Don Anderson, Karen Clayton,
Kathryn Kozora, Chuck Knakal,
Kevin Lowry, Alan Moen,
& Michele Spainhour**

Phage Press

Dedicated to
Kate,
my one true love

Thinking of writing to Phage Press?

Please do!

We love getting mail. Especially mail with large checks in them, but any kind of mail will do. If you write to us we will put you on our mailing list so that you may receive notice of our upcoming books, information about *Amberzine*, the magazine devoted to *Amber* Role-Playing, and about the yearly (since 1989!) **Ambercon** role-playing convention. Plus details on T-Shirts and other neat things.

We are very, very, very good at responding to our mail. We pride ourselves upon it.

We are, however, weak in two very human ways.

First, we are sometimes slow. Sometimes we are slow because a piece of mail loses itself in a stack of the wrong stuff.

We are also slow when we receive letters filled with intelligent, perceptive and insightful questions. Then we have to actually *think* about how we might respond. Large numbers of intelligent, perceptive and insightful questions in a single letter cause us great consternation, and send us into such flurries of concentration as takes us near forever to actually respond. If there were but one intelligent, perceptive and etc. question in a single letter then we might have a somewhat quicker burst of insight.

Another weakness we have is that we are customers of a rather large organization, the Post Office. This institution is known for it's amazing timeliness and accuracy. Most of our mail is delivered promptly, accurately, and with great pleasantness.

However, we daily return to our Postmistress items destined for other parties. Sometimes an item appears in the wrong zip code area, or the zip is right but the person is one unknown to us. We also receive the occasional letter intended for Nigeria. Upon these occasions we wonder if somewhere, some lonesome piece of mail intended for us, has found itself elsewhere (in Nigeria?).

And so, if you have not heard from us in some unreasonable length of time, please try again. Thank you.

Library of Congress Cataloging-in-Publication Data

Wujcik, Erick.
      Amber diceless role-playing : diceless role-playing system / designed & written by Erick Wujcik ; based on the Chronicles of amber by Roger Zelazny ; cover painting by Stephen Hickman ; trump portraits & interior illustrations by Michael Kucharski ; editing help from Don Anderson ... [et al.].
          p.   cm.
      Includes index.
      ISBN 1-880494-00-0 : $22.95
      1. Amber (Game)   I. Zelazny, Roger.   Chronicles of amber. II. Title.
      GV1469.62.A42W85   1992
      793.93--dc20   92-4529
                                                  CIP

Third Printing - April, 1993

Phage Press™ • P.O. Box 519 • Detroit • Michigan • 48231-0519 • USA

Phage Press Product Number 0100

Table of Contents

INTRODUCTION

"What would another generation have been like?"
"How can such a question be answered? I have no idea."
               The Hand of Oberon

That's where you come in, the players and Game Masters of *Amber*. If there is one thing that we all have in common, all who would use this book, it's trying to answer that question, "what would another generation have been like?"

For players, this means becoming part of that new generation, playing a character whose parents are as powerful as gods, a character who is inheriting those powers in full measure. For Game Masters, it is the challenge of unfolding the grandest of all settings.

The next generation of *Amber* ...

In the beginning there was nothing at all.
Then there was *Chaos*.
A spinning, complex maelstrom of change, Primal Chaos. Pure destruction, pure fury, the essence of change.

At the fringes of Chaos, where it interacts with the great *Abyss*, the remnants of the original nothingness, are small regions of habitability. Shards of worlds, shreds that are not stable, but constantly changing, appearing and disappearing.

Into these primal fragments of reality came life, and then sentient life.

Only *Shape Shifters* can survive in these places, those who can continually adapt to ever-changing worlds.

And the Shape Shifters formed the *Courts of Chaos*, and they learned to tame the *Logrus*, the representation of Primal Chaos, and use it as a tool. They discovered arts and powers, including *Magic*, the casting of spells, and *Trump*, creating artistic images used to communicate between worlds.

They are the *Lords of Chaos*, they were the masters of the universe.

Were, until, for some unknown reason, one of their number rebelled against them. He is named *Dworkin*.

Dworkin somehow found, forged, or stole a very special artifact. A device of unknown properties called the *Jewel of Judgement*.

With the blessing of the mythical *Unicorn*, against the violent raging of his elders, Dworkin remade the universe. He inscribed a force that opposed Chaos.

In his own blood Dworkin drew, or wrote, or conjured, the *Primal Pattern*.

The Pattern was an imposition of a different kind of power onto the universe. Where all is fluid and changing near Primal Chaos, order and stability dwell in Pattern realms.

Between these two great symbols, of chaos and order, of Logrus and Pattern, there came to be an infinite number of *Shadows*. Each a possibility, a different mirror image. Each Shadow, an entire physical universe, is a reflection of the Primal Pattern, but each is unique, each warped in its own special way by the distant fluctuations of Chaos.

Surrounding Primal Pattern is a region of great power.

And, following those apocalyptic days, it came to pass that a kingdom was formed, to be named *Amber*.

Its ruler, *King Oberon*, son of Dworkin, initiate in the ways of Pattern, fought wars, built a dynasty, fathered heirs, and founded a Castle, known also as Amber.

Oberon had sons and daughters, each of whom had a measure of their father's power. In each was born the power to walk the Pattern.

These *Princes and Princesses* of Amber, each carry in their veins the blood of Amber. Having walked the Pattern they can move through the Shadow universes, and shape Shadow itself, to their own ends. Theirs is the power of life and death, creation and destruction.

And then the children of Oberon bickered and fought among themselves...

Was it because they were children of a rebel's child? Or were they manipulated, set one against another, by their father? We'll probably never know.

They plotted against each other, and against their father. And one of them even dreamed of remaking the universe.

That evil, or insane, brother calculated that the Primal Pattern could be erased. He predicted that blood, the blood of any of Oberon's children, or Oberon's children's children, could be used to black out the Pattern, damage it, and ultimately destroy it. With the original Pattern gone, it would be possible to draw a new Pattern, a Pattern more to his liking, a Pattern where he would have absolute power.

For the full story, you'll need to read Roger Zelazny's *Chronicles of Amber*. Suffice to say, the evil brother did not succeed, the Pattern was restored, and the threatening forces of Chaos were beaten back.

There was something new in the air, the ground, the sky. This was a new place. A new primal Pattern. Everything about me then was a result of the Pattern in which I stood.

I suddenly realized that I was feeling more than refreshment. It was now a sense of elation, a kind of joy that was moving through me. This was a clean, fresh place and I was somehow responsible for it.

The Courts of Chaos

I loved reading Zelazny's saga. And *I didn't want it to end*. Not then, not now, not ever! For me, the designer of this system of role-playing, Amber is alive, and pregnant with opportunity.

Why can't we go back to Amber, again and again, to make our own stories there? Why not? There are an infinite number of Shadows yet to be explored. Each eternal Prince or Princess, must surely have daughters and sons. Why not role-play one of these new immortals?

Starting in 1985 Amber became a place where players burst out with enthusiasm, as their characters came to life, lived and loved, and made Amber live for a few hours each month. Other Game Masters picked up the torch, and the role-playing of Amber spread across the world.

Now, gentle players, and cunning Game Masters, Amber is no longer a private domain. What gave birth in Roger's brain, what took unsteady steps behind closed doors, what bold and rash scenarios graced the privileged few, that is now before you. Throw open the gates of Amber!

Let the role-play begin...

INTRODUCTION TO ROLE-PLAYING

If you've never played a role-playing game, we sure hope you start with this one. Here's how it works, in four steps:

1. Role-Play Set-Up. Before role-playing someone has to do some pretty elaborate planning. There are two kinds of preparation.

Players, most of the folks, have to create fictional *characters*. In *Amber* you start by bidding in an **Attribute Auction** that sets up the character's basic Attributes.

So what's an attribute? It's an artificial way of measuring things that the characters might be able to do. For example, the attribute *Strength* determines how strong the character is. Which means it's a way of figuring out how big an object the character can pick up, or how much damage they can do with a punch. The other three attributes are *Warfare*, fighting ability, *Psyche*, mental power, and *Endurance*. The higher the ranking (first place is highest) the better the attribute.

Once the auction is over, it's time to spend the remaining points. Then the character can get powers, like the ability to walk the Pattern, or the talent to draw Trump Cards. Spending the rest of the points, and finishing the character can take quite awhile.

Let's say, just to help you picture things, that you've put together a character, and that you've named him Farley. Assume that Farley is pretty well rounded, with good attribute rankings, and with the power of the Pattern.

Meanwhile, the person running the show is called the **Game Master**, also called the GM. While each player is in charge of a single character, the GM is in charge of just about everything else. Just like the author of a book, the GM takes care of creating and presenting the setting, villains, the rest of the Amber family, monsters, Shadow Storms, etc...

The Game Master is also in charge of coming up with an interesting story. While a Game Master can use a prepared adventure, most GMs like to create their own version of Amber, with their own problems and encounters.

2. Campaign Set-up. You have your character, as do the rest of the players, and the Game Master has the universe figured out, so it's time to play ball.

It's time for the game to start. The Game Master asks you, and the other players, "If your character, Farley, wanted to spend a couple of years in a single Shadow, where would he be? And what would he be doing?"

Having created the character, you've now got to start getting into his head. What would Farley be doing?

He could be anywhere, and he could be doing anything. It's up to you, the player, to make that decision.

Let's say, just to keep things moving, that you decide that Farley will be here on Earth, which Amberites call *Shadow Earth*, visiting Los Angeles, and studying to be a movie director. When the GM asks for a more specific place, say on a Wednesday afternoon, you decide that Farley is working part-time on a studio back lot, helping film a famous Television sitcom.

The GM gets different starting points for each of the players, most of whom are scattered all over the place. One character is in Amber, another out in the Courts of Chaos, and the rest in Shadows where they're either learning something, fighting something, or just having a good time.

3. Role-Play Time! "Okay," the Game Master starts, "you're standing around while the light people are fiddling with a burned out bulb. Where would you be standing?"

Again, it's up to you to picture the place, where you'd be, and to fill in other details. Finally the Game Master is satisfied with your explanations, and tells you, "Suddenly, one of the back doors bangs open, and you see silhouetted against the bright California sun, a large winged figure! It flies up and into the studio. It's no bird, even though it has huge feathered wings, and the body has a human shape, possibly female. What are you doing?"

Now things are left up to you. Gather information, grab for a weapon, get out a Trump, run, or whatever. From this point on, you tell the Game Master what you're doing, the Game Master tells you the results, and so it goes...

4. Role-Playing Together. Up to this point the game has mostly been between you and the Game Master.

What changes things is the addition of the other players. Each of them is playing another character, someone your character has to deal with one on one.

So, let's advance the action, and the Game Master says,

"You're up-side-down, with the winged valkyrie trying to haul you away by your right leg. You've been trying to kick with your left leg, but that hasn't done much good, and your right hand grip on the scaffolding is starting to slip. On the other hand, the left one, ha ha, a little GM joke there," At which point all the players groan, "you've managed to get your Trump out and you're looking at Dorell's picture. Are you trying to contact her?"

"Yes!" you yell, understandably annoyed at how the Game Master, not unlike some computers you have known, keeps asking stupid questions.

"Fine," says the Game Master, "I'll get back to you."

Waiting your turn is also part of the game. Even though the player controlling Dorell is sitting right across the table, listening to every word, that doesn't mean that Dorell knows what's going on. In fact, Dorell may still be several minutes, days, or even years behind, still taking care of business that "happened" long before Farley got himself in his predicament.

Finally the Game Master turns to you and says, "Farley has the contact with Dorell."

Now, even though you'd like to get this over with, you can't just tell the Game Master that you want Dorell to extricate Farley from his little problem. No, you've got to talk directly to Cindy, you speaking for Farley, and she for Dorell.

"Dorell, pull me out of here!" you might say.

"Um. Actually Farley, I find I am presently occupied. I'll call you back shortly."

You see? Once again it is up to you to play your character, but not just with the Game Master. This time it might be a matter of begging, threatening, shouting, or whatever it takes to get Cindy, aka Dorell, moving to assist you.

So, role-playing consists of four basic elements. First, the creation of a character and a setting. Second, putting the character into the Amber universe. Third, interacting with the Game Master for information and to manipulate the character. And, fourth, dealing with your fellow players, and their respective characters.

Amber Role-Playing System Terms

Attributes: These are the ratings of your character's advantages, and weaknesses. There are four Attributes in the Amber RPG; *Psyche, Strength, Endurance* and *Warfare*. They are measured in two ways. First, by the *rank*, which is where the character fits in compared with everybody else (1st being best). Along with rank, each Attribute has a numerical score, starting in the negative numbers (-25 for Human rank), and increasing to represent improvement.

Campaign: Most games, run by a Game Master, for a regular group of players, is considered an Amber campaign. It can range anywhere from a short campaign of a couple of sessions, to a long-term epic campaign lasting many years.

Character: Every player (except for the game master) has a character, also called a *Player Character* or *PC*.

Cross-Over: A cross-over is a special kind of *Amber* tournament, where characters from different campaigns (based on different Game Masters), get together. Usually a cross-over is based on the idea of there being many different versions of Amber, and that occasionally characters can slip from one version to another.

Death: Just as in real life, characters can die. In the Amber RPG death is unlikely but always a threat to every character. The player's main job is to keep their character

alive. As long as your character is alive you can consider yourself a "winner."

Diceless: Doing without dice, and any type of chance or random number generator. Diceless also means going without coins (2-sided dice), card shuffling (52-sided dice), spinning wheels (flat dice), Electronic Number Crunchers (infinite dice), Yarrow Sticks (multi-dimensional dice), or anything else other than character interaction.

Game Master (GM): This is the person who controls the "world" and runs the game. All the non-player characters (NPCs), including guards, innocents, and villains are controlled by the GM. The GMs control even extends to things like weather, cross-universe politics and natural disasters. Amber Game Masters have more power, and more responsibility, than those who run more traditional role-playing games.

Scenario: This is a specific mission assigned to the characters in a role-playing game. A scenario is usually a story with a beginning (Julian is missing from Forest Arden), a middle (using the blood stained trump of a strange woman, after discovering it in Julian's bedroom), and an end (rescuing Julian). Most campaigns are developed around a number of scenarios.

Tournament: The flip-side of a Campaign is a *Tournament*, where the role-playing usually lasts no more than one or two sessions.

Playing a Princess (or Prince) of Amber.

A player's view of *Amber*.

In our example Cindy is playing *Dorell*, a Princess of Amber. Of the four Attributes that make up a character, Dorell is *ranked* second (45 points) in Warfare *and* fourth (9 points) in Strength. For other two, Psyche and Endurance, she is Amber rank (0 points), which means she has no weaknesses. Dorell has one Power, that of Pattern.

Like any character who buys the power of Pattern, Dorell automatically became a descendent of Oberon and one with the legacy of the Blood of Amber. The Game Master has decided that Julian is Dorell's father.

Notice that the player, Cindy, usually describes the setting. Infinite Shadow is Dorell's playground. Any place that Cindy can imagine, any world, any world at all, any world she has ever read about, thought about, or dreamed, exists somewhere in Shadow. As in many campaigns, the Game Master starts by letting Cindy describe Dorell's situation.

GM: Okay, has everybody got their situation in order? Remember, when I get to you, I want you to tell me about your character and where you'll be at the start of the campaign.

Cindy: Anywhere we want?

GM: Pretty much. If I've got a problem, it will be because you've thought of something I missed. In that case, I'll let you know. Cindy, we'll start with you.

Cindy: My character is named Dorell. She's 5'9" tall, 130 pounds, with bright blue-green eyes, and fiery red hair. She prefers to wear black or dark brown, with a blouse of light blue. She is also arrogant and opinionated. She knows she is among the finest blades anywhere, and she'll test her sword, "Dragon-Tooth," against anyone.

GM: Good. Where will she be starting the campaign?

Cindy: Dorell is in a Shadow I'll call Mith. Spelled with an "i" instead of a "y." It's a fairly modern world, with everything except electricity and gunpowder. So it has good plumbing, steam engines, and a code of chivalry. It's also a place where swordsmanship is considered a high art, as if you combined a Victorian Age with something like the Bushido devotion to martial arts. It's dawn, and I'm pacing the wet grass, waiting for my dueling opponent to show up.

• Cindy could "buy" a Shadow, an entire world, with points. Mostly characters just wander out into Shadow and find a convenient one. It could be a world where the character is worshipped as a god, a place of interstellar spaceships and galactic empires, or a paradise filled with nothing but sunny beaches and friendly natives. Cindy could have picked a world described in any book, movie, television show or just a variation of Shadow Earth. "Mith" is a place that Cindy invented, a place she thought would be neat for her character.

GM: I can't help but ask, but what's the reason for the duel?

Cindy: My lover has been accused of treason by his jealous ex-wife, and I intend to have satisfaction.

GM: Ho ho! How about your seconds? Who are they?

Cindy: I hadn't thought about that... Let's just say that they are friends of mine. If it turns out to be important I'll work out the details later.

GM: No problem. You are waiting in the pre-dawn light when you suddenly feel something change. As if you were moving through Shadow, but you didn't cause it.

Cindy: I'll look around, has anything changed?

GM: Yes, for some reason one of your seconds is holding an open case containing what seems to be a pair of dueling pistols.

Cindy: Pistols? I was looking forward to a duel with swords. Besides, gunpowder isn't supposed to work here.

GM: Well, the guns are here. What are you doing?

Cindy: I'll pace across the field, using my Pattern to Shift Shadow.

GM: No problem. What are you doing to the Shadow?

Cindy: I'll subtract the guns, and change everything back to the way it was.

GM: A few minutes later, the field is just as you left it. Your opponent's carriage pulls up.

Cindy: Fine. I'll proceed with the duel.

GM: The ex-wife turns out to be no great challenge for you.

• The Game Master made an assumption here. Since the Shadow is one of Cindy's choosing, he just figured that her opponent would be pretty easy. At this point Cindy could have objected, saying something like, "Wait a minute, this Shadow is supposed to provide me with challenging duelists, I think she should be almost as good as Dorell."

GM: What injury do you wish to leave her with?

Cindy: I want to wear her down to exhaustion, then put a small cut on her nose.

GM: You want to disfigure her?

Cindy: No, just enough for a tiny scar. Just enough to teach her a lesson.

GM: Twenty minutes later the duel is over, the doctor is tending to your rival's hurt nose, and your seconds are congratulating you on your victory. What are you doing?

Cindy: I'd like to find out about what caused that earlier shift. Just who was behind adding those guns into my Shadow?

GM: How are you going to find out?

Cindy: I'll seek out the source of the problem by walking through Shadow.

• Amberites can find anything in Shadow, including other Amberites, and even things as vaguely defined as "whatever is causing my problem."

GM: What details are you changing?

Cindy: Hmmm... For now, just a few tiny things. I'll start by making the clouds a little more blue.

GM: No problem, the clouds gradually turn light blue, and you feel like you're making progress toward whatever caused the shift.

Cindy: As I walk along, I'll think about finding a horse. A pretty white mare, with a tan bridle and matching saddle, grazing in a meadow.

GM: A few minutes later, as you come out of the woods, you see a meadow with exactly the horse you described.

Cindy: Excellent. I'll take a few minutes to make friends with the horse. Then I'll mount and continue.

- Cindy could just as easily have "found" a sports car, a bycicle, a space ship, or a magic carpet. As it happens, Cindy's character Dorell is comfortable on a horse. Horses also have the advantage of working in Shadows where technology or magic might not.

GM: Okay, riding the horse, what are you changing?

Cindy: I'll start with flowers. I want a lot of sweet-smelling lilacs to cover our path.

GM: The path changes to lilacs and you notice the sky is taking on a familiar hue. It looks like you are heading towards Amber.

Cindy: Hmmm... I'm not sure if I like that idea, but I'll keep going. Let me know when I get to Forest Arden.

GM: A bit later you see a figure seated on a tree stump, huddled under a dirt-brown blanket. He holds his hand out to you, and you can see he is wearing some kind of gauntlet. What are you doing?

Cindy: I'll stop my horse and look at him.

GM: He throws back the blanket and you can see the features of your father, Julian. His long black hair is tangled and dirty, his white armor stained with blood and muck, his right eye is black and blue, and he looks exhausted.

- Just as player characters can find anyone or anything in Shadow, so other Amberites can arrange to be found. Of course, it is possible that Dorell's father, Julian, is the source of her problem...

Cindy: Father! What has happened?

GM: "Daughter, I have a mission for you." He takes a small pouch from his belt, tears it away, and holds it out to you. Are you dismounting?

Cindy: No, I'll stay on my horse and reach for the pouch.

GM: You take it. He says, "Now daughter Dorell, you must Hellride away from here, and away from Amber, until the stars fall."

Cindy: Until the stars fall? Father, what does that mean?

GM: You will know, you will know. Then, and only then, take out the cards I have given you, and gather together the young children of Amber.

Cindy: Then what?

GM: "I don't know, and I have no more time.." Julian says, and then, "Go! Now!" What are you doing?

Cindy: Father, why should I be riding away from Amber? If it is in danger, isn't that where we should be going?

GM: He turns away from you, drawing his sword, and you see something moving, back in the forest. Are you doing anything else?

Cindy: I'm waiting for his answer.

GM: He doesn't say anything. What are you doing?

Cindy: What is it?

GM: You can't tell, but it looks awfully big. Julian looks back at you, angry as you've ever seen him, his mouth drawn into a tight line. What are you doing?

Cindy: Father, I want to help you!

GM: He whirls, says something that sounds mystical and magic and also like a curse. At the same time he whacks your horse's flank with the flat of his sword. Your horse rears straight up! What are you doing?

Cindy: I'll try to hang on.

GM: Your horse is galloping like crazy! Now what?

Cindy: I'll try to calm her, and slow her down.

GM: By pulling on the bridle?

Cindy: I'm strong enough, aren't I?

GM: You're probably strong enough to tear the head right off the horse. What are you doing?

Cindy: Putting my hand on her neck, I'm going to reach out with my mind, and try to make her feel calm and soothed.

- Amberite Attributes, even Dorell's Amber rank in Psyche, are enormously powerful. True, Cindy has high rankings in Warfare and Endurance, but she is smart enough not to neglect her other options.

GM: The feeble mind of the horse easily bends to your will, and she slows down to a canter. Right then your sword cries out, 'duck!'

Cindy: I duck down!

GM: A split second later, an arrow whizzes overhead. You hear shouts, about two hundred feet away, and the sound of horses. What are you doing?

- Cindy had used a couple of points to buy "Sensitivity to Danger" for Dorell's sword. Obviously, it just paid off!

Cindy: What do I see?

GM: Looking behind and off to your right, you see at least two dozen armored men on horseback, all wearing pink. They're accelerating toward you. What are you doing?

Cindy: Could I take them?

GM: One at a time, probably, but you're out in the open where you can't stop a mob from getting behind you. What are you doing?

Cindy: Do I still have mental contact with my horse?

GM: Yes.

Cindy: I'll tell the horse to start running, and I'll start shifting Shadow as we go.

GM: Dorell shifts Shadow as she gallops away. I'll get back to you...

- At this point, the Game Master shifts away from Cindy, over to some other player. The outcome of the story depends on how the players respond and react.

Playing a Lord of Chaos, a Trump Artist, a Shape Shifter, or a Shadow Sorcerer.

While Dorell is a Princess of Amber, with the Power of the Pattern, there are many other choices available. You can make characters based on any of the Powers. For example, a player might create a character with Logrus, the sign of Chaos, and an heir to the Courts of Chaos instead of Amber. Such a Lord of Chaos, instead of walking through Shadow, might reach out with the tendrils of the Logrus.

Another possibility is a character based on Trump Artistry. Such a character can paint pictures of people, places or things, and use the images as gateways across Shadow.

Characters based on Shape Shifting base their powers on what they can do to their own bodies. For example, Shape Shifters can take on the abilities of Shadow-travelling creatures. Finally, one can take the more laborious route of mastering spells and/or magical artifacts.

CREATING PLAYER CHARACTERS

The first step to role-playing in *Amber* is creating characters for each of the players. Here's the simple version, in five steps for creating your Amber character.

STEP ONE: THINK ABOUT IT.

Dream up the best character you can imagine. How do you want that wonderful new person to look? In your dream of dreams, what kind of person would you really like to be? Use your imagination first, and then take a look at the character sheets and start figuring out how to make that person come to life. Ask yourself what most interests you. Combat? Mystical Arts? Power Politics? Exploration? Or some combination?

Remember, this is an immortal character. One you'll want to play for years and years.

STEP TWO: BID FOR ATTRIBUTES

There are four Attributes that define every character; Psyche, Strength, Endurance and Warfare. Bid for your character in the Attribute Auction using some of your one hundred (100) points. Pay attention to the other players, and remember who your rivals are going to be. Save enough points for the rest of the character creation process.

STEP THREE: BUY POWERS

Use the points left over from the auction to buy Powers. You can choose basic or advanced forms of either Pattern, Logrus, Trump, or Shape Shifting. There are also three forms of magic available.

STEP FOUR: BUY EXTRAS

Add extra details to your character. These include magical creatures or artifacts, personal worlds (Shadows) that you create, or Allies who will help your character.

STEP FIVE: BALANCE THE POINTS

See how many points you have left out of your original 100. If you come up short you can chat with the Game Master about getting more points for reducing your Attributes, or by pledging player contributions to the game. If you have to, go back and change your powers and extras. Any points you have left over become "Good Stuff" and make your character lucky. If you have a point shortage, your Game Master can give your character "Bad Stuff" that will make your character unlucky.

STEP FIVE: TAKE THE FREEBIES

Work up your characters looks, background, skills and personality.

So much for the simple version. Let's start with the Attribute Auction.

> Decide on a "focus" and then stick with it. Develop your character along the lines of your specialty. Explore and use *your* talents. Don't keep changing your character's skills.
>
> - Chuck Knakal

ATTRIBUTE RANKING

All characters start out with Amber level Psyche, Strength, Endurance and Warfare. Points can be spent to raise these to match rankings in the auction. Or, if a player has made no bid in an Attribute, then that attribute can be sold down to Chaos or Human level.

Amber Rank. You don't bid, and you don't sell the attribute down. Which means you end up with an Amber level attribute. That's great, because Amberites are so very much better than all those wimps out in Shadow. An Amber Rank means you have no weakness. Amber Strength makes you stronger than the Earth's strongest human. Amber Warfare puts you ahead of every Earth swordsman or strategist. Same with Psyche and Endurance.

Ranked. When it's not enough to be better than Shadow mortals, you spend extra points on an Attribute, and you enter into a contest with other Amberites. Any kind of ranking, even just a point, puts you way ahead of the do-nothings with Amber level.

Dominant Rank. First place holders in the auction hold a very special place in the rankings. Even if they're only a point ahead, even if somebody buys up with enough points to almost pull even with them, they're still way out in front when it comes to any kind of contest. Those with a first place Rank aren't just better, they're *superior*.

Secret Rank. You don't have to settle for Amber, or for your last Auction bid, you can spend even more points on any Attribute. This is a secret advancement, one that the other players will not know about. However, remember that when you match points with someone who made an official bid, say 3rd Rank, you'll always be slightly behind, what we'd call "3.5" - better than 4th, but inferior to 3rd Rank.

Chaos Rank. The common level of dwellers of the Courts of Chaos. Also called "Below Amber." Equals the very high end of the human spectrum. Shadow Earth's strongest man, or the world's best fencer. Pretty respectable by Shadow standards, but no match for any of Amber rank. Selling down to Chaos is good for an extra 10 Points.

Human Rank. Wimp out. Human level means the Amber equivalent of a couch potato, which means something equal to an ordinary person of Shadow Earth. You become one of the masses of inferior Shadow folk. Taking any human attributes is making yourself vulnerable. Human level Rank is always dangerous because it makes a character vulnerable. Selling an attribute down to Human is good for an extra 25 Points.

THE ATTRIBUTE AUCTION

Auction Time. Otherwise known as the "Bidding War." All the players get together and participate in four consecutive Attribute Auctions. In each auction the ranking of the Attributes will be determined. Just as Gérard is the strongest Prince of Amber, and Benedict the best warrior in creation, so too will player characters be the best of their generation of young Amberites.

Once the entire group has been assembled for the event, there will be separate auctions for each of the four Attributes. The first auction is for Psyche, the next for Strength, then Endurance, and finally the bidding on Warfare.

THE AUCTION RULES

Here are the hard-and-fast rules of running an Attribute Auction. You, the Game Master, should be sure to explain each rule to the players before the bidding starts.

1. All players start with 100 points.

You can get more points in three ways; (a) by "selling down" Attributes, (b) by taking "Bad Stuff," and (c) by "contributing" to the Game Master.

(a) Getting more points by reducing Attributes.

- Amber level in Psyche, Strength, Endurance and Warfare is **Free.**
- Chaos level is just below Amber. Going down to Chaos level is worth an extra ten (10) points.
- Human level is below Chaos level. Going down to Human level is worth an extra twenty-five (25) points.

For example, if you plan on getting ten (10) more points by "selling down" Psyche to Chaos level, you can't make **any** bids in the Psyche Attribute Auction. **You can't "sell down" an attribute after you bid on it!**

(b) Getting points for Bad Stuff.

You can take as many points of Bad Stuff as your Game Master will allow. Every point of Bad Stuff you take gives you another point you can spend on your character. However, every point of Bad Stuff is also used by the Game Master to arrange for bad things to happen to your character.

Bad Stuff characters will be naturally unlucky, and will enter the universe with enemies, problems, or misfortunes.

(c) Getting points for making contributions.

Contributions are some kind of regular, on-going chore that the player promises to do for the Game Master. Examples include keeping a diary for your character, taking notes for the whole group, or drawing a picture every game session.

2. Every bid is permanent. Each player's highest bid is paid and cannot, ever, under any circumstances, be reduced or refunded.

3. No Attribute can be sold down after it has been bid. If a player doesn't bid on an Attribute, they can later adjust it up or down. Once a player bids on an Attribute, that Attribute can only be raised, never lowered.

4. Players can only bid for themselves. It's just not fair to the other players. If missing players want to participate, it's up to the Game Master to enter bids for them.

5. The winner of an Attribute Auction is unbeatable. Whoever wins for an Attribute is that Attribute's ultimate winner (among the younger generation). First place is the only safe place. If you get first place in any Attribute Auction you will *know* you are the best of your generation. Anyone, after the Auction, can buy up, even spending as many points as the first place winner, but they can never beat, or equal, whoever gets first place.

6. Ranks are based on the Attribute Auction. Each group of players really sets up their own standard, and their own levels for advancement. With your bids you are *making* the rank system for this campaign. What you bid will determine how the campaign will be set up. The final results also determines how characters advance later on.

Here's how it works. Assume the final results for Warfare turn out to be fifth place for ten points, fourth place for eighteen points, third place for thirty-five points, second place for fifty points, and first place for fifty-one points. That means buying up Warfare later will have to match those point levels, just as if they were rungs on a ladder. You couldn't possibly spend twenty points in Warfare. You have to "match" the bidders who bid in the Auction, so you can only spend eighteen, or thirty-five, but nothing in between.

7. Bad Stuff Hazard. Too much Bad Stuff, which you get when you spend too many points, can result in losing some control over your character's creation. If you have a great deal of Bad Stuff, the Game Master may end up changing your character's Powers, Items, Shadows, and background.

How an Auction Works

Just in case you've never seen an auction in real life, this will explain how to conduct an *Amber* Attribute Auction.

Step One. You describe what is up for Auction. At that point the Game Master's job is to be a salesman, to get people excited about the "product," and to make sure everyone understands how important it is.

Step Two. Everyone writes down their opening bids. These bids, like any others, can't be taken back. The main point is to allow everyone an equal chance to get in a bid before things go crazy. It's possible that someone could just shout out "fifty!" for the first bid, freezing out all the other players. So the opening bid gives everyone a fair chance. a chance to make a modest bid.

Step Three. When everyone is done with their opening

bids, the Game Master checks each one, and announces the results.

Step Four. After the opening bids have been reviewed, say, "Open bidding starts now!"

Each bid must be higher than the last. So, for example, the first bid must be higher than the highest opening bid. Pay close attention to who says what. If things get confused, then the Game Master should stop the bidding and review the current bids, saying, "Stop for a minute! I've got Jill in first place with twenty-one, Tim's in second with twenty, Tony last bid fourteen, and Vince is still at twelve. Hugh and Diane both made opening bids of five, and they haven't raised them yet. Is that right? Okay, we'll start again. We're at twenty-one, do I hear twenty-five?"

Step Five. Ending an Attribute Auction is always done with the following formula, "Elaine, with forty, has top bid in Endurance. Endurance is going once... Going twice... Going... Going.... Going... Gone! Sold to Elaine for forty points!"

At the end of each Attribute Auction the results are compiled by the Game Master, and the rankings announced. Winners, the top bid of particular auctions will be absolutely best in that attribute (at least among player characters), and will be Ranked First. The next highest bid is Ranked Second, then Third Rank, and so forth. Players who have not bid should be announced as "No Bid."

Tips on Bidding. Hardly anyone is ever satisfied with the results of an Attribute Auction. If you don't bid, you'll regret not getting the good Ranks. If you bid, you'll regret the points you spend. Still, what do you expect? You're competing with the other players. No matter how unhappy you are with the results, remember that every bid you make throws a monkey wrench in somebody else's private plan.

Bid Points are Spent Points. It's not just the "winner," the player who gets first place, who pays. Every time you bid, you spend points. And, at the end of an Auction, your highest bid "buys" your ranking.

Warning! Once you make a bid, you've effectively spent the points! There's no "taking it back," no "I didn't mean it." All Attribute Auctions are "Final Sale" with no returns, no credits, and no exchanges. *You can't take the points back after they've been "bid" in an Auction!*

Bid According to Plan. Start high if you really want to have first place. Start low if you just want a ranking. Budget your points in advance. It's a good idea to figure out how many points you're willing to spend in all the auctions, usually fifty points maximum. Then break it down for each of the four attributes.

Bid for What You Need. If you want your character to be a hotshot in mental powers, then bid heavily in Psyche, but don't neglect Endurance. Combat is determined by Warfare, the last auction, but a good warrior should have some ranking in Strength and Endurance as well.

Stay Flexible. Sometimes opportunities open up unexpectedly. Don't pass up the chance to be highly ranked in a slow-moving auction, even if you haven't planned on spending quite that many points.

PSYCHE

This is the rating of the character's mental strength. Psyche also shows the character's force of will, and their agility in manipulating Pattern, Logrus, Magic or Trump. In addition, a character with a high Psyche will be sensitive to danger and to other things.

Most important, the relative Ranking in Psyche will determine all contests of the mind. These Psychic battles, often fought through Trump contacts, or by touch, but also through mystic connections, will be always judged by a character's Psyche. The loser in a Psychic battle can, if overwhelmed, be killed by the power of the mind alone.

Of the children of Oberon, it seems likely that Fiona is number one in Psyche in the Amber family. She has used her power to hold off her brother Brand, when he was at the height of his power, and she is, by all reports a sorceress and a student of both Pattern and Trump.

Why Bid on Psyche? Mental power, no matter what your power base, is the battery that runs things. Whether it be Pattern, Logrus, Trump, or Magic, it's the character with the highest Rank in Psyche who will be most feared.

SAMPLE ATTRIBUTE AUCTION - PSYCHE

Here's the Attribute Auction for a typical group. After running a practice auction, the Game Master has assembled all the players, Alex, Beth, Cindy, Kevin, Mick, Ted, Peggy and Willy, into a room and is ready to begin.

GM: Okay folks, settle down, it's time for the bidding war. The first item of The Bidding War is Psyche. The winner will be the person with the strongest psychic sense and will. Is everybody ready?

Beth: Almost. Can I ask a question?

GM: Yes, please! Remember everyone, it's for real this time. Any bids will be permanent. You can't get your points back, so make sure you ask your questions before we get rolling. Beth?

Beth: How important is Psyche compared to the other Attributes?

GM: It's the most important Attribute.

Ted: Can you give us some kind of idea of the usefulness of Psyche?

GM: Sure. A person with a strong Psyche can dominate someone else through the Trumps, or through other means, engaging in a struggle of wills. High Psyche also helps contact someone difficult to reach, or to sense trouble, like some slight disruption in shadow. The classic use of the ability is to use it as the mental force to drive the major powers of Pattern or Psyche. It's also important in any kind of Magic.

Beth: When you say a struggle of wills, do you mean like when Corwin's Fleet is engaging Caine's, and Eric is trying to distract Corwin?

GM: Exactly!

Willy: Five points!

GM: Sorry Willy, you've got to write down your initial bid.

Mick: But thanks for the information, Willy!

GM: Come on folks, write down your initial bids. If you're not going to bid, then write down zero or "no bid..." Everybody ready? Good.

Ted: Wait a minute. Who was the strongest in psyche in the original books?

Beth: Fiona!

Peggy: Hardly! It was Brand.

GM: I think the point is debatable but evidence points to either Brand or Fiona. Fiona shied away from taking Brand on, but she did manage to hold him in place once or twice. We'll start with your opening bids now. Hold up the paper with your number, then announce your bid.

Alex: No bid for me.

Mick: Five.

Willy: Zero also.

Cindy: No bid.

Peggy: I've bid one point. And I still say it was Brand.

Kevin: Irrelevant anyway. I bid ten.

Beth: Five.

Ted: Zip.

GM: For starters we've got Peggy with one. Mick and Beth both bid five, so they'll be tied unless one of them bids again. Next bid will have to beat Kevin's ten. Do I hear twenty?

Alex: Eleven.

Kevin: Twelve.

Peggy: Twenty!

GM: Alex with eleven, and Peggy with twenty. Good! Do I hear thirty?

Beth: Twenty-one.

Kevin: Twenty-two.

Beth: Twenty-three.

Kevin: Twenty-four.

Willy: Twenty-five.

Kevin: Twenty-six.

Willy: Thirty!

GM: Wow! We've got some action. Do I hear forty?

Alex: In your dreams...

Kevin: Forget it!

GM: We've got a measly thirty points threatening to take first place in Psyche? That's it?

Beth: What do you mean measly? Thirty is a lot of points!

GM: Not when you consider that Willy can burn right through your mind... Peggy?

Peggy: Let me get this straight. We've only got one hundred points to work with. Right?

GM: Yes.

Peggy: That's not like one hundred points for Attributes and then another hundred points?

GM: No. It's hundred points total. So how about thirty-one for

Psyche?

Peggy: Too rich for me.

GM: Anybody?

Kevin: Let me get this straight. I've already bid twenty-six, so that means I've already spent the 26 points, right?

GM: Bid 'em and lose 'em, that's the way it works.

Kevin: Then I bid thirty-one.

Willy: Thirty-five.

Kevin: Ahhh, the heck with it. Forty!

GM: Willy?

Erick Wujcik here. The game designer. Speaking to you, the first-time Amber player. I want to give you the best advice I can manage before you start building your very own Amber character. Aside from designing the system, and game mastering a couple of hundred player characters, I've also played a character or four. So here are my own words of wisdom on character design, in two laws of Amber creation:

Always Get Pattern Imprint. Do what you have to. Spend the 50 Points. No character is complete without Pattern.

Never Sell Down Attributes. No matter what, don't sell any of your attributes. If you absolutely have to sell attributes, stop at Chaos rank. If you need the points, selling down three attributes to Chaos is better than selling down one to Human level.

Willy: Fifty. So there!

Kevin: That's crazy!

GM: I hear fifty, do I hear sixty?

Kevin: That's nuts! Crazy! Insane!

GM: You already spent forty points Kevin, and all you've got is a distant second place.

Kevin: I'm happy.

GM: Anybody else? I've got Psyche at fifty points. Fifty going once...

Willy: Yes!

GM: Going twice...

Kevin: Wait a minute.

GM: What?

Kevin: Well, fifty-one.

GM: Willy?

Willy: Oh, fifty-two.

GM: Kevin?

Kevin: Willy, are you going to keep this up forever?

Willy: You bet.

GM: Kevin, he's probably bluffing. Two more points and you beat him again.

Kevin: NO!

GM: Hey, just another couple of points...

Kevin: NO!

GM: Anybody else?

Whole Group: NO!

GM: Sigh... Well, Going Once. Anybody?

Willy: Come on...

GM: Going Twice...

Willy: Yeah!

GM: Hey, Willy, want to go for fifty-five?

Willy: What?!? I'm already the top bid!

GM: Just thought you might want a round number. Going once, going twice, going... going... Gone! Top bid in Psyche goes to Willy for 52.

Willy: Yes!

GM: So, here's the final tally for the Attribute Auction for Psyche. Top bid is fifty-two points from Willy, followed closely by Kevin with fifty-one...

Mick: Nuts, both of 'em, nuts...

GM: We'll see how nuts it is, the first time you end up in a Trump contact. Third is Beth with twenty-three points, twenty points makes Peggy fourth, Alex is fifth with eleven, and in last, sixth place, for a chintzy five points, is Mick.

Cindy: What about Ted and I?

GM: Currently you are considered average. That's far better than any normal human but no match for any of the bidders.

Ted: Are we stuck with that?

GM: No. Anytime between now and when we actually start playing you can change your Psyche.

Cindy: You mean we can buy a better placement?

GM: Exactly. You do it secretly.

Ted: Does that mean one of us could become better than Willy?

GM: No, that's one thing you can't do. In fact, nobody can spend more that fifty-two total points on Psyche.

Cindy: What if we spend the fifty-two?

GM: Then you'll be just behind Willy in ability, but ahead of everybody else.

Peggy: I've already spent twenty points. Can I decide to spend more points?

GM: Yes, but you can only match a higher bidder. In other words, you'll have to spend either another thirty-two, making you fifty-two, to match Willy, or thirty-one, to match Kevin, or just three, to match Beth's twenty-three. And even then you'll be just below them.

Beth: Does that mean I can get three points back, and go down to Peggy's position?

GM: No!

Kevin: Since I'm only second, does this mean that anybody can buy their way over me?

GM: Yeah, but if you're worried you can also buy your way up. And everyone who buys a higher level, at the same number of points, is considered dead even.

Ted: That's a lot of decision making.

GM: There's even one other option. Ted, since you and Cindy didn't bid you can sell some of your Psyche. It'll make you much weaker but you might find it worthwhile.

Peggy: Can I buy just four points to get over Beth?

GM: Nope. You have to spend either fifty-two, fifty-one, twenty-three, the twenty you already spent, or, if had you started lower, eleven or five, the same amounts as the bidders finished with.

Here are the results after the first Attribute Auction:

PSYCHE

Willy	1st place rank [52 points]
Kevin	2nd place rank [51 points]
Beth	3rd place rank [23 points]
Peggy	4th place rank [20 points]
Alex	5th place rank [11 points]
Mick	6th place rank [5 points]
Cindy	NB (No Bid)
Ted	NB (No Bid)

The Potential of Psyche.

Fiona, probably the elder Amberite with the greatest Psyche, can perform quite a few "tricks." Here are the limits of what Fiona can do. Player characters can strive for any or all of these abilities.

- In moving through Shadow, using the Pattern, Fiona can sense the disruption caused by another Hellrider, and follow that disturbance, tailing the leader.

- Fiona has a basic feel for Shadow paths, and knows when one has been altered or artificially Barred.

- Fiona can sense the gathering of magical energies accompanying use of a Sorcerer's spells, warning her of their imminent usage. Even if the Sorcerer is unseen, she can detect the magical aura and have some indication of its intended target.

- Fiona can detect the location of any other Psychic presence in the immediate area, including the minds of those invisible or hidden.

- The "stench of Chaos," as Amberites call it, is obvious when someone nearby summons the Logrus. Likewise any strong use of Logrus, such as reaching through Shadow, or manipulating objects with Logrus tendrils, will be quite noticeable. Note that Logrus Masters of high Psyche have the same ability to detect characters summoning or using Pattern.

- Fiona can pick up on the physical condition of those in range of her Psyche, seeing their hidden weaknesses, pain, and injury.

- Fiona can sense the presence of objects of power, whether the personal artifacts or creatures of another Amberite, or items powered by Pattern, Logrus, Trump, or other strong energies.

- In sensing characters and creatures, Fiona can tell whether they've reached the limits of their Endurance, and whether they are tired, hungry, thirsty or otherwise in need.

- Fiona, when opening up her mind to the environment, is intuitively aware of any potential hazards or dangers. If a character or creature is plotting or preparing an attack, then the source of that mind will be obvious. Traps and innately dangerous situations are simply recorded by Fiona as uneasy feelings, without specific details other than a vague sense of some locations or movements being more dangerous than others.

- Fiona notices it whenever someone is touching her Trump.

- When in the company of others, the general mood and emotional state of each individual will be quite visible to Fiona. This works through Trump contacts, and other long-distance sensations using powers.

- When in mind to mind communication, or even while attempting contact, Fiona can read a number of things about the other character. She can get their name, though, if the character is sensitive (Amber or better Psyche), they will be aware of the transferral of the knowledge. As the contact continues, unless the other character takes steps to block out the data (mind to mind combat), Fiona can draw out other pieces of information.

- In a Shadow, Fiona can read its composition and resonance, making it possible to identify much about the place. This basically translates into a description of all the Shadow's features, including time differential, and the degree of Pattern and Logrus.

- She can also pick up an impression of anyone who has a strong impact on the shape or form of the Shadow.

- Meeting, or sensing, any individual or creature, will mean Fiona can detect the presence of Pattern in their blood, and just how activated that Pattern may be. This allows for identifying someone with the blood of Amber who has not yet walked the Pattern, or telling whether someone has recently walked the Pattern. The potential for Logrus, Trump Artistry, Shape Shifting, and Magic are also noticeable.

- In conversation Fiona has a feel for the truth. This means that a direct lie is obvious. Likewise, it's possible for her to tell when one is skipping around the truth, or evading a question.

- Although someone gifted in Shape Shift can cloak their identity from most, Fiona can, if she has the mental contact, and the time, find the exact identity of the changed one.

STRENGTH

Strength rates the character's muscles, and therefore the damage that a character can inflict in hand to hand combat, as well as the character's resistance to damage. Normal Strength, in an Amberite, is sufficient to heft and toss around small automobiles.

Once characters lay hands on each other, the contest, no matter what it may have been before, becomes one of sheer Strength. Unlike other forms of combat, where a character can usually run away, there is likely no escape from a wrestling match with a character of superior Strength.

In the Amber universe Gérard is Ranked number one in Strength. Once Gérard gets a good grip on anyone of lesser Strength, they are doomed. Never forget Gérard's words to Corwin, "I can kill you, Corwin. Do not even be certain that your blade will protect you, if I can get my hands on you but once..."

Why Bid on Strength? If there's any such thing as a "sure thing" in Amber, it's having a superior ranking in Strength. Most combat can go either way, influenced by any number of factors, but once you've got your hands around your enemy's throat, all that counts is Strength.

SAMPLE ATTRIBUTE AUCTION - STRENGTH

Continuing with Strength, the players, Alex, Beth, Cindy, Kevin, Mick, Ted, Peggy and Willy, move on to their contest for Srength.

GM: Our next Attribute Auction will be for Strength. Strength determines the winner in hand to hand combat, which is just about anytime you can grab somebody. A character

with higher Strength can literally break a weaker character.

Beth: How does Strength compare with the other Attributes?

GM: Easy, Strength is the most important Attribute.

Beth: Didn't you say that for Psyche?

GM: You must be mistaken. Strength is far more important than Psyche.

Cindy: He's lying!

Beth: Yeah, you said Psyche was the most important.

GM: Don't be ridiculous, the Game Master never lies. Strength is the most important Attribute. Gérard is master of Strength in Amber, and he is feared by everyone.

Willy: I think I see where this is going...

Beth: Yeah, and I don't like the looks of it.

GM: If you're done arguing, we'll start with your opening bids for Strength. Take a minute to write down your starting number... Ready?

Alex: No bid for me.

Mick: I bid Five.

Willy: None.

Cindy: No bid.

Peggy: Two points.

Kevin: Zero.

Beth: Five.

Ted: Zip.

GM: Only three opening bids? Two points for Peggy, and a tie at five for both Beth and Mick. Top bid is five, do I hear twenty? (silence) Come on folks!

Willy: Hey, Strength doesn't seem all that important.

GM: Just remember, Strength is unbeatable, there is no escaping someone who is superior to you in Strength.

Cindy: I don't know... I guess I'll jump in with six.

Beth: Seven.

Mick: Eight.

Cindy: Nine!

Mick: Twelve.

Peggy: Thirteen.

Beth: Twenty.

GM: Okay, good start! Who's next?

Mick: Not me.

Cindy: Uh uh. No way.

Beth: Yeah, I knew they'd wimp out...

Peggy: Don't be so sure.

GM: Peggy, do you have a bid?

Peggy: I'm thinking.

GM: Well, don't take forever. Strength going once.

Peggy: Wait a minute. Tell me again what Strength is good for.

Beth: She's stalling!

GM: I think so. Strength going twice. Ted, want to jump in now?

Ted: No.

Peggy: Ah, come on Ted, it'll be fun.

GM: I guess that's it, Strength is going...

Peggy: No! Twenty-four!

Beth: What?

GM: Back to the race. Beth, your bid?

Beth: She's crazy.

Peggy: Hey, I didn't jump all the way from seven, over thirteen, to twenty, the way you did. Why didn't you just bid fourteen?

GM: Settle down. I've got a Strength bid of twenty-four. Do I hear thirty?

Beth: No, and I'm not bidding any more.

GM: Anybody? Strength going once... twice... going, going, gone! Gone to Peggy for twenty-four points. Sounds like a bargain to me.

Mick: You always say that!

GM: Okay, here's the final results. Peggy in first place with twenty-four, Beth in second with twenty, Mick takes third with twelve, and nine gives Cindy fourth place.

Ted: Cheesh, what a bunch of muscle-bound women!

Peggy: Yeah, well let your character get in close and say that...

After the Strength Attribute Auction, the results are as follows:

	PSY	STR	TOTAL POINTS
Willy	1st [52]	NB	52
Kevin	2nd [51]	NB	51
Beth	3rd [23]	2nd [20]	43
Peggy	4th [20]	1st [24]	44
Alex	5th [11]	NB	11
Mick	6th [5]	3rd [12]	17
Cindy	NB	4th [9]	9
Ted	NB	NB	0

The Potential of Strength.

Gérard, the strongest creature known, is the yardstick of Strength's possibilities.

His prowess comes in three forms, Hand-to-Hand Combat, Exertion and Resistance.

1. Hand-to-Hand Skill. Aside from sheer physical brawn, Gérard is trained in a wide range of hand-to-hand combat skills.

- It is possible that a single obscure move, from some specialized form like Aikido or Atemi, might catch him unprepared. However, this will be a one-time advantage, as Gérard will learn it and compensate for it in all future conflicts.

2. Exertion. This is the pure muscle end of things.

- Gérard can bend iron bars, break boards of hardwood, smash individual bricks or cinder blocks, and snap light chains and ropes.

- He is also capable of tearing open things like car doors, conventional door locks, and ripping plate armor off an opponent (snapping the bindings that hold it on).

- If Gérard shoves, taking the time to lean against a solid object, he can topple stone walls, or turn a well-constructed building on its side.

- Gérard can punch through solid brick, cinder block, or wood. His punches against flexible objects, like a living opponent with the freedom to flex or give, are sufficient to break bones and rupture internal organs.

- Gérard can break black iron manacles or handcuffs, bend thick pieces of metal, and, holding something solid (a hammer, a rock), crack solid rock with just a few blows.

- Gérard can also uproot any tree small enough for him to grasp around the trunk.

3. Resistance. Covered with a layer of muscle, just as a prize fighter conditioned to absorb punches, Gérard can withstand an awful lot of damage.

- Gérard can resist any physical blows delivered from anyone of Amber Rank Strength or less.

- Gérard can withstand the impact from falling from thirty feet or less, or crashing with an impact of twenty miles per hour or less.

- Properly braced, Gérard can withstand the impact of a fully armored mounted knight, at full gallop, without budging more than a couple of inches.

- Gérard can strike, block and parry with enough force to break steel swords and other metal weapons and armor.

ENDURANCE

Where normal mortals tire after a few minutes of fierce combat, Amberites can keep fighting, fencing, or partying for days on end. In addition, Amberite Endurance includes the ability to heal all wounds rapidly, even to the extent of (eventually) regenerating lost limbs and organs. In a fight of any kind, whether Warfare, Strength, or Psyche, the character with the greatest Endurance always has a chance of holding out longer than an opponent with lesser Endurance, and winning by default. In addition, anyone with less than Amber level Endurance will not be able to walk the Pattern unassisted.

Endurance is the only Attribute that comes into play in every situation, involving either physical combat or the use of arcane powers.

Corwin, the main character of the *Chronicles*, is Amber's champion in Endurance. Why? Mostly because he could just keep going, in spite of the wear and tear of his adventures. In addition, Corwin regenerated his eyes, after they were burned from his head, far faster than any other Amberite expected.

Why Bid on Endurance? Frankly, you need it for everything. And, even if you are somewhat inferior in some battle, a higher rank in Endurance, combined with a little patience, virtually guarantees victory. Endurance is the ultimate Amber tie-breaker.

SAMPLE ATTRIBUTE AUCTION - ENDURANCE

GM: Time for the third phase of the Bidding War, the auction for the Endurance Attribute. Endurance is really the battery that drives all the other Attributes and Powers. Where each of the other Attributes is directly useful,

Psyche for mental battles, Strength for wrestling, and Warfare for combat, Endurance is just a measure of how long you can hold out against anyone else.

Cindy: Which of the guys from the books had the best Endurance?

Peggy: Wouldn't it be Corwin?

GM: Yes, Corwin! He surprised everyone in the family by growing back his eyes, after he'd been blinded, in record time. Which brings me to one of the other main things about Endurance, it measures how fast you can heal, and how long it takes to grow back any body parts you happen to lose.

Ted: What if you've got just a normal Endurance?

GM: If you don't bid, and if you don't sell it down, then you'll have an Amber Endurance. That means you'll be able to fight, or run, or push yourself somehow, for a full day without falling over. It also means you can heal a lot quicker than a human, and you can recover from anything eventually.

Alex: What if you go down in Endurance?

GM: There are a lot of disadvantages. Healing is slower. Even worse, if you don't have at least an Amber level of Endurance you can't walk the Pattern unassisted.

Ted: What do you mean "unassisted?" I thought you could only walk the Pattern by yourself.

GM: It's possible for you to get help, say from one of the elder Amberites. So someone with a lesser Endurance could gain the Pattern in the first place, but wouldn't be able to walk it again without help. And they'd always know the Pattern could kill them...

Beth: I don't suppose you'll tell us how Endurance compares to the other Attributes?

GM: Sure! Endurance is, of course, the most important of all the Attributes.

Willy: I told you, he's going to say that for everything!

Alex: No bid for me.

Mick: Zero.

Willy: None.

Cindy: No bid.

Peggy: Three points.

Kevin: Do I see a trend here, Peggy? Zero for me.

Beth: No bid.

Ted: Zip.

GM: Only one bid? Yikes! Peggy, you're way out in front with three points in Endurance.

GM: Ted, if you're saving your points for Warfare, this is the perfect complimentary Attribute. And it's going awfully cheap...

Ted: I'm not bidding.

GM: Not even on Warfare?

Ted: Hey, I don't have to say!

GM: True, but either way, a good rank in Endurance will serve you well. Either as a complement for Warfare, or, if you're saving all your points for powers, Endurance is what drives the powers.

Ted: I'll pass.

GM: Kevin? Don't you want to get in on this?

Kevin: I've already spent fifty-one points on Psyche. Why would I want to waste any of my remaining forty-nine on Endurance?

GM: Simple. You're second place in Psyche, but in a really close race with Willy. A superior ranking in Endurance could put you out front.

Kevin: Hmmmm. I hadn't thought of that.

Peggy: I thought of it. That's why I bid.

Willy: Hey, I thought you said I couldn't be beat for first place?

GM: You can't. But if somebody has a really good Endurance, and it's a close contest, they might be able to wear you down...

Kevin: An interesting idea, but no.

GM: Are you sure?

Kevin: Just can't afford the points.

GM: Well, Endurance is at three. Going once...

Mick: Hey, this seems way too cheap. I bid four.

Peggy: Eight!

Mick: Nine.

GM: Better! You're catching on. Ten?

Alex: Oh, ten.

Peggy: Rats!

GM: All right! It's finally picking up. Do I hear twenty?

Peggy: No, but I'll go eleven.

Alex: Twelve.

Peggy: Thirteen.

Alex: Fourteen.

Peggy: Fifteen.

Alex: Sixteen.

GM: Okay. Peggy?

Peggy: Nope. I can't go any more points.

GM: Mick? You've already spent nine.

Mick: I'm thinking about it...

GM: Endurance is at sixteen. Do I hear twenty? No? How about seventeen? Peggy? Mick?

Peggy: Sorry.

Mick: Nope.

GM: Well this is pathetic. Endurance going once at a crummy sixteen. Going twice... Beth, how about you?

Beth: Why would I want Endurance?

GM: You've already got good scores in Psyche and Strength, why not put yourself ahead all the way?

Beth: Maybe because I've already spent forty-three points. No bid.

GM: Last chance people! Endurance is going... Going... Going... Gone for sixteen points!

Alex: Whew! I got it!

GM: Yup, and pretty cheap too. Here's our totals for Endurance. First place goes to Alex for sixteen, second to Peggy for fifteen, and third to Mick for nine.

	PSY	STR	END	Total Points
Willy	1st [52]	NB	NB	52
Kevin	2nd [51]	NB	NB	51
Beth	3rd [23]	2nd [20]	NB	43
Peggy	4th [20]	1st [24]	2nd [15]	50
Alex	5th [11]	NB	1st [16]	27
Mick	6th [5]	3rd [12]	3rd [9]	26
Cindy	NB	4th [9]	NB	9
Ted	NB	NB	NB	0

The Potential of Endurance.

Corwin has probably the highest Endurance of all Amberites. He seems to be able to keep going, in spite of overwhelming exhaustion and fatique. He also heals quickly.

- Corwin heals from any serious wound in less than twenty-four hours. Someone with Amber Rank would take five days to heal completely and a character with Chaos Rank might need two weeks. Such a wound would mean bed rest of a month or more for a normal human.

- Scratches and nicks seem to heal in less than an hour. That's five times as fast as healing would be for someone with Amber Rank. Chaos Rank characters would need two days to heal such a wound and Human Rank characters would not heal for a week or more.

- Fully exerting himself, Corwin seems to be able to keep going for days without rest. Anyone with Amber Rank Endurance can fight for twenty-four hours straight. A Chaos Rank character could last for perhaps two or three hours. Humans wear out in ten or fifteen minutes.

- Corwin can regenerate anything in his body in less than four years, even his eyes. A character of Amber Rank would need twenty years to heal anything as delicate as the eyes. Chaos Rank folk might heal from such an injury in a century or two. Human Rank characters never regenerate lost body parts.

WARFARE

All Amberites are trained in Warfare. Use of any weapons, from daggers to machineguns, requires Warfare. All duels of swordplay, the most common way of settling disputes, are judged according to the relative Warfare of the participants. Warfare also determines a character's "knack" for tactics and strategy, for everything from the placement and leadership of troops on a battlefield, to a quiet game of chess.

Weakness in Warfare is more dangerous than in any other Attribute. That's because the relative Rank in Warfare determines how long the combat will last. If the Rank is too low, either Chaos or Human, the character can be hurt or killed before having a chance to flee.

Benedict is the family master of Warfare. In combat, whether with swords in the dueling arena, or massed armies on the field of battle, he is unsurpassed. Never underestimate his skill. Never fail to fear him.

Why Bid on Warfare? Face it, most conflict boils down to Warfare. There's nothing faster, nothing more decisive. Powers are too slow, compared with the flash of a blade. In Amber, the most feared character is the one with the fastest blade, the one with the highest Warfare Rank.

SAMPLE ATTRIBUTE AUCTION - WARFARE

GM: Now to the close of the Bidding War, and the last Attribute Auction, for Warfare. Any combat that involves tactics or strategy, including fencing with swords or shooting with guns, is based on Warfare. The better the Warfare, the better the character will be as a soldier, sergeant, captain, general, or admiral. Unlike the other Attributes, where rank is pretty much a measure of raw power, Warfare is a measure of the character's skill.

Beth: I don't know why I'm bothering to ask, but I suppose Warfare is the most important Attribute?

GM: All together now...

Everybody: Warfare is *the* most important Attribute...

GM: I'm glad you agree. Any other questions?

Willy: I guess it's pretty obvious that Benedict was the best in Warfare, right?

GM: Exactly. Benedict's rank in Warfare is so high, the rest of his brothers assume that he is unbeatable.

Beth: Let me get this straight. You say Warfare determines who will win in a sword fight, or some battle, but you also said that someone with better Strength would win battles. Which is it?

GM: It depends on the situation. Most cases, where the characters are using weapons, or leading a bunch of guys, Warfare is the Attribute we use. On the other hand, if the fight actually turns into a hand-to-hand melee, then Strength comes into play.

Beth: I still don't get it.

GM: Okay, here's an example. Let's say your character gets into a fight with Alex's character.

Alex: I'd have better Endurance!

Beth: But my Strength and my Psyche are both better than Alex.

GM: Right! Now, let's say that Alex ends up with a higher Warfare than Beth. The two of you start fighting with swords. So long as you keep it that way, relying on your skill with the blade, Alex, having a better Warfare, would probably win.

Alex: What do you mean, *probably*?

GM: Other things can come into it. The biggie would be if Beth managed to grab Alex. Then she could use her superior Strength and win based on that.

Beth: What if it were the other way around? I'm already better than Alex in Strength and Psyche. What if I had a higher rank in Warfare as well?

GM: That would put Alex in a really tough place. Not only could you kill him with your sword, but you could also start bashing at him, using your Strength as well. His only chance then would be if you were only a little bit better in Warfare. Then, fighting defensively, he could try to stretch out the combat, hoping to wear you out because of your inferior Endurance.

Beth: So what you're really saying is that it depends on how well we role-play, right?

GM: Yes!

Willy: Hey, can we get started?

GM: Okay. For the last time, let's get your opening bids. Again, be sure to write them down. Ready? Okay, let's start with Alex...

Alex: Ten!

Mick: Five.

Willy: No. None.

Cindy: Eleven.

Peggy: I wanted to, but... zero.

Kevin: Zero.

Beth: No bid.

Ted: Zip.

GM: Looks lively! Mick's got an opening bid of five, Alex with ten, then Cindy way out front with eleven. Do I hear twenty?

Mick: Twelve.

Alex: I'll go fifteen.

GM (after a period of silence): People? Are we stuck at a crummy fifteen for Warfare? You've got to be kidding!

Alex: This is fine. Let's stop here.

Mick: Okay by me. I'm happy.

GM: Hey, Willy! How about you?

Willy: What? I already blew fifty-two points in Psyche. Why would I want to mess around with Warfare?

GM: Think about what an awesome combination it would be! First place in Psyche **and** Warfare! You'd be unbeatable.

Willy: Forget it.

GM: Willy, think about it, this is really a bargain!

Beth: Hey, I think it's cheap. Seventeen!

Cindy: Well, I think that's cheap. So twenty.

Alex: Twenty-one.

Cindy: Twenty-five.

Alex: Twenty-six.

Cindy: Thirty.

Alex: No way. Thirty-one.

Cindy: I can keep this up a lot longer than you! Thirty-five.

Alex: What makes you think that? Thirty-six.

Cindy: 'Cause you've already spent eleven in Psyche and sixteen in Endurance. So I bid forty.

Alex: Argh!

GM: Well, Alex? Is that grunt a bid?

Alex: Give me a minute.

GM: Ted? Here's a great time for you to jump in. You've got the points to blow these guys away!

Ted: No way. Leave me out of this madness!

GM: Well, I've got forty in Warfare. Forty going once.

Cindy: Yes!

GM: Forty going twice.

Alex: Forty-one. Cindy, I don't care what you bid, I can always go a point higher than you.

Cindy: Really? Forty-two.

Alex: What happened to your five point leaps? Forty-three.

Cindy: Oh, I'm sorry. Forty-five!

Alex: Uhn...

GM: Well? Alex?

Alex: I hate this. I hate her!

GM: We've got forty-five going once.

Alex: You know, Cindy, you can be really irritating.

GM: Forty-five going twice.

Cindy: Especially since I'm going to be number one in Warfare.

GM: Going...

Alex: No! Forty-six!

GM: Very good. I now have forty-six. Do I hear fifty?

Cindy: I'm thinking...

GM: Don't take too long...

Cindy: No. I pass.

GM: Are you sure? You're only a couple of points away from a solid first place... No? Anybody else? Warfare is going once for forty-six points. Going twice... Ted? Here's your big chance to get a real bargain for all those points you've been hoarding!

Ted: Leave me out of it!

GM: Well, looks like Alex is going to get it... Warfare at forty-six is going... Going... Gone! Sold to Alex for forty-six points!

Willy: Is that it? Are we finally done?

GM: We're really just getting started, but we're done with the Attribute Auctions. I'll tally the points everyone spent in all four Attribute Auctions...

Beth: I don't understand how some people are going to be able to afford Pattern. I'm just barely going to be able to squeak it in.

Willy: Don't assume all of us want Pattern. Some people might have other plans.

Cindy: Really? What are your plans Willy?

Willy: None of your beeswax...

GM: Here's the totals, from highest to lowest. Alex spent more than anybody, seventy-three points, and is ranked first in both Warfare and Endurance, and third in Psyche, with no bid in Strength.

Mick: I think I see his weakness...

Willy: Yeah, like not being able to buy any Powers!

Alex: Don't be so sure...

GM: Next in points spent is Beth with sixty, and she bid in everything but Endurance. Peggy is right behind her with fifty-nine, bidding in everything but Warfare.

Willy: Well, if they do a diary or something, they'll at least be able to buy Pattern.

GM: Cindy spent fifty-four, bidding just in Warfare and Strength. Then there's Willy and Kevin at fifty-two and fifty-one, and both of them bid only in Psyche.

Mick: Looks like Willy and Kevin are going to be at each other's throats!

GM: Which brings us to you, Mick, spending thirty-eight points, but the only one who bid in every Attribute Auction.

Kevin: Wow! Look at all the points that Mick wasted! He didn't get first place in anything.

GM: It's not necessarily a waste. After all, he bid in every category, but he ended up spending fewer points than you. Which means he's ranked in everything, has no real vulnerable points, and doesn't have to go into debt with Bad Stuff in order to get Pattern.

Peggy: Yeah, Kevin, you're the one who screwed up! Fifty one points for Psyche, and you didn't even get first place.

Kevin: Hey, I didn't screw up! I stopped before the whole Psyche auction got out of hand. Besides, overall I spent fewer points than you did!

Willy: Pretty slick, Mick! You spent less than anybody and you're strong across the board.

GM: Less than anybody except Ted, who didn't spend a single point.

Peggy: Boo! Hisss...

Willy: Yeah, Ted, how are we supposed to trust you when you've got all those points saved up?

Ted: I've just got a little restraint!

GM: Bear in mind folks, that Ted may have kept his options open, but any Attribute he buys will be more expensive for him.

Beth: What do you mean, more expensive?

GM: Well, for example, you spent twenty-three points to get third place in Psyche. If Ted spends the same number of points, twenty-three, he'll be a half rank behind you. That's true of anything he buys up.

Willy: So Ted really ends up paying more than any of us?

GM: True, but let's move on to the next step. Start working up the rest of your characters, and spending your remaining points. I'll start seeing you privately as soon as you're ready...

	PSY	STR	END	WAR	Total Pts
Willy	1st [52]	NB	NB	NB	52
Kevin	2nd [51]	NB	NB	NB	51
Beth	3rd [23]	2nd [20]	NB	3rd [17]	60
Peggy	4th [20]	1st [24]	2nd [15]	NB	59
Alex	5th [11]	NB	1st [16]	1st [46]	73
Mick	6th [5]	3rd [12]	3rd [9]	4th [12]	38
Cindy	NB	4th [9]	NB	2nd [45]	54
Ted	NB	NB	NB	NB	0

The Potential of Warfare.

Early on, in one of the play-test sessions, one of the characters, cloaked with a magical invisibility, snuck up on one of the player characters who was a master in Warfare.

"I'm going to plunge my sword into his back," said the invisible one.

"He turns," replies the Game Master to the invisible character, "and parries your blade. As your swords clash, it's obvious he knows you're there. What are you doing?"

"How? I'm *invisible*!" protests the player, "How could he possibly parry my blade if he can't see me?"

"Well, he's awfully good," says the Game Master, "good enough so he assumes there are always invisible opponents around, and he always moves to counter them."

"What? Even if he thought there was somebody invisible around, how could he parry my invisible blade? How could he know where I was standing?"

"Tactics. When he imagines invisible opponents, he tactically figures out where they'd be standing, how they'd strike, and what would be the best, most efficient block for invisible blades."

Ridiculous?

Not if you think about in terms of *Amber's* best warrior. We know Benedict could, and probably would, be good enough to parry invisible blades. Any other player character who invests in Warfare is moving up to Benedict, gradually absorbing the skills that Benedict has mastered.

Any character, given sufficient Warfare, and advancement, *can* take on any of Benedict's special abilities.

Not every character would want to.

After all, living every minute of every day as if in the midst of a battle can be pretty tiring after awhile. And what's the point of being an Amberite if you can't kick back, relax, and let the good times roll?

So there will always be characters with high Warfare who can perform brilliantly in one-on-one combat, but who might never develop any "other" abilities.

Benedict's Warfare can be broken down into five categories. Tactical Vision, Leadership, Weapon Skills, Reaction Time, and Strategy. Here are some of the possible capacities a character can pick up in each category.

1. Tactical Vision. Benedict sees the world around him tactically. He automatically picks out the best spot for defense, attack, ambush, or flight. Knowing these areas, he then *automatically* moves to avoid being vulnerable.

- This works out as a resistance to surprise. Gradually the character comes to view all situations as potential combat situations, constantly reviewing the current tactical position. A character would say, "the guacamole dip is over there, the mariachi band is playing over there, those six windows, and these three doors, which means the trap door would be over there, a skilled warrior would pop in over there, and I'd be best off standing right about here."

- Benedict reacts, without surprise or pause, to any and all threats, no matter how well hidden. That's because Benedict goes through life looking around and expecting ambushes, snipers, and other nasty things. He even expects things out of nowhere, like from below ground, from out of the air, and he is always ready.

- Benedict also anticipates even invisible and intangible opponents. In other words, Benedict just naturally adjusts for possible enemies that can't be seen, blocking or parrying when it seems "right," even when there is no one around.

- All Benedict has to see is the stance, the outline of someone standing still, in order for him to spot a familiar person, or to recognize an opponent's style of combat.

- In one on one combat, Benedict can instantly size up a stranger, figuring out movements and reactions, their attitude, along with any hidden weapons or secret abilities. This doesn't mean that Benedict will know that the stranger is carrying a .45 automatic pistol, but he will know that the stranger has some kind of ranged weapon, probably to be used with the right hand, and that its effective range is somewhere between 10 and 50 feet.

- Based on scouting, troop placement, or even the arrangement of supplies, Benedict can recognize familiar opponents, see the strengths and weaknesses in any army, and make very accurate "guesses" about what they will do and how they will react.

- Benedict can look at an opponent's tactics on the battlefield, and figure out any and all secret weapons that may be held in reserve, their probable effects, where they are kept and where they anticipate being used, and how the enemy's strategy will change.

2. Leadership. Benedict knows how to lead soldiers. "The moral is to the physical as three to one," said Napoleon, something he might have learned from Benedict.

- Benedict can at least triple the effectiveness of any force using group psychology and loyalty. He can make rousing speeches, convince troops of the superiority of their cause, and even rally badly beaten armies back to the fight.

• Just as Benedict can manipulate his own army, his mastery of the soldierly mind is such that he can subvert an enemy force. He knows how to break their morale, and drive them from the field of battle, using psychological tricks, playing on their fears, and timing advances and attacks.

• Raising and training armies is second nature to Benedict. He can recruit troops instantly, send propaganda rippling through a population and, given enough fresh candidates, he can build an army of any size. He also is expert at picking out and training instructors, so that any army can be trained and ready for battle in a matter of days.

3. Weapon Skills. Benedict, when using a rapid fire long-range weapon, like an automatic rifle, can pick off targets as fast as he can look and pull a trigger.

• Benedict simply knows every weapon that has, does, or ever will exist.

• Anything can be used as a deadly weapon by Benedict. A pencil, a butter knife, or a sharpened twig can be used to thrust. Anything from a piece of sharpened rock, to the edge of a piece of paper can be used as a cutting weapon. Thrown projectile weapons can range from pebbles to objects as large as Benedict's Strength can handle. A piece of cloth, a length of rope, or virtually anything else can be used by Benedict as a means of damaging or killing the enemy.

• Benedict visualizes all possible new weapons, so he is never surprised, and never at a loss at how to apply a new technology.

4. Reaction Time. means that Warfare serves as a measure of how fast a character can be.

• Benedict is too fast for any spell casting that takes more than a moment. In other words, if the spell takes even one lynchpin, Benedict can, if within arm's reach, or knife's toss, interrupt the casting of any spell.

• Benedict, unless he's somehow restrained, trapped, or hit with an inescapable barrage, can dodge all incoming long-range fire.

5. Strategy. Warfare doesn't just cover sword battles and rifle shots. It also covers strategy and tactics, for everything from chess games to continental wars.

• Benedict can master any strategic game. It is possible that a lifetime master, someone who has devoted every waking hour to the game, could beat Benedict at chess or Go. It's just not likely.

• Benedict can cheat the "fog of war," collecting superior information, flanking any enemy, surprising any enemy, and always outdoing the enemy in the battle of will or morale. This means Benedict can defeat armies vastly larger than his own.

AUCTION RECORD WORKSHEET

Player/Character	Psyche	Strength	Endurance	Warfare	Total Points
1					
2					
3					
4					
5					
6					
7					
8					
9					
10					

CREATING A CHARACTER

Once the Attribute Auction is complete, it's time for the players to put together the rest of their character. Among other things, you can "buy up" your Attributes and obtain personal artifacts, creatures and Shadows. However, the most important decision now is which of the Powers you'll get for your character.

POWERS

The most expensive point purchases are Powers. Each Power provides the character with a basis for manipulating Shadow and the forces of the Amber universe.

There are two *Primal Powers*.

The great symbol of Order is **Pattern**. To have its power means being born with the Blood of Amber, to be a descendent of Oberon, and to be a member of the ruling family of Amber. It also confers immortality and a few other trifles. Basic Pattern Imprint is 50 Points, and the advanced form is 75.

In opposition to Amber there is the Courts of Chaos, and their sign, the writhing sygil of the **Logrus**. To gain this power is also to become one of the sprawling family that reigns in the dark realm. Logrus Mastery costs 45 Points, plus the power of Shape Shifting (another 35 points). Advanced Logrus Mastery is 70 Points.

If there is a third major power, then it is that of the **Trump**, the art of mystic images. It costs 40 Points. Advanced Trump Artistry, actually just the first step on a stairway of great power, is 60 Points.

Shape Shifting is not exactly a power, for it enables a character self-control, but not control of the wider universe. Shape Shifting is required for any Logrus user. Basic Shape Shifting, such as that practiced in the Courts of Chaos, is 35 Points. The advanced form costs 65 Points.

Of lesser powers there are many, of which Magical Powers are the most common. A character may choose from three types of Magic. **Power Words**, used mostly for defense against other magic, are but 10 Points. Spell-casting is known as **Sorcery**, and costs 15 Points. Creating creatures and things of magic requires **Conjuration**, a 20 Point power.

Caution on Advanced Powers. Amber is a learning game. It takes many, many hours of play time to master even the basic forms of power. There is a danger from jumping directly into the advanced forms, in that you may become power-heavy, and never learn to use the subtlety of the power. It's been proved over and over that an experienced player with a basic power can always overcome advanced powers wielded without skill.

Caution on Logrus Mastery. Taking Logrus in an Amber campaign has its hazards. Amber, and the Shadows of Amber, are hostile to the forces of Chaos. Becoming a Lord of Chaos in Amber is to accept the role of a stranger in an unfriendly land. A mere prick of Pattern, at the wrong time, can cause you to flare up and die in blue flame. Take Logrus if you must, but be forewarned that your Game Master has no obligation to make your life comfortable.

CHARACTER ITEMS

A sword of power, a steady horse, an enchanted hawk, a cup of magic and power. Anything you like can be "built" for your character. The only limitation is the number of points you wish to expend. For more details, consult the section on "constructing artifacts and creatures."

However, any item that you can imagine is also just sitting somewhere out in infinite Shadow. To get it, all you have to do is spend the time to Hellride, or, if you've got Logrus, reach for it.

So, why spend points at all?

First, when you spend points on an item it becomes an integral part of your character. Just as Greyswandir, Corwin's sword, is a part of him. So, having spent the points, you can always get it back if it's lost, stolen or destroyed.

Second, remember the insidious nature of the Amber universe. Anything worth having is likely to have connections with someone, or something, else. By spending points on an item, you are making sure that the *hooks* on an item are your hooks.

Shared Items. As with Shadows, players can join forces to create items. Since the multiplier, *Named & Numbered*, creates a number of items for merely double the base cost, it may be desirable to share some prized objects or beasts.

CHARACTER SHADOWS

In a universe of infinite Shadow, an Amberite can do more than just locating a perfect world. The world is as much created, or defined, as found. Although Shadows can be found in play time, without cost, prepared Shadows are often valuable. Note also that characters without Pattern may have no other way to obtain a Shadow of their desires, short of buying it with points. Each Shadow is constructed using the Shadow Construction Guide. There is no limit to the number of Shadows you can buy.

Shared Shadows. It's possible for any group of two or more players (otherwise known as a "cabal") to pool their points and construct a joint Shadow. Shadows created in this way require that each participant submit a separate description sheet to the Game Master. Points are contributed in which ever way the players wish. For example, for a six Point Shadow, one player could chip in one point, another two points, and the third would make up the remaining three points. Of course, the heavier a character's investment, the more control over the Shadow relative to the others.

> Develop your dream character: the one you have always wanted to play - the one you have always wanted to be.
>
> - Ron Miller

CHARACTER ALLIES

Your character enters the Amber Universe complete with background, family, and a history. For a few points you can arrange to have some of the character's prior history, there is the possibility of having staunch allies, friends and even doting relatives.

An Ally in Amber. While not a family member, an ally will be a familiar face in Amber. Examples are Court retainers, the royalty of nearby kingdoms, servants, sea captains, or soldiers in the forces of Amber. In any case, an Ally is always someone with the power to influence affairs. The Ally will be loyal to the character no matter what the current political situation and will always be willing to share information, and assist in any reasonable plans. **1 Point.**

Family Friend. These are members of the family of Amber, or of the Courts of Chaos, ranging from old Dworkin, down to young Merlin, who will regard your character as a friend. The friend is someone likely to go out of their way to help your character, to always look out for your welfare, and to tell you things they think you should know. **2 Points.**

Chaos Court Devotee. Although a Chaos Devotee is just as loyal as one from Amber, they are unlikely to be able to help the character quite so much, because of their lesser influence and power in places of Pattern. On the other hand, a Chaos Court Devotee is an outstanding asset should the character ever visit the Courts of Chaos. As with a devotee from Amber, this gives the character the heritage of the Logrus. Therefore, a character with a Chaos Court Devotee will be able, when the points are available, and the opportunity comes up, to become a Master of the Logrus. **4 Points.**

Amber Court Devotee. Same category of relatives as "Family Friends," except they don't just like you, they love you. Which means they go to an awful lot of trouble for you. For example, they will be likely to observe you secretly, and may appear when your character's life is in danger. Devotees will risk their own lives, betray others, or even change sides in a conflict, if necessary, to help your character. An Amber Court Devotee also means that your character is a direct descendent of someone with the blood of Amber. This is important as a "place holder" for a character without Pattern. It means that the character may eventually be able to Walk the Pattern, once the required points become available. **6 Points.**

Naming Court Friends and Devotees. Who will it be? Sorry, it's not up to you. Choosing your Family Friend is up to the Game Master. Worse yet, the Game Master isn't going to tell you who your character's friend really is. Amberites always fear that their friends and loved ones may be used against them, so they will be reluctant to ever reveal their true feelings.

If, by some chance, the Game Master does mention the name of your Family Friend, remember, sometimes the Game Master lies.

GOOD STUFF, BAD STUFF & ZERO STUFF

In Amber, characters make their own luck.

Or, to be more accurate, they *buy* their own luck.

As a player, you use "Stuff" to determine the character's luck, and karma. Stuff works in three ways.

First, Stuff can be used by players to adjust their character's points. Extra points can be spent on Good Stuff, while point shortages can be made up by accepting Bad Stuff.

Second, Stuff determines the character's luck. Good Stuff makes the character lucky, Bad Stuff unlucky, and, with Zero Stuff, the character has no particular advantage or disadvantage.

Third, Stuff is used as a way of declaring the "karma" or "alignment" of the character's personality. Good Stuff characters are in harmony with the universe, and tend to be friendly and loyal to their families. Bad Stuff characters are self-centered and tend to use those people who care about them. Zero Stuff characters are neutral, walking a middle way.

Stuff Quantity. What is a large amount of Bad Stuff or Good Stuff, how many points? It's relative, and there is no absolute answer. It depends on how many points all the players spend on Good Stuff and Bad Stuff. If you take all the players' Stuff scores, and end up with an average of six points of Good Stuff, then five points of Bad Stuff will be a lot, but eight points of Good Stuff might be only a relatively small amount.

Good Stuff. Characters with Good Stuff will tend to have good luck. In combat, when characters take risks, Good Stuff will mean they are more likely to succeed. The more Good Stuff, the luckier the character. Anytime an event has a "chance" element, Good Stuff will push the result to the advantage of the character.

Good Stuff Attitude. First impressions count, and when you're holding Good Stuff, people will tend to take a shine to you. Taking Good Stuff means declaring that your character is basically a "good guy."

Good Stuff Encounters. As the Amber game is played, the Game Master is always designing "encounters" for each of the characters. Good Stuff means that the events, contacts, and other encounters will tend to be helpful and beneficial for the character.

Why some players love Good Stuff! No more bad rolls! They never have to worry about things turning out badly just because some regular solids roll badly on a flat surface. Having Good Stuff means being naturally lucky, being instantly liked, and always getting the benefit of the doubt.

Bad Stuff. The only advantage of Bad Stuff is that you get points for taking it.

How Bad is Bad? It's really bad. Every time something is left up to chance, your Game Master is going to have a look at your Stuff. Any Bad Stuff, and the Game Master is going to assume that "chance" just took a turn for the worst.

Bad Stuff Attitude. Anyone with Bad Stuff is going to have to overcome their first impressions. There's something sinister about a character with Bad Stuff, something vaguely repellent. The more Bad Stuff the worse it is. Of course, it isn't permanent. You can always talk your way into a better reception, it's just an uphill fight when your Bad Stuff puts a sour note into your first introduction.

Bad Stuff Encounters. When the Game Master is looking around for a fall guy, somebody to find out something unpleasant, or for someone to experience the first attack from an unknown enemy, guess who the Game Master is going to pick? Right, the one with Bad Stuff. This isn't as bad as it sounds. Since bad things happen to people with Bad Stuff, it can be fun. After all, the more points of Bad Stuff you've got, the more exciting your character's life will become.

BAD STUFF WARNING! First, you never run out of Bad Stuff. As long as the character "owes" points, the Bad Stuff will just keep coming and coming. It doesn't matter that ten minutes ago your character just had three outrageously bad breaks, because Bad Stuff doesn't get used up and the Game Master can apply it all over again. Second, don't forget that Bad Stuff is also an indicator of how bad a wound can be. Enough Bad Stuff, at the wrong time in a combat, and the character could end up dead.

Bad Stuff Combat. Ooh. The worst aspect of Bad Stuff is in combat. You won't even know about all the bad breaks you'll be getting. The mistake the other guy could have made, but didn't. A split decision that turns against you because of a run of misfortune. Yes, you've got to be pretty skilled to make up for a heavy dose of Bad Stuff.

Why some players like Bad Stuff! Sure, they've got to take a few falls from time to time, and they do end up with hard knocks, but Bad Stuff has three advantages. First, the game is more exciting, because the Game Master throws more surprises, traps and attacks at folks with Bad Stuff. Second, a character with Bad Stuff has a sinister, evil appearance, just the right kinda guy for some players! Third, those extra points can be used for making a more powerful character.

Zero Stuff. Neutrality is kind of a strange topic, isn't it? But, if you've avoided getting Good Stuff, or Bad Stuff, that makes you Neutral, a Zero Stuff character. Zero Stuff characters tend to be very well balanced, and are often the most Power-oriented characters in the game. All well and good, but remember that Zero Stuff characters have also set themselves up with a certain attitude, a kind of "wait and see" approach.

Why some players like Zero Stuff. There is an air of mystery about a character with Zero Stuff. Also, characters with Zero Stuff are built just right, with no points missing, and no points left over. Plus, with Zero Stuff, every encounter starts from neutral, and the character has the optimum range of choices.

How Good Stuff, Zero Stuff & Bad Stuff Works. The secret is for the Game Master to picture any situation and figure out whether a character's luck might play a role. In the following scene, a character is performing a difficult maneuver.

GM: The hounds are baying behind you, and the hunters with the spears are getting pretty close. Looking down over the edge of the cliff, you see a tiny ledge, obviously ice-covered like the rest of the mountain side. What are you doing?
Player: I'll take the chance, I jump down.

What happens next can depend on the character's "Stuff." a variety of factors. Here, ranging from lots of Good Stuff, to lots of Bad Stuff, are five options.

To Yvonne, a character with a lot of Good Stuff:
GM: Tripping forward, you stagger, miraculously slipping under the spear point. That bit of fortunate missed footing has put you in easy range of the enemy's belly. You can strike with your sword, punch him right in the gut, or toss him over the edge. What are you doing?

To Farley, with a little bit of Good Stuff:
GM: You land on the ledge, sliding a bit as one foot goes off the edge, but you catch hold of a solid bit of rock. What are you doing?

To Dorell, a character with Zero Stuff:
GM: You sprawl down on the ice-covered ledge, your feet going out from under you, one boot goes over the edge, but your hand brushes a tip of stone, catching just a finger on the slippery ice. Your grip won't last another second! What are you doing?

To Garvin, a character with a bit of Bad Stuff:
GM: You land on the ledge, but your feet slide over the edge, twisting you and knocking you flat on your stomach. You are continuing to slide off, your legs flailing in empty space, watching the ice slide by. Frantic, your fingers find no purchase on the slippery ice. You're a man on his way to oblivion, what are you doing about it?

To Harick, a character just loaded with Bad Stuff:
GM: Whoops! You land hard on your left, and a spasm of pain from your ankle reaches you just as you slide clear off the ledge, your palms and fingers go bloody from trying to get a handhold on the jagged rock edge. You are making the long fall down toward the noisy water. What are you doing before you hit?

How Good Stuff, Zero Stuff & Bad Stuff Affect the Attitude of Others. Here's another example of how the universe reacts to a character's "Stuff," where the setting is a first meeting with an unknown elder relative.

GM: Stepping through the mysterious Trump, you feel that there was some kind of barrier in the place, a barrier that may have just been dispelled by the use of the card. Looking around, across the room you see an armored man seated on an iron chair.

Player: I'll cautiously step closer.

GM: Without opening his eyes, the man brushes away cobwebs from his face. He uses the back of his left hand to do this, and you get a good look at the palm side of the hand. Strong, blunt fingers, badly in need of a manicure, and the center of his palm is stained with a blotch of something dark and brown and spidery. What are you doing?

Player: I'll say, "Who are you?"

GM: He opens his eyes, which are bright and hard, and young and green. He looks you over. His voice is deep and scratchy when he repeats your question. "Who are you?" he says.

Player: I give him my name, "I'm a scion of Amber," I say, "and I would have your name."

Now the response to the character's question would vary, and here are the variations, ranging from a lot of Good Stuff, to a lot of Bad Stuff.

To Yvonne, a character with a lot of Good Stuff:

GM: He chuckles. Green eyes twinkle. "It has been a long time since I've seen such a brash youngling. Yvonne, I am Finndo, your kinsman. If you would have my favor, tell me of Amber."

To Farley, with a little bit of Good Stuff:

GM: His features ease a touch, the green eyes lose a bit of their hard glare. He says, "Long ago I was known as Finndo. You may call me that, if you wish."

To Dorell, a character with Zero Stuff:

GM: He answers slowly, cautiously, his green eyes neutral. "I am an old man, an Amberite. Why do you ask?"

To Garvin, a character with a bit of Bad Stuff:

GM: The green eyes grow hard. "Never heard of you," he says. And he says nothing more.

To Harick, a character just loaded with Bad Stuff:

GM: He squints. His sword slides noisily out of its scabbard, green sparks flickering as metal scrapes metal. "You go too far, and you remind me of ancient days, and the betrayal of a kinsman."

How Good Stuff, Zero Stuff & Bad Stuff Colors the Character's Perceptions.

We each look at the world differently, filtering everything through our attitudes and beliefs. There's an old saying that optimists will see a glass as half full, which a pessimist would see the glass as half empty. The more Good Stuff a character has, the more a glass will seem to be full, and the more Bad Stuff, the emptier the glass appears. Here is an example of what our five sample characters might see in the same situation, this time where the non-player character only reacts to the player character's actions.

GM: The smells, sights and sounds of this part of Amber all tell you that the sea is nearby, and that here is a place for the common sailor to relax informally. In other words, there are more brothels and bars than you can shake a stick at.

Player: This looks like the right place to ask about the *Goldenrod*. I'll walk into a nearest bar.

GM: Okay, once inside, after the moment it takes your eyes to adjust to the gloom and the tobacco smoke, you see three sailors seated at a table, an old man in merchant's robes off by himself in the far corner. A woman sits behind the bar, her face hidden by the book she seems to be reading. In a back room you can hear the sounds of water sloshing around. What are you doing?

Player: I'll walk up to the bar, and throw a coin on the counter, and ask for a beer.

Now what each player character sees varies according to their perceptions. In this case the woman behind the bar is a customer, and someone of importance, just filling in for the real barmaid who is doing some washing in the back room. The sailors are no threat, just a jolly bunch out for a good time. They'll fight if insulted, but will be friendly to anyone who treats them well. The merchant in the corner is a mage, hostile to Amber, and a danger to any player character. The more Good Stuff a character has, the clearer the true picture of what is happening.

To Yvonne, a character with a lot of Good Stuff:

GM: The woman puts down her book, and looks at you perceptively as she reaches for a glass. She gives a little smile and you realize that she is as out of place here as a thoroughbred champion in a glue factory.

The sailors seem jolly and friendly, but you get the impression that the robed figure off in the corner is sizing you up.

"Light beer or dark?" the woman asks, in an educated tone. The woman obviously has a resemblance to the Amber royal family. What are you doing?

To Farley, with a little bit of Good Stuff:

GM: The woman puts down her book, looking at you thoughtfully as she reaches for a glass. She seems kind of high class for this joint.

Behind you the sailors are sharing some joke, and the robed figure off in the corner is looking you over.

"Light beer or dark?" the woman asks, in a pleasant tone of voice. What are you doing?

To Dorell, a character with Zero Stuff:

GM: The woman puts down her book, looks at you, and reaches for a glass.

The sailors are acting a bit rowdy, and the robed figure off in the corner is quiet.

"Light beer or dark?" the woman asks. What are you doing?

To Garvin, a character with a bit of Bad Stuff:

GM: The woman lowers her book, looks coolly at you, and reaches for a glass.

A couple of the sailors look over at you, point, and the group erupts in noisy laughter.

"Light beer or dark?" the woman asks, disrespectfully.

You notice she didn't bother calling you sir. What are you doing?

To Harick, a character just loaded with Bad Stuff:
GM: The barmaid, still holding her book, gives you a surly look, and slowly reaches for a glass. Obviously the service around here leaves something to be desired.

"Light beer or dark?" she asks, in a haughty tone, as if you were some scum off the street. As she says it the sailors erupt into laughter, obviously sharing a joke at your expense. What are you doing?

PLAYER CONTRIBUTIONS

Player Contributions. If your character is becoming part of a campaign, as opposed to a game that's just going to take a session or two, then there is another way to get some extra points. You can do something for the game.

The Game Master of an Amber game has to do a lot of work, a lot of intense planning, and has to engage in challenging game mastering skills. So, it's only fair that the players be set up with the opportunity to do their fair share.

Here are some of the most common forms of Player Contribution, the "rules" for applying them, and how many points they're worth:

Diary. This is a write-up of the game, written by the player, from the point of view of the player's character. Not only does this provide excellent documentation, but it serves to "fix" the character in the mind of the player, allowing for a better handle on the persona. This doesn't have to a big deal. You don't have to do more than scribble a page or two of handwritten notes in each play session. On the other hand, many players have done marvelous things with their notes. 10 Points.

Campaign Log. The idea is to write up notes for everybody, and then handing out copies for the whole group. This is usually a public account of the game's events. 10 Points.

Trumps. For the artistic player, the assignment is to draw "Trump" of each of the player characters in the game. GM and players alike enjoy having actual trump of the various player (and non-player) characters. Of course, the more the merrier, and each talented player is encouraged to come up with his or her own version of the trump. Usually the player will have to do one drawing for each game session. 10 Points.

Amber Stories. If the player likes writing, especially fictional writing, then this is the opportunity to write Amber stories. The idea here is to write about your character, or about other characters in the game. Either a story, or a chapter, is required for each game session. 10 Points.

Poetry. Poems, based on the events in the game, or on the characters' quirks. About a page, or about twenty lines, is required for every game session. 10 Points.

Other than the contributions listed above, the Game Master can also offer points (say, 5 Points) for things like typing, photocopying, or record-keeping.

No matter what the Player Contribution, the Game Master should get something on a regular basis, once per gaming session. If a player is late, the Game Master is entitled to start turning the bonus points into Bad Stuff for the character.

NOTE: No player can get more than twenty (20) points for Player Contributions, no matter how great the effort! Repeat, there is an absolute maximum of 20 points!

SAMPLE CHARACTERS

Example of a fatally flawed character: Kevin's Character Roderick

We'll start by looking at what happens when a player shows the Game Master a character with problems. Here's what Kevin first presented to the Game Master:

Kevin - **RODERICK** - First Version

PSYCHE: 2nd [51 Points]
STRENGTH: Human [+25 Points]
ENDURANCE: Amber Rank
WARFARE: Human [+25 Points]

    1 Point Total in Attributes
    Advanced Logrus Mastery [70 Points]
    Shape Shift [35 Points]
    STEVE - The Demon Skateboard [4 Points]
        Double Normal Speed [2 Points]
        Shadow Path [2 Points]
    Trump Contribution [+10 Points]

Kevin: So, what do you think?
GM: I think it looks really bad.
Kevin: What's wrong?
GM: Well, for starters, you're going down to Human in both Strength and Warfare. As far as I'm concerned, that's suicide for your character.
Kevin: Why?
GM: First off, you can't protect yourself. If you get in either a hand-to-hand fight, or any kind of fight with weapons, with just about anybody really good, you'll be dead. Instantly.
Kevin: What about my Psyche, I've got 51 points and second place. Won't that protect me?
GM: If someone gets their hands on you? No. And your Psyche won't help you when someone levels a sword or a gun at your belly.
Kevin: So I'll avoid getting into fights.
GM: That's just the problem. I don't see how you can avoid it. You're taking Logrus!
Kevin: What's wrong with Logrus?
GM: Well, most of my campaign is going to take place in and around Amber. Where they're not too fond of Logrus Masters. You'll be setting off alarm bells and getting into

fights all the time.

Kevin: So you think I should switch my Endurance and Warfare? So I'll be able to fight better?

GM: Just a second. Let's talk about your character. What kind of guy is he?

Kevin: I see him as really easy going. He just wants to go hip swinging on his board.

GM: So he's not really interested in combat?

Kevin: No way. That's not Roderick's scene.

GM: Then I don't think you should just switch around your Endurance and Warfare. Warfare isn't something that interests you. Likewise, I question your picking Advanced Logrus.

Kevin: Why, isn't it a lot of power?

GM: Yeah, but it's also a lot of work. Taking Advanced Logrus would make sense for a studious, hard-working character. You're describing a guy who just wants to have fun.

Kevin: So what are you saying? What do you want me to do?

GM: Kevin, it's your character, I can't tell you what to do. I just want to point out your options. Think about this. If you leave your Psyche at fifty-one, and drop all three of your other Attributes to Chaos...

Kevin: That seems okay, 'cause I'm from Chaos...

GM: That means you'll get thirty points back, which means you'll have spent twenty-one points total on Attributes. Now, buy Shape Shifting for thirty-five, and regular Logrus Mastery for forty-five, and you end up spending eighty for Powers. Eighty plus twenty-one equals one hundred and one. With your Trump Contribution you still have nine points left over for other things.

Kevin: I see what you mean. That makes sense.

GM: How about Pattern. How come you didn't take Pattern?

Kevin: Well, I would have, but Logrus cost too much.

GM: Does that mean you'd like to have Pattern someday?

Kevin: Yeah, sure. Eventually.

GM: Then you better think about getting an Amber Devotee. It's only six points. You'll still have three points left, and that way you can pick up Pattern sometime in the future.

Kevin: But what about Stevie, my skateboard? It costs more than three points. And I don't want to get any Bad Stuff.

GM: Hmmm. Why don't you tell me what you really want from your skateboard. Not in terms of points or qualities. What do you want it to be able to do?

Kevin: Mostly I want to be able to skate in really neat places.

GM: But don't you really want to be able to skate with neat tricks, like over ramps and curves and the like?

Kevin: Yeah! That's the idea.

GM: Okay, here's an idea. Why don't you put the power, *Able to Mold Shadow Stuff*, into your skateboard? It's only one point, and you could set it up so that the board would warp the ground where it's travelling.

Kevin: Oh cool! Like everything would get smooth, and there would be half-ramps, and neat curls! I wouldn't need extra speed in the board, because there would always be inclines for accelerating. And it would work everywhere. I like it.

GM: Sounds like we've talked about everything. Go away, think about what we've talked about, and rework your character. We'll get together again later.

After thinking it over, Kevin returns with the following, revised, character description:

Kevin - **RODERICK** - Final Version

PSYCHE: 2nd [51 Points]
STRENGTH: Chaos Rank [+10 Points]
ENDURANCE: Chaos Rank [+10 Points]
WARFARE: Chaos Rank [+10 Points]

    21 Total Points in Attributes
    Logrus Mastery [45 Points]
    Shape Shift [35 Points]
    STEVE - The Demon Skateboard - Creature [2 Points]
       Psychic Sensitivity [1 Point]
       Able to "Mold" Shadow Stuff [1 Point]
    Amber Court Devotee [6 Points]
    Good Stuff [1 Point]
    Trump Contribution [+10 Points]

GM: Hey, one hundred percent better!

Kevin: I figure I'm covering all my bases. I've still got Logrus, which has always intrigued me. I'm covered if I ever get the points to buy Pattern, because I've got the blood of some close relative with the blood of Amber. As far as Warfare and Strength are concerned, I'm not planning on getting that close to any enemies. Which, by the way, is why I spent the point on Good Stuff, so I'll have a break to get away from any inconvenient sword play.

GM: Good going!

Example of a player working through some options: Willy's Character Garvin.

In this case Willy has to make some critical decisions before working up his character. The Game Master points out some of the possibilities of Creature Creation.

GM: So where's your character sheet?

Willy: I'm not quite ready. I want to ask you about something first.

GM: Go ahead.

Willy: What if I want a bunch of items...

GM: Named and Numbered?

Willy: Yeah, like a half dozen...

GM: Yes, that would be Named and Numbered.

Willy: Well, can I make them different?

GM: Different how? Like each one is a different color? I don't have a problem with that. There's nothing that says each item has to look exactly the same...

Willy: No! Can I give them, like, different qualities and powers?

GM: You mean spend the points differently?

Willy: Yes. So long as they're all the same number, like, eight points, can I just pay double, sixteen points, for named and numbered?

GM: Let me get this straight, you want to put together a bunch of items, each different, and then say they're Named and Numbered?

Willy: Yeah. Like I'll have these six cats, and they'll each be eight point items, and then...

GM: NO!

Willy: Why not?

GM: To use Named and Numbered, or any multiplier, the items have to have the same number of points, spent in exactly the same way.

Willy: So I can't have Named and Numbered cats that do

different things.

GM: I didn't say that.

Willy: Yes you did. I just heard you...

GM: I said you can't use Named and Numbered on things that are different. That doesn't mean Named and Numbered can't be changed afterwards. You can make Named and Numbered items, artifacts or creatures, with different points, as long as you do it in the right order.

Willy: So how do I do it?

GM: Okay, first you figure out what things, what qualities and powers, that you want all the items to have. For example, well, was there something you wanted all the cats to have?

Willy: Yeah. I wanted them all to be able to communicate mentally.

GM: You mean Psychic Communication?

Willy: Uh huh, but presence, not just touch.

GM: That's a two pointer. Anything else?

Willy: Hmmm. It would be good if they were all, hmmm, Resistant to Firearms. That's another two points.

GM: Okay, so now we've got your basic cat, which costs four points. Double that, to eight points, and you've got Named and Numbered cats. How many did you want?

Willy: Oh, six would be okay.

GM: There are these six Named and Numbered cats, all the same.

Willy: And they've each got Psychic Communication and Resistance to Firearms. But they're all the same. That's not what I wanted.

GM: Wait a minute. Now you can customize them by spending points on each individual cat. Give me an example, what might one of them do?

Willy: I'd like one to be able to follow through Shadow.

GM: Okay, you'd spend the points for just that one cat. It wouldn't help any of the others, but that one cat could Follow Through Shadow.

Willy: Which means I'll have spent eight for all the cats, plus another two points, for a total of ten.

GM: Right! And, since you were talking about spending sixteen points anyway, that would leave you another six points to spend on the individual cats.

Willy: I'd have to spread the points out over the other five?

GM: No, it's up to you. You could spend all the rest on that first cat. What the heck, take as many points as you want!

Willy: As many as I want? Great!

GM: Yeah, all you've got to do is trade for Bad Stuff...

Willy: No thanks!

Willy goes off to work on the character, eventually returning with the following version:

Willy - GARVIN

PSYCHE: 1st [52 Points]
STRENGTH: Chaos Rank [+10 Points]
ENDURANCE: Amber Rank
WARFARE: Amber Rank

    42 Total Points in Attributes
Advanced Pattern Imprint [75 Points]
Personal Shadow [1 Points]
GREYMER - Lynx - Creature [2 Points]
    Sensitivity to Danger [2 Point]

Bad Stuff [+1 Point]
Campaign Log [+10 Points]
Personal Diary [+10 Points]

GM: You sure you want to commit for 20 Points of contributions? The Campaign Log will probably be a lot of work.

Willy: I want heavy duty power. There's no point in spending 52 points for first place in Psyche unless I've got something to power with it. From my point of view, Pattern is the most potent power, and I want the most advanced form of it. Strength is no big deal to me.

GM: And the point of Bad Stuff?

Willy: My way of telling the universe to get out of the way, Garvin is coming! I know it means I'll have a few unlucky breaks, but I like the idea of everybody treating me with respect and a bit of fear.

GM: What happened to your cats?

Willy: I changed my mind. When I decided on Advanced Pattern it turned out I didn't have enough points.

GM: So you've got a Lynx?

Willy: I'm calling him Greymer. I like the idea of something nasty, but not too big, and able to warn me about trouble. I figure a Lynx already has great natural senses, so adding the Psychic stuff should make him really hard to surprise.

GM: Sounds reasonable. What are you doing with your Shadow?

Willy: I'm not sure yet, I just know I want one. Do I have to tell you now?

GM: No, take your time. Just make sure I get the write-up on it soon, before we actually start playing.

Willy: How much detail do you want?

GM: That's up to you. The better you describe your Shadow, the closer it will be to what you want. After all, any details you leave out are things I can fill in for my own purposes.

Example of Giving in to a Player Concept: Alex's Character Harick.

In spite of the Game Master's best advice, some players will continue to stick with their original conceptions for a character. There's nothing wrong with that, just as long as the player knows the character's limitations and weaknesses.

Alex - HARICK

PSYCHE: 5th [11 Points]
STRENGTH: Amber Rank
ENDURANCE: 1st [16 Points]
WARFARE: 1st [46 Points]

    73 Total Points in Attributes
Shape Shift [35 Points]
Power Words [10 Points]
Bad Stuff [8 Points]
Personal Diary [+10 Points]

GM: Okay, here's how I see it. You could drop your Strength to Chaos, reducing your Attribute Points to sixty-three, the Diary would take it down to fifty-three, add fifty points for Pattern, and you're at one hundred and three. Then, either take three points of Bad Stuff, or we can figure out something else you can do for me.

Alex: Nope. I want Shape Shift. Non-negotiable.

GM: Well... Same plan, only take Logrus. That's forty-five instead of fifty for Pattern. Plus the thirty-five for Shape Shift. Makes your total, um, ich! one hundred and thirty-three! It's trouble, but you could still take Strength down to Human, and take Bad Stuff...

Alex: Naw. It's not bad, but it's not my idea.

GM: Why not?

Alex: I don't want my Strength going down. I'm mainly a fighter, and I don't need that kind of weakness.

GM: Alex, I still think you'd be better off with Pattern or Logrus. Look, take your seventy-three for Attributes, subtract the ten for the diary and that leaves sixty-three. Add Pattern for fifty and you've only got one hundred and thirteen. You're already taking eight points of Bad Stuff, so that means we're only, hmmm, five points apart here. Then you'd be more likely to survive.

Alex: Hey, I'm first in Warfare, and Endurance. I have no weaknesses. Shape Shift plus Power Words. Sounds like a really strong combination to me.

GM: Yeah, I'll have to admit, Harick is pretty formidable. But what about your Bad Stuff? Eight is a lot of Bad Stuff!

Alex: First, I don't want Harick to be a good guy. He's one mean customer. And second, let them try. I think Harick is tough enough for all comers!

GM: How about getting around in Shadow? Won't Harick be pretty limited?

Alex: Well, I've thought about that. For the time being, I guess I'll have to depend on other people. However, when I can, I'm planning on getting Advanced Shape Shift, so I can Shape Shift into a form that can move through Shadow.

GM: Have you thought about getting a Court Devotee, either Amber or Courts of Chaos, so you can be sure of getting Pattern or Logrus later on?

Alex: I've thought about it. And I don't care. I want to choose a different path for Harick.

Example of Sound Character Conception: Cindy's Charcter Dorell.

Cindy has a solid character concept. The Game Master just questions her on a few details.

Cindy - **DORELL**

PSYCHE: Amber Rank
STRENGTH: 4th [9 Points]
ENDURANCE: Amber Rank
WARFARE: 2nd [45 Points]

    54 Total Points in Attributes
    Pattern Imprint [50 Points]
    DRAGON-TOOTH - Sword [6 Points]
        Doubling Damage [2 Points]
        Sensitivity to Danger [2 Points]
        Can Speak and Sing [2 Points]
    Zero Stuff [0 Points]
    Personal Diary [+10 Points]

GM: Looks good. Why don't you tell me a little about your ideas for Dorell's personality.

Cindy: I see Dorell as a classic Amberite. Bigger than life and twice as loud. She's a fighter, but also a romantic.

GM: You sure you want to spend six whole points on a sword? A couple of points of Good Stuff can make your life a lot easier...

Cindy: I think that Dorell is basically a good person. But she's arrogant and egotistical, so I think that makes her a Zero Stuff character. Besides, I'm willing to make my own luck.

Example of Sound Character Conception: Peggy's Character Iresa.

Peggy walks in with her character perfectly thought out. While the Game Master might question some decisions, there's no real reason to change anything.

Peggy - **IRESA**

PSYCHE: 4th [20 Points]
STRENGTH: 1st [24 Points]
ENDURANCE: 1.5 [16 Points]
WARFARE: Chaos Rank [+10 Points]

    50 Total Points in Attributes
    Pattern Imprint [50 Points]
    Zero Stuff [0 Points]

GM: I see you're buying up on Endurance.

Peggy: It only cost a point, and I want to be able to surprise Alex.

GM: Even though you spent the same number of points, sixteen, he's still going to be ahead of you in Endurance. After all, he won the auction.

Peggy: Well, at least nobody else is going to sneak over me. The best they can do is match me.

GM: What about your selling down in Warfare? With your Strength and other scores, don't you think you should keep up the Warfare?

Peggy: Yeah, I know. But I needed the points.

GM: Why don't you just contribute a character diary or something? Then you would...

Peggy: Sorry, I don't want to do any extra work. I just want a solid character. Besides, I like Iresa! Simple, but functional. Like a female Gérard.

GM: Well, I guess the characters don't have to all be complicated.

Example of Helping a Player's Ideas: Mick's Character Farley.

Mick comes in with a solid character. No weaknesses or problems. It's just that what Mick says about his character Farley doesn't match what he's done with the points.

Mick - **FARLEY** - First Version

PSYCHE: 6th [5 Points]
STRENGTH: 3rd [12 Points]
ENDURANCE: 3rd [9 Points]
WARFARE: 4th [12 Points]

    38 Total Points in Attributes
    Advanced Pattern Imprint [75 Points]
    Bad Stuff [3 Points]
    Personal Diary [+10 Points]

GM: Well, except for the Bad Stuff, I don't see any problems here.

Mick: Yeah, putting Farley together was pretty easy.

GM: So, tell me about Farley...

Mick: I kinda see Farley as young and inexperienced. Sort of a kid in Amber. He's got a lot to learn, but he doesn't have any real weaknesses.

GM: Young? Does that mean you want Farley to be carefree? Sort of casual?

Mick: Yeah, the kind of guy who could go out looking for the Shadow with the perfect beach.

GM: I don't get it. You bought Advanced Pattern. That means the character is going to be a lot more studious, somebody serious and intent on gaining power.

Mick: So? Doesn't it also give him a lot of power?

GM: Mick, you're telling me you want a fun character. How does Advanced Pattern fit with Farley being a barrel of laughs?

Mick: Hmmm. I guess I see what you mean.

GM: Also, it sounds like you want Mick to be innocent and nice. Three points of Bad Stuff is telling the universe that Farley is rotten and evil. Look, you could just get regular Pattern, trade in your Bad Stuff for a few points of Good Stuff, and still do some interesting things.

Mick: I don't know...

GM: All I'm asking you to do is think about it. If you want Farley as it stands okay. For now take your character sheet and come back later.

Mick comes back later, with the following version:

Mick - **FARLEY** - Final Version

PSYCHE: 5.5 [11 Points]
STRENGTH: 3rd [12 Points]
ENDURANCE: 3rd [9 Points]
WARFARE: 4th [12 Points]

    44 Total Points in Attributes
    Pattern Imprint [50 Points]
    Courts of Chaos Devotee [4 Points]
    Good Stuff [2 Points]

GM: Wow! I guess you changed your mind.

Mick: I'm happy we changed Farley around. Now he seems a lot younger. A little lucky, but not a lot. And a nice guy. I figure he can develop big deal powers later, when we see how Farley turns out in play.

GM: So what's with spending four Points for a Courts of Chaos Devotee?

Mick: Easy to explain. I've read the books, I know how important an ally in the right place can be! Also, I want to keep my options open in case I have the chance to explore new powers.

GM: Fair enough. Looks like a solid character to me. Why did you knock out the diary? Don't you want the extra points?

Mick: Nah. I like Farley just like this. But I'll probably do a diary for him anyway, just because.

GM: Mick, you're too good to be true.

Mick: No, not really, if I like writing a diary for Farley, then next time, when we get a character advancement, that's

when I'll expect the extra points.

Example of Negotiating Player Contributions: Beth's Character Yvonne

Beth comes to the Game Master with a pretty solid character. She just wants a few more points to work with.

Beth - **YVONNE**

PSYCHE: 3rd [23 Points]
STRENGTH: 2nd [20 Points]
ENDURANCE: Amber Rank
WARFARE: 3rd [17 Points]

    60 Total Points in Attributes
    Pattern Imprint [50 Points]
    Shadow Coza [3 Points]
        Shadow Type: Personal Shadow [1 Point]
        Shadow Barrier: Restricted Access [2 Points]
    Good Stuff [7 Points]
    Personal Diary [+10 Points]
    Poetry [+10 Points]

GM: Yvonne looks like a fine character. Why you want to do so much work just to get seven point of Good Stuff?

Beth: I want tons of good luck. In dice games I always roll lousy. In Amber, I want to have everything fall my way, all the time!

GM: You want to write poetry? Exactly what is that going to involve?

Beth: My idea is to write a poem for every session, either a piece on something that happens in the campaign, or else something on Amber in general.

GM: Okay, but you'll have to supply copies to every player.

Beth: I can manage that.

GM: Now, what about the Shadow?

Beth: I see Coza as being Yvonne's private vacation land. Hmmm. That reminds me, can I make Coza a heavy gravity planet?

GM: Sure, it's your Shadow, you can do anything you want.

Beth: Good! Because that way anyone with lesser Strength than mine is going to have a real problem there.

Example of a Player with Secrets: Ted's Character Ariel.

Since Ted didn't bid in any of the Attribute Auctions, he starts out being able to shift his Attributes around at will.

Ted - **ARIEL**

PSYCHE: 6.5 [5 Points]
STRENGTH: 4.5 [9 Points]
ENDURANCE: 3.5 [9 Points]
WARFARE: Amber Rank

    23 Total Points in Attributes
    Pattern Imprint [50 Points]
    Trump Artistry [40 Points]
    Good Stuff [2 Points]
    Personal Diary [+10 Points]
    Game Master Assistance - Recording the Sessions [+5 Points]

GM: You didn't bid in any of the Attribute Auctions, but you bought up in Psyche, Strength and Endurance.

Ted: Yeah, I'm really pleased. I get rankings in three of the four Attributes.

GM: If you had bid in those things you'd be ranked higher. For example, if you'd bid nine points in Endurance you'd be third instead of just behind it.

Ted: Well, I wanted to keep my options open.

GM: Okay, so what about your Powers?

Ted: I've got both Pattern and Trump. That's what I really wanted. I've got no weaknesses, and I've still got two points left over to spend on Good Stuff.

GM: Which gives you two extra points if I okay your getting five points for recording the sessions. Otherwise you're three points into Bad Stuff.

Ted: Aw, come on. You know that Willy needs a tape if he's going to do a good job with the campaign log.

GM: Okay, but you'll have to pass the tapes on to me when Willy is done with them. If you're going to get points for taping, the tapes will have to be a permanent part of the campaign collection.

Keeping Track of the Numbers.

Aside from being responsible for running the whole universe, the Game Master also has to keep track of player's character numbers.

To make sure that nothing gets lost or confused, and to help during character advancement, it's recommended that the Game Master and the player each keep a duplicate set of character sheets.

Another useful tool for the Game Master is an Attribute Table. For example, once the players in our example got done changing their points around, their Attribute table would look like this:

G/B	PLAYER	CHAR	PSY	STR	END	WAR
0/1	Willy	GARVIN	1st [52]	C	A	A
1/0	Kevin	RODERICK	2nd [51]	C	C	C
7/0	Beth	YVONNE	3rd [23]	2nd [20]	A	3rd [17]
0/0	Peggy	IRESA	4th [20]	1st [24]	1.5 [16]	C
0/8	Alex	HARICK	5th [11]	A	1st [16]	1st [46]
2/0	Mick	FARLEY	5.5 [11]	3rd [12]	3rd [9]	4th [12]
0/0	Cindy	DORELL	A	4th [9]	A	2nd [45]
2/0	Ted	ARIEL	6.5 [5]	4.5 [9]	3.5 [9]	A

The first column of numbers, from "0/1" to "1/0" track the character's Good Stuff and Bad Stuff. "0/0" means Zero Stuff, "2/0" means two points of Good Stuff, and "0/1" means one point of Bad Stuff.

Each Game Master is responsible for keeping these numbers absolutely secret! The best way of keeping the secret is by not talking about it. To anyone. Ever. When players "forget" their Attributes, let them know that you can be depended on to remember.

CHARACTER FREEBIES

Some things don't cost any points. You get to invent most of the details about your character, according to the following guidelines.

Physical Description. Choose your own Physical Characteristics. This includes height, weight, build, hair color, eye color, physical description, clothing styles and colors, and a personal symbol.

Get comfy.

After all, you're going to be wearing this body for hundred, perhaps thousands, of years. Fulfilling every fantasy, dreaming every dream, meeting every challenge.

You want to look good for the occasion.

Absolute Age. More than just a number, picking your character's age really says a lot about the character's persona. Picking a young age, under thirty, is best if you want young and brash. Middle age, between thirty and seventy, is a time of seasoning and learning for Amberites.

The older the character, the more cynical, more cautious, and more worldly they become. Older characters are less likely to have elder Amberites make allowances (a pretty major concern, given the legendary tempers and powers of that bunch). A young character has a "puppy" factor, and is likely to be treated better and even protected by the elder generation.

"Absolute" means this is the amount of time that the character has actually lived through. What with fast and slow time Shadows, a character may be a lot older, or younger than their "Amber" birth date would indicate.

Skills. The simple answer is, in Amber, if you want skills, you can get them.

Why? Corwin has pointed out that every one of his brothers and sisters is the equivalent of a skilled surgeon. They're all interested in learning whatever skills are necessary for survival. And they all have had the time to learn.

So time becomes the important factor.

Time is always available. Any character with the power to travel through Shadow can always take a side trip to a "fast time" Shadow. For example, someone can spend eight years in a Shadow medical school and surgery residency, while only a lazy weekend passes in Castle Amber.

On the other hand, if you want your character to be young, and fresh, and inexperienced, it means limiting the number of skills that you've picked up.

Also, you may not *want* to pick up a lot of skills. After all, a millionaire may have tons of money, yachts, fast cars, polo ponies, and all the toys he can imagine. That doesn't imply wanting to learn how to keep books, trim sails, repair engines, or groom horses. Why should Amberites waste their time doing boring things, when they can do any *fun* thing they can imagine?

Rather than trying to list all the various skills you've picked up, it's better to list the experiences that the character has lived through. Here's a few examples:

Two Years of Touring Shadow. A little time as a Shadow tourist gives a character a knack for learning new languages and cultures, getting an overview of how things work in different places. For example, after a couple years of just wandering around, the character will have a "feel" for how payment works in different places, from the sharing rituals in hunting and gathering societies, to bartering with nomadic tribes, to weighing gold and precious metals in city states, to money exchanges in kingdoms and empires, to the use of checks and credit cards in industrialized societies, all the way to purely abstract "work credit" transactions of futuristic societies. Likewise, the character learns how to get around on horseback, with wagons and carriages, arranging for rail and steamship passage, and driving everything from cars and helicopters to spaceships and teleport platforms.

Five Years of fantasy adventures. What if your character travelled to a fantasy Shadow, a place where there were wondrous adventures, evil mages, brave knights, dark dungeons, and beautiful maidens in need of rescuing? Obviously the character would pick up all the weapon skills, horsemanship, and even a bit of that Shadow's magic system. Need to pick a lock? A character with this kind of background probably has the experience. Other fantasy adventurers might involve spending five years as a pirate, a cowboy, a wandering samurai, or anything else you can imagine.

Ten Years of Civil Engineering. Aside from a long period of study, followed by apprenticeship to a Master Engineer, the character has engaged in a number of major public works. Note that any culture, at any level of technology, provides certain standard knowledge, so learning how to build Egyptian-style Pyramids, Roman-style Aquaducts, Bavarian-style Castles, Cross-Continental Railroads, Urban Skyscrapers, or Orbital Factories, all give the same kind of understanding of the problems involved. Means and machinery may change, but the basics of design and implementation are universal. Once accomplished, the character would have mastered dozens of special skills. And, just as an engineer who worked on China's Great Wall could see how to apply a bulldozer or backhoe, the character can quickly learn related skills. Medicine, law, philosophy, and meditation are all variants on this course of study.

Twenty-five Years shipboard. Can be on a sailing ship, a modern battlecraft, or a spaceship. Starting with a short stint as a trainee, and graduating to junior officer status, and eventually becoming a captain. Aside from the technical stuff, which varies with the type of ship, the character learns the ins and outs of absolute rulership and moral command of a well-organized and highly trained crew. The character would also come to know that Shadow very well, and end up with fluency in virtually all of its languages. Other 25-year possibilities are the roles of Medieval Lord, Renaissance Great Artist (painter, sculptor, etc.), Napoleonic Military General, or Modern Corporation President.

In general, you need about two years for any technical or mechanical specialty (jet engine mechanics, automatic weapon design and repair, locksmith, or pharmaceutical chemistry). More formal schooling, say in sculpture, mathematics, or to gain skill in a musical instrument, can take four years for the basics, and another ten or twenty for really detailed specialization. Really difficult skills, like the aforementioned surgery, or quantum physics, can take ten years or more of exclusive, intensive study.

Equipment. Material possessions come ridiculously easy to Amberites. Start off stark naked, and at the end of a two-hour stroll through Shadow a character can have accumulated enough gear to put any billionaire to shame. Feel free to make life in your home Shadow as opulent as you like. Objects are just objects, so help yourself.

OUT OF A PLAYER'S CONTROL

There are some things that players can't choose. These are background things that the Game Master must decide in setting up the game.

Your Parents. There's an old saying, "you can choose your friends but you can't pick your relatives." It's all too true in Amber. The Game Master will choose your Amber (or Chaos) parent. And, unless you are very lucky, it's entirely possible that you won't even know who it is. It's also a good idea to remember that your Game Master is coming up with a unique version of all those elder Amberites. Which means your "favorite" Amberite may turn out to be a real snot.

Your Amber Friends and Devotees. Again, it's up to the Game Master. The way things work in Amber, the person who is most friendly is not likely to be the person most devoted to your character's well-being.

Your Character's History & Background. This is a strange case. Sometimes the player's ideas will work out, and every detail will be allowed by the Game Master. On the other hand, the Game Master has the right to completely change the character's history, in order to fit things into the story.

The Effects of Your Good Stuff and Bad Stuff. Not only can't you choose, but the Game Master can actually switch the effects around from session to session. Good Stuff and Bad Stuff is a tool of the Game Master, a way to plan adventures, encounters and other exciting things. Sometimes the Game Master will assign a particular effect, say, a bitter enemy resulting from a point of Bad Stuff, that will follow the character for years. Other times the Game Master may shift the effect of the point of Bad Stuff to something different every time circumstances change.

Your Age in Amber. Players do get to pick their characters age. Twenty, or two thousand, it's up to you. However, a twenty-year-old character could have been born two thousand years ago, having been raised in a slow time Shadow. Or, just as easily, a character aged two thousand could have been born just twenty years back in Amber years, and have spent all that time in fast time Shadows out near the Courts of Chaos.

Items with "Real" Power. Any creature or person, no

matter what their origin, who comes to have "real" Power, such as the Pattern, or Logrus Mastery, automatically become Non-Player Characters, and the sole property of the Game Master. If points were spent, they'll be returned, but children, clones, or other living creations of player characters, will become independent, and even, perhaps, hostile to the "parent." Likewise, objects that contain Pattern, like the Jewel of Judgement, or Greyswandir, are not available to a starting character (though they may come to "buy" such things later on).

Two Great Characters

As a Game Master, I've got two favorite characters.

Back in 1985, when the first Attribute Auction was held, and I was the first novice Amber Game Master, there were many interesting characters. Some players came up with powerful characters, in fact, with the most powerful Amber characters I've seen, or ever expect to see. I love those powerhouse guys!

But they weren't my favorites.

My favorites were the two who were flawed, yet not fatally. Two characters very much of Amber, yet misfits. Were, and came to be, people totally apart from their players.

Don Woodward came up with Carolan, noble and forthright, an honest innocent among the cynics of Amber.

Mike Kucharski came up with Morgan, vile and underhanded, a rat among men.

These two characters, one based on Good Stuff, the other on Bad Stuff, were wonderful.

Because they were tragic.

They made mistakes.

Not little piddling mistakes.

No, Morgan and Carolan made world-beating, gasping, horrific and apocalyptic mistakes.

They were like the two sides of the Amber character.

Carolan was a Good Stuff kind of guy. Trusting, honest, and earnest about Amber. As Game Master I stomped all over him, abused his trust, and sent the worst of the elder Amberites to manipulate him shamelessly.

Morgan invested in Bad Stuff and loved it. No good guy he. He enjoyed back stabbing, murder and mayhem. In response to his Bad Stuff his luck was always out, and behind every door the Game Master placed enemies seeking his blood. Morgan, in his haste to eliminate opponents, also killed his share of innocents.

Six years later, I know why they were, and are, my favorites.

It's because they honestly grew.

Carolan, embittered by fate and his own gullibility, managed to maim a feared Uncle, and kill a beloved Aunt. He experienced betrayal of his every honest emotion. And turned to denial, denying responsibility for his own actions.

Don, the player, complained bitterly about a game where nothing was "fun" and where he found pain everywhere. Worse, he seemed to bring pain to everyone he loved.

Eventually he got through it. Full circle, Carolan faced his guilt, and conquered it.

Morgan left a trail of bodies and saw nothing gained by it. Those he killed were no challenge, and were, in retrospect, blameless. He found fulfillment where he least expected it.

He, Morgan, sought a team of killers. He found them, but in them he also discovered his own weakness. He turns out to be a better father than he thought possible.

And, Mike, playing Morgan, had plenty of cause for complaint, when I turned his lovely children into a new batch of argumentative player characters.

Now tempted by violence, Morgan finds reasons to turn the other cheek.

Each of them, Carolan and Morgan, had to question their motives, and their character.

With each turn of fate they grew.

Now, six years into the campaign, they each remain among the hunted, banished from Amber for their crimes, their lives forfeit by the order of the King.

Why are they my favorites?

Because they, unlike any characters I've ever witnessed, have been through the baptism of fire. They've been hammered, and bent, and hammered again, like swords in a forge, until their characters have emerged in glory.

Nowadays, when Morgan, or Carolan, speaks, there is no confusion. The characters have become real, and deep, and there is no simplicity left within them...

So, why go on about these guys?

Well, maybe it'll tell you something about putting together an Amber character.

Neither character was frugal in the Attribute Auction, they both made strong bids. Mike bid on three of four Attributes. Don, bidding for Carolan, was one of the only players to bid in every Attribute.

The main thing, though, is that they both saw that they were making a beginning for their characters. They didn't try to do everything, or be all-powerful. Instead, they recognized, maybe subconsciously, that they were in the game for the long haul.

They took the time to grow.

GLOSSARY OF AMBER TERMS & CONCEPTS

The Abyss. There exists a place, somewhere at the border of the Courts of Chaos, where the known universe simply ceases to exist. This void, which we assume is contained by the combined effect of the Logrus and the Pattern, appears as a huge bottomless pit, where material land just stops and emptiness begins. The contents of the Abyss, that which falls in, would seem forever lost. This must not be an absolute, since there is evidence that the Unicorn is capable of retrieving items from the Abyss.

Amber. The center of all things. The only true reality. Here's a passage from pages 99-100, of *Nine Princes in Amber*,

Now, it is written that only a prince of Amber may walk among Shadows, though of course he may lead or direct as many as he chooses along such courses. We led our troops and saw them die, but of Shadow I have this to say: there is Shadow and there is Substance, and this is the root of all things. Of Substance, there is only Amber, the real city, upon the real Earth, which contains everything. Of Shadow, there is an infinitude of things. Every possibility exists somewhere as a Shadow of the real. Amber, by its very existence, has cast such in all directions. And what may one say of it beyond? Shadow extends from Amber to Chaos, and all things are possible within it. There are only three ways of traversing it, and each of them is difficult.

If one is a prince or princess of the blood, then one may walk, crossing through Shadows, forcing one's environment to change as one passes, until it is finally in precisely the shape one desires it, and there stop. That Shadow world is then one's own, save for family intrusions, to do with as one would. In such a place had I dwelled for centuries.

The second means is the cards, cast by Dworkin, Master of the Line, who had created them in our image, to facilitate communications between members of the royal family. He was the ancient artist to whom space and perspective meant nothing. He had made up the family Trumps, which permitted the willer to touch his brethren wherever they might be. I had a

feeling that these had not been used in full accord with their author's intention.

The third was the Pattern, also drawn by Dworkin, which could only be walked by a member of our family. It initiated the walker into the system of the cards, as it were, and at its ending gave its walker the power to stride across Shadows.

The cards and the Pattern made for instant transport from Substance through Shadow. The other way, walking, was harder.

The Amber Universe. Our familiar Earth is nothing but a Shadow of the one real world. That real world is Amber. It is the only world of substance, and Shadows stretch out around it like images in a room full of mirrors, warped by the distant shifting Logrus. Our Earth is Shadow Earth, a remote and twisted image, and each person we have ever known, everything we have ever seen, is a little piece of Shadow stuff. And out among the infinite Shadows, there is any world you can imagine.

Castle Amber. The room where the Pattern is located is inside of a castle, a castle on top of a mountain called Kolvir, looking over the sea. The castle is called Amber, and it is the home to that family which has power over the Pattern. Ancient, powerful, rambling, filled with arcane secret passages, mysterious halls, and built atop a cavernous maze of tunnels.

The Courts of Chaos. Beyond the Shadows of Amber, beyond the point where "rational" things like the laws of physics and nature have any real meaning, there is the Courts of Chaos. Occupied by Lords of Chaos, the users of the Logrus, they are hostile to Amber, but involved in their own affairs. It would seem to pre-date Amber, and it would seem that Dworkin, the creator of Pattern, was born in the Courts of Chaos. Sadly, there's no room in this book to give a fair treatment of the Courts, so we'll reserve a full discussion of the Courts for a future supplement.

Dworkin. There is no better explanation than Corwin's, on page 106, of *Nine Princes in Amber:*

I fingered my cards, weighed the deck in my hand. I could try a contest of wills through them, with either Eric or Caine. There was that power present, and perhaps even others of which I knew nothing. They had been so designed, at the command of Oberon, by the hand of the mad artist Dworkin Barimen, that wild-eyed hunchback who had been a sorcerer, priest, or psychiatrist - the stories conflicted on this point - from some distant Shadow where Dad had saved him from a disastrous fate he had brought upon himself.

The details were unknown, but he had always been a bit off his rocker since that time. Still, he was a great artist, and it was undeniable that he possessed some strange power. He had vanished ages ago, after creating the cards and tracing the Pattern in Amber. We had often speculated about him, but no one seemed to know his whereabouts. Perhaps Dad had done him in, to keep his secrets secret.

Nine Princes in Amber

Hellriding/Shadow Shifting.
There are an infinite number of different worlds. The problem is getting from one to the next. That process, getting from Shadow "A" to Shadow "B," we call Shadow Shifting. It's a mental process, a matter of picturing possibilities, and gradually adding or subtracting details. Hellriding is more of the same, just faster, and usually a matter of riding a horse, a mount, or a vehicle.

The Jewel of Judgement.
The main mystic artifact of the Amber universe, and most prized possession of the throne. Look within its depths and you will see a three dimensional version of the Pattern. It is said that Dworkin inscribed the Pattern using the Jewel of Judgement, although it is not known if the Pattern within the Jewel was there before Dworkin made the Pattern. Certainly the Jewel of Judgement now possesses strange powers, among them the absolute control of weather in Amber, and throughout Shadow.

The Logrus.
As Pattern is the representation of order, so Logrus is the representation of chaos. Taking the Logrus, navigating it's multi-dimensional twists and turns, is the acceptance of madness, but the only way to gain its power. Initiates of the Logrus can use its tendrils to reach through Shadow, for defensive and offensive combat, and as a container for magical spells.

Magic.
Wandering through Shadow, one may come upon many kinds of magic, from the tittering enchantment of pixies, to curses cast of man, to the mighty pronouncements of dragons. However, most of these magics are special only to their own place. They are confined to a single Shadow, a single reality, and work no where else.

Greater magic works everywhere. It comes in three varieties.

Power Words are a simplified form of magic, useful primarily as a defense against the magic of others. They are few in number, although specific to each individual. Their effect is instantaneous, and their power is very limited. However, they require no preparation prior to use.

Sorcery is the ability to weave and cast spells. A time consuming process, each spell must be made specifically for a unique use. Once triggered, a spell is expended, and cannot be repeated without another lengthy process of spell creation.

Conjuration allows for the creation of magic in a more permanent form. Either by inserting magic into some creature or item, or by creating something innately magical. It is difficult and time consuming, but the results are long lasting.

Pattern.
Picture a circular pathway, with the entrance on the outer edge of a large circle, maybe the size of a baseball diamond. The path winds between glowing lines of power arranged to spiral around and around, toward a small clear zone at the very center. Along the way are strange twists and curves and knots. Walking the Pattern is a dangerous and difficult task, and a weak or fatigued character could easily die before completing it. But, having walked it, one can transport oneself, instantly, anywhere one can imagine. Again, Zelazny can explain this best—

I strode forward, setting my left foot upon the path. It was outlined by blue-white sparks. Then I set my right foot upon it, and I felt the current Random had mentioned. I took another step

There was a crackle and I felt my hair beginning to rise. I took another step.

Then the thing began to curve, abruptly, back upon itself. I took ten more paces, and a certain resistance seemed to arise. It was as if a black barrier had grown up before me, of some substance which pushed back upon me with each effort that I made to pass forward.

I fought it. It was the First Veil...

...Each raising and lowering of my foot suddenly required a terrible effort, and sparks shot from my hair.

I concentrated on the fiery line.

Suddenly the pressure was eased. The Veil had parted before me, as abruptly as it had occurred.

I was well into the Pattern now, and the sparks flashed continually about my feet, reaching to the height of my knees. I no longer knew which direction I faced... The currents swept through me, and it seemed my eyeballs were vibrating. Then came a pins-and-needles feeling in my cheeks and a coldness on the back of my neck. I clenched my teeth to keep them from chattering.

I took six more rapid steps, reaching the end of an arc and coming to the beginning place of a straight line.

I set my foot upon it, and with each step that I took, another barrier began to rise against me. It was the Second Veil.

There was a right-angle turn, then another, then another.

Another curve began, and it was though I were walking in glue as I moved slowly along it.

One, two, three, four... I raised my fiery boots and let them down again.

It was tricky, so devilish tricky... Instinctively, I knew that to leave the Pattern before I'd completed it would mean my death. I dared not raise my eyes from the places of light that lay before me, to see how far I had come, how far I had yet to go.

I emerged from the filigree and marched along the Grand Curve.

I walked three more curves, a straight line, and a series of sharp arcs.

Ten turns which left me dizzy, another short arc, a straight line, and the Final Veil.

It was agony to move. Everything tried to beat me aside. The waters were cold, then boiling. It seemed that they constantly pushed against me. I struggled, putting one foot before the other. The sparks reached as high as my waist at this point, then my breast, my shoulders. There were into my eyes. There were all about me. I could barely see the Pattern itself.

Then a short arc, ending in blackness.

One, two... And to take the last step was like trying to push through a concrete wall.

I did it.

Nine Princes in Amber

The Patternfall War. The conflict between Amber and the Courts of Chaos, seemingly engineered and encouraged by certain forces within the family Amber. Chaos was urged into the conflict by a number of signs. Corwin's curse against his brother Eric and the partial damage done to the Primal Pattern opened the Black Road, the invasion route from the Courts of Chaos to Amber. Forces in the Courts of Chaos were also encouraged by the warring factions of Amberites, especially the rebel cabal lead by Clarissa's children, Bleys, Brand and Fiona, who implied that they might be willing to ally themselves against Amber. In the end, the war proved disastrous to Chaos, and nearly as damaging to Amber. Peace was eventually achieved through Amber's military victory, but also through the influence of a third,

"peace party," headed by Dara and Merlin, both of whom have a mixed heritage from the two courts, and Martin, the son of Random.

Corwin's Chronicles record the events of the Patternfall War, from his viewpoint of course. Published on Shadow Earth, they consist of five books, *Nine Princes in Amber*, *The Guns of Avalon*, *Sign of the Unicorn*, *The Hand of Oberon*, and *The Courts of Chaos*. Read them, and be enriched.

A later series, of another five books, are based on the adventures of Corwin's son, Merlin. Although very little of that material is covered in this book, they are invaluable resources. They are *Trumps of Doom*, *Blood of Amber*, *Sign of Chaos*, *Knight of Shadows* and *Prince of Chaos*.

Rebma & Tir-na Nog'th. The Pattern itself is reflected strongly in only two places. The first, is in the water under the seas bordering Amber. This is a pattern that is the center of a duplicate castle and city, a place structured like Amber, but in an underwater world known as Rebma. The other great reflection appears only by moonlight, in the clouds over Mount Kolvir. When the moon is bright, and the clouds are firm, it is possible to climb a set of shimmering stairs and wander about in Tir-na Nog'th. In that strange place one finds weird pieces of reality and alternate reality, a mix of visions and prophecies.

Shadow. Shadow is that which Amber creates, which seems real, which comes in infinite profusion, but which is not real. Everywhere between Amber and the Courts of Chaos is Shadow. Worlds of fantasy, dotted with faerie forests and orcish castles. Worlds of science fiction, bristling with technology, spacecraft and maybe even Bio-Implants. Worlds like Earth, but variants based on different histories; or like Earth, but inhabited by different creatures; or like Earth, but of a different age. Amberites can walk to any Shadow that they can imagine, and have it be exactly the world of their dreams. Here are two important passages that describe Shadow:

...There had been fifteen brothers and six were dead. There had been eight sisters, and two were dead, possibly four. We had spent much of our time in wandering in Shadow, or in our own universes. It is an academic, though valid philosophical question, as to whether one with power over Shadow could create his own universe. Whatever the ultimate answer, from a practical point we could.

Nine Princes in Amber

After a time, I stopped at a hollow tree that had to be there. I reached inside and drew forth my silvered blade and strapped it to my waist. It mattered notthat it had been somewhere in Amber. It was here now, for the wood that I walked was in Shadow.

The Guns of Avalon

Shadow Storms. Although a Shadow Storm has some of the aspects of a normal storm of cloud and wind and lightning, it is different in that it cuts across Shadows, ripping things, living and otherwise, from one world to another. A fearsome thing, even for Lords of Amber, a Shadow Storm is capable of destroying entire Shadows, and can easily rip apart an army, or a fleet, or anything travelling through the boundaries between Shadow.

Shape Shifting. The ability to shape one's body into other forms. Can range from something as simple as a change of appearance, to something as involved as the taking on of the shape of an entirely different form of creature. Shape Shifting, at its most advanced, can extend to changes in personality and even imitation of Powers and abilities.

Thari. The language of Amber, as well as the Courts of Chaos. It is said that all other languages are simply distortions of Thari, the one true tongue.

Trump. Another power, used by both courts, is that of the cards of Trump. In shape they are no more than ordinary cards, of the kind used by mischievous fortune tellers, but with images of extraordinay detail. Trump containing the portraits of people have the power to reach the minds of their subjects, across infinite reaches of Shadow. There are also Trump of places and objects.

Those who devote themselves to the study of Trump, learning to create the cold images, tap into yet another source of power in the Amber universe.

Trump Scrying. Amberites have their own superstitions. They believe they can take a deck of Trump, shuffle the cards, and deal out an arrangement that will tell them something of the future or of importance. A harmless enough pursuit, and sometimes helpful in pointing out the hidden meaning of a situation.

Trump Travel. Using a Trump connection, it's possible to move from either point of the contact to the other, instantly moving across the intervening distance. Aside from using character Trump, those illustrated with peoples' images, it's also possible to use place Trump, which picture a particular scene, and use them to transport to the place in the image.

Units of Measure. Fortunately, a "foot" on Shadow Earth is roughly equal to an Amber Imperial Foot (based on the absolute measure of Oberon's foot when he paced out the streets of the city. Even today cast versions of Oberon's foot imprint are found in the streets of the city...). Likewise, a pound is comparable to the Amber Imperial Pound used throughout the trade routes of the Golden Circle Kingdoms, and is said to be one twenty-fifth of one tenth of Oberon's true weight.

The Unicorn. As much as Amber is built on the mystic lines of the Pattern, or the stone and mortar of the Castle, it has a spiritual foundation. Even a fleeting glance of this magical creature, pure and dainty, whether in the Grove of the Unicorn, or anywhere in the realm of Amber, is taken by Amberites as a sure sign of good fortune. Although the nature of the unicorn is unknown, some say unknowable, it is the family's connection with the divine forces shaping the universe.

POWERS

The Attributes of player characters define their physical capacities, but it is their Powers that get them true control over the surrounding universe. The primary Powers are Pattern, Logrus, and probably Trump. Shape Shifting is definitely secondary, and Magic is the least of all Powers.

PATTERN IMPRINT

50 Points.

Located in the depths of family castle, behind well-locked doors, lies the great Pattern of Amber. This is actually the reflection of the true Pattern, the Primal Pattern, that lies in a mystic realm dominated by itself. There are also two other 'reflections' of the pattern, one in Rebma (the underwater reflection of Amber), and the other in Tir-Na Nog'th (the moonlight's reflection of Amber).

Walking the Pattern was the greatest trial in your life, and also your greatest triumph. Only those with the blood of Amber can walk the Pattern and live.

Pattern is the most powerful and most useful of all Powers available. At only fifty points, it's the bargain of the *Amber* role-playing system. Even more so than the other Powers, Pattern takes time and experience for players to master. In the hands of an expert, Pattern use is subtle and precise, wielded with an understanding that Pattern is a very large tool that can be used to do very fine work.

Attribute Tips for Pattern Imprint. Although Psyche is important for manipulating Pattern and Shadow, remember that your Endurance must be at least Amber level in order to actually walk the Pattern. Those with lower Endurance, either Chaos or Human, must get some kind of assistance, or they will die in the attempt. The higher your Endurance, the more damaged or exhausted you can be and still be able to walk the Pattern.

Pattern Imprint Abilities

Travelling through Shadow. Sometimes called the Hellride. This is done by changing minor things that lay in your path. You, the player, decide that over the next hill will be a patch of blue flowers, around the next corner is a jagged rock face, in the next village is a country fair. Down that alley is a shop dealing in obscure weapons. By such minor changes you move closer to what you seek in Shadow. Certain places are easy to manipulate, such as those closest to chaos. Manipulating shadow is harder the closer you get to Amber, until, in Amber itself, it cannot be changed at all. Some places are difficult to get to, barred by monsters, hazards or whatever. Finally, following a traveller through shadow is always easier than taking your own path.

Since the shadows are without number, anything you wish, anything at all, can be found somewhere in shadow. Whether these shadows were always there, or whether they have no existence until you arrive, or whether you create them by your desire, is a question that has not been answered.

Moving Through Shadow. In Shadow, if you stand unmoving, you are in the middle of a vast universe. However, each step forward is an opportunity to visit another "next door" universe, another of the infinite number of possible variations on a world. In Shadow, so long as an Amberite can move, walking, riding, sailing, swimming, flying, driving, or crawling, it is possible to move through Shadow.

Shadow Walking is the easy way and is accomplished by changing very subtle, very small details. So, for example, in a ride through the woods of Shadow Earth, around the corner you might expect to see a greenjay instead of a bluejay. That one detail is an easy change. Shadow Walking is easy, but slow.

The Royal Way. Another way of walking or riding through Shadow is by travelling through only Shadows whereverythin is to the character's liking. For example, one can choose to walk through the worlds where there will always be welcoming faces and rosebuds strewn on the roadway. Or, if you choose to travel in a luxurious train, you can stick to those Shadows that contain rails. The Royal Way is slow, and the changes are gradual, but it's very comfortable.

Hellriding is the harder way to go. In Shadow walking you changed one detail, a bluejay into a greenjay. In a Hellride you focus on a detail, and change the whole world around it. Spot a bluejay and you decide to enter another Shadow, completely different from Shadow Earth, but where bluejays still fly. Hellriding is hard, but fast.

Character Objectives. Anythin and anyone can be found somewhere out in Shadow. What you find depends on what you describe to be your objective. Here are some of the possibilities.

- *A Shadow of the player's description.* "I'm going to a version of Shadow Avalon, but where Morgana is Queen and where they use steam engines instead of horses."

- *A character that the player is seeking.* "I'm going to out to find Uncle Bleys."

- *An object that the player describes.* "I want to track down a couple of SCUBA outfits and some other diving equipment."

- *A situation defined by the player.* "I want to find a place with a good bar fight, but where I won't seriously hurt anybody, and where I can't be hurt too badly."

Most of the time the Game Master will simply respond with the results, taking the player to the alternate Avalon, the Shadow where Bleys is at present, just outside a SCUBA shop, or right into the bar. The usual response from the Game Master is, "after walking through Shadow for a time, you find yourself in the place you were seeking."

Infinity. There are an infinite number of versions of every Shadow. For example, there are countless versions of Shadow Earth. Sometimes this gets a little confusing. After all, when two characters arrange to "meet at the Savoy Hotel on Shadow Earth," how do they avoid each going to different versions?

Strengthening Shadow. Even though there are an infinite number of each Shadow, they all pale compared to those few Shadows that have been "strengthened" by a touch of something real. Whenever someone from Amber visits a Shadow they make it more solid than all of the surrounding Shadows. For an Amberite character an overnight stay makes the particular Shadow into that "place where I slept the night."

Stay for years, and the Shadow becomes correspondingly real and vivid. Shadow Earth was made very real by Corwin's lingering for four hundred years, *plus* the occasional visits of other Amberites. Then, when Corwin left the Jewel of Judgement on Shadow Earth for a time, the Shadow became even more real.

Leading Others Through Shadow.
In the following quote Bleys loses many of the troops he leads through Shadow. Had he been travelling alone he would still have had to pass through the Shadows of the centaurs, of the jungles and of the killer machines, but he would have passed through these places so quickly that there would have been no time for anything to attack him. Since he has such a long column of troops, he must move through Shadow slowly.

Bleys by now was marching across the plains of the worlds. Somehow, I knew that he would make it, past whatever defenses Eric had set up. I kept in touch with him by means of the cards, and I learned of his encounters along the way. Like, ten thousand men dead in a plains battle with centaurs, five thousand lost in an earthquake of frightening proportions, fifteen hundred dead of a whirlwind plague that swept the camps, nineteen thousand dead or missing in action as they passed through the jungles of a place I didn't recognize, when the napalm fell upon them from the strange buzzing things that passed overhead, six thousand deserting in a place that looked like the heaven they had been promised, five hundred unaccounted for as they crossed a sand flat where a mushroom cloud burned and towered beside them, eighty-six hundred gone as they moved through a valley of suddenly militant machines that rolled forward on treads and fired fires, eight hundred sick and abandoned, two hundred dead from flash floods, fifty-four dying of duels among themselves, three hundred dead from eating poisonous native fruits, a thousand slain in a massive stampede of buffalo-like creatures, seventy-three gone when their tents caught fire, fifteen hundred carried away by the floods, two thousand slain by the winds that came down from the blue hills.

Nine Prince in Amber

Shadow Time.
A most useful aspect of Shadow is the way it can be used as a handy time machine. It's impossible to go backwards in time. That leaves skipping forward or holding back time.

'He will probably be tired. That should slow him some.'

'No. Put yourself in his place. If you were Brand, wouldn't you have headed for some shadow where the time flow was different? Instead of an afternoon, he could well have taken several days to rest up for this evening's ordeal. It is safest to assume that he will be in good shape.'

The Hand of Oberon

Skipping Forward in Time. The character goes to a place where time moves slowly. While the character is spending a few minutes in the "slow time" Shadow, time is speeding hour by hour along in Amber. So a character can leave, spend a few minutes out in Shadow, and return days, or weeks later. It's even possible to find a Shadow that moves so slowly that it becomes almost like a stasis, and where the character can wait for years to pass in Amber.

Holding Back Time. The other handy trick of Shadow time is going out to a "fast time" Shadows, where hours pass relative to minutes ticking by in Amber. A character could go to Shadow Earth for a ten hour sleep, and return to Amber where only four hours have gone by. In faster Shadows it's possible to spend years in preparation, while only a day or so passes in Amber.

As I sailed into Shadow, a white bird of my desire came and sat upon my right shoulder, and I wrote a note and tied it to its leg and sent it on its way. The note said, 'I am coming,' and it was signed by me.

A black bird of my desire came and sat on my left shoulder and I wrote a note and tied it to its leg and sent it off into the west.

It said, 'Eric - I'll be back,' and it was signed: 'Corwin, Lord of Amber.'

Nine Prince in Amber

Shadow Resources.
In the passage above, what is Corwin doing?

As he moves through Shadow he is also seeking certain things. In this case he first seeks a white bird that will alight on his shoulder, and that will, after he has tied a note to its leg, travel through Shadow to the place where he is destined to arrive.

Then he seeks another bird, a black one. The black bird will travel through Shadow to Amber and to someone who will take his note to Eric.

Sure enough, later on in the story the white bird landed on his shoulder, a signal that he had found the place he was looking for. Likewise Eric got the message carried by the black bird.

Corwin could have easily have found two fish, each destined to be caught, and cut open, at the two destinations. Or two ships, each of which would happen to end up where he wanted them to be.

Using Pattern near Amber.

Using Pattern near Amber. Pattern, in Amber, or anywhere near the center of things, is nearly worthless. Amber is just too fixed, too real, to be manipulated. Characters can use Pattern to get away from Amber, gradually moving out into Shadow, but it may take many hours before even subtle changes start to take effect.

Power over Shadow.

Power over Shadow. In shadow, the shadings of one shadow to the next are vague. That means that small changes can be made without travelling. Short of money? Imagine a loose floorboard, under which is hidden a treasure, look and it will be there. Those with the imprint of Amber's pattern are endlessly lucky. What is possible can, with time and effort, be made probable, and what is probable can be made certain. While this power is absolute, it always involves some movement, a shifting of the variety of Shadow for a particular shade. Any Shadow manipulation is also slightly hazardous because of your involuntary movement in shadow.

With enough time it's possible to shift just about anything in Shadow. However, there are two important limitations.

First, in order to shift Shadow you must move. Subtle changes don't require much movement, but the more drastic the change, the more you must move around. The problem here is that you can easily move right out of one Shadow and into its neighbor. Be wary that, while shifting, you don't start moving away from your objective.

Second, shifting Shadow is fairly slow. Far slower than combat time. If you wish to make use of some Shadow shifting it's best if it be prepared before a battle. A minor change, such as the color of a coin, might take as much as a half a minute. Something useful, such as weakening a wall, or making a weapon brittle, could take a minute or more.

Using Pattern to Affect Probability.

Using Pattern to Affect Probability. Out in Shadow, if there is any reasonable chance at all of something happening, then a character with Pattern can make it happen. The main limitations are time, since using Pattern takes a few moments of concentration, and probability. A character can't make something unlikely happen, not without going into an adjacent Shadow.

Example of Using Pattern to Affect Probability. Mick's character Farley has been fleeing from an odd set of enemies, but is still attempting to stay in the same Shadow. Here is how things might go:

GM: You see a carnival show. There's a barker out front, and a couple of scantily-dressed women. It looks like you have to pay a cashier for a ticket to get in.

Mick: Do I have the right change?

GM: I don't know. Does Farley want to have the right change?

Mick: Yeah, I'd appreci... Oh, I get it. Farley can use Pattern to affect things. All right, I'll pull the right change out of my pocket.

GM: Son of a gun, that seems to be just the right amount.

Mick: I'll buy a ticket, and go on inside.

GM: You're taking in the sights and smells of the carnival, when you notice three guys in pink suits...

Mick: Again? When am I going to lose these guys? I'll head toward the exit.

GM: No problem, you get to the exit. Before you can leave, you notice a bus has pulled up in front, and dozens more of the people in pink suits are milling around.

Mick: Ick. I really want to get rid of them, so I'll use Pattern to make them not be here.

GM: That doesn't seem to do anything. What are you doing?

Mick: I know, I know. You've got to move in order to really affect Shadow. I'll start walking around the Carnival, gradually subtracting everybody with pink suits.

GM: Good going! In a few minutes the crowd is still there, and it feels like a slightly different Shadow, but the guys in pink are gone. What are you doing now?

Mick: Well, I'm here, so I might as well enjoy myself. What is there to do in this carnival?

GM: Let's see, you've got a choice of a sleazy-looking sideshow filled with freaks, a cafe that's serving food and drink, a carousel with carved horse-like things, and a row of little booths, each with a different game.

Mick: What kind of games?

GM: Oh, things like the wheel of fortune, a weight-guesser, ring toss and a shooting gallery, but with crossbows.

Mick: That sounds neat, I'll walk up to the shooting gallery and try a game.

GM: What are you using for money?

Mick: I'll just reach in my pocket and pull out the right change.

GM: Problem! You already emptied your pocket, so it's very unlikely that there are any coins left in there. Now what?

Mick: Well, people are always dropping coins, so I suppose there's one right about... here, where I'm standing.

GM: Right you are! Now are you going to the shooting gallery?

Mick: Not yet. I think I'll build up my stake first. You said there was a roulette wheel?

GM: A Wheel of Fortune, almost the same thing.

Mick: I'll put my coin down on the winning number.

GM: Using Pattern again, you influence events and end up winning a fistful of coins. And right about then you notice a guy in a pink suit standing right next to you.

Mick: Oh no!

Shadows of Desire.

Shadows of Desire. Any character can find exactly the Shadow they desire. When this is done by a player character, the Game Master should get a written description of the Shadow from the player. There's really no limit to what a character's Shadow might be, except that the more elaborate it is, the longer it's going to take to get there. Shadows sought by players can have any of the features of "bought" Shadows, except, of course, that their Shadows run the risk of being influenced by elder Amberites, or even by other player characters.

Game Masters should also be aware of the unconscious aspect of a player character's desires. The character's problems, hopes, dreams, fantasies, and fears, all the things that can show up in quizzes or in role-playing, can also show up in a character's Shadow. For example, if a character is having a problem with a particular elder Amberite, then one, or more, Shadow versions of that character might be found in the Shadow.

Pattern Defense.

Pattern Defense. Picture the Pattern, walk it in your mind. When it is complete, concentrate on the image. This takes a few minutes, depending on your Psyche and how well you can concentrate, but it strengthens your mind, and lends strength to your existence. The main use is defensive, so a character with Pattern to mind is pretty much immune to the Logrus, and other Chaos-generated forces. Likewise, it's difficult or impossible to use invasive Magic on a character who has Pattern brought up like this. On the other hand, keeping the Pattern in mind requires real concentration, so the character can't run or fight effectively. If the character's attention is broken, then the Pattern instantly flickers out. Characters with Human rank Psyche are incapable of picturing the Pattern.

Walking the Pattern. Should you want to get somewhere quickly, there is a shortcut available. Just walk the Pattern, any of the Patterns, and when you reach the middle, you can instantly teleport yourself to anyplace you can imagine. Walking the Pattern is tiring, it strains your Endurance, and even your Strength, but it also purifies your mind, repairing gaps in your memory, and shaking off any lingering curses, magical enchantments, or demonic possessions.

Dying on the Pattern. Be cautious. Walking the Pattern in a weakened state is fatal! If your Endurance falls to under Amber level then you won't have the energy to get through the Pattern. Even if you normally have the required Endurance and Strength, if your Attributes are temporarily weak, you can die attempting to walk the Pattern. Even then, even when you are at the height of your power, it is still possible for you to be attacked as you wind your way inward. And, having walked the Pattern, unless your Endurance is exceptional, you will be weakened and drained.

The Blood Curse of an Amberite. The rarest, yet most feared, of all the powers of those of the blood of Amber, is their ability to deliver a *Blood Curse*. Usually delivered when dying, the curse is an evoking of a character's personal relationship with the Pattern, a directing of their life energies towards some terrible goal.

In game terms a character is trading away ten points (or more) of "Stuff." If, after the curse, the character's life continues, this means they will have lost as many Good Stuff points, and gained as many Bad Stuff points, as is needed to make up the ten points of the curse. Of course, if the character actually dies, well... Then it's not a problem that concerns the character any more.

The subject of the curse, whether person, place, or thing, is then invested with the Bad Stuff, all of which will be directed toward fulfilling the curse. It is also, theoretically, possible to perform a "Blessing," such as that bestowed by Oberon at the end of the *Chronicles of Amber*.

Once performed, there is no known way of removing a Blood Curse.

ADVANCED PATTERN IMPRINT

75 points. Or, if you've already got Pattern Imprint, an additional 25 points.

Your character has illuminated the reality of the Pattern far beyond that of mere imprinting. With this power you understand more deeply how the pattern works, how it affects shadow, how it can be manipulated to serve your own ends.

Attribute Tips for Advanced Pattern Imprint. As with other powers, the degree of your control over Pattern is governed by your Psyche.

Advanced Pattern Abilities

Pattern Mindwalking. Once the pattern is summoned to your mind you can use it as a great lens, to peer into shadow. Anyone, anything can be observed in this way. And, having found them, you may reach through the Pattern to touch mind to mind, or even physically.

This power also allows you to open shadow to others, putting the changes of the Hellride in the way of those who know nothing of it. Those already in Hellrides can be obstructed and sidetracked as long as you devote your attention to it. Shift the Shadow that lies in their path, and you can make passage smooth or rough, as you see fit.

Finding a particular item, or person, is, however, not as easy as it might be. Frankly, there's a lot of ground to cover. Just scanning all of Castle Amber could take hours. On the other hand, if your Psyche is sharp, and if object of your search is not disguised, then things can go much more quickly. Likewise, if Pattern is being disturbed, by a Hellride, or by any manipulation of Shadow, then the source of the disturbance should be easy to locate.

Walking the Pattern of the Mind. Almost all of the abilities of Advanced Pattern Imprint require that the character imagine walking the Pattern, all the while visualizing the Pattern coming into existence. Then, upon completing the imaginary walk, the character must continue to focus on this vision of the Pattern. As soon as the character's attention is turned away, the Pattern will go down, and the whole process starts all over again. The effort of holding the Pattern is draining, and how long the character can keep it up depends on the character's Endurance. Likewise any strenuous mental activity, such as a Psychic battle, will cause the character to lose the Pattern image.

Sensitivity to Logrus. If the Pattern contacts the Logrus, or Primal Chaos, then it tends to flare up, causing an even greater drain on the character, and shoving the offensive Chaos away. When the Pattern flares up, it's all the character can do just keeping it up in the mind.

Looking Through the Pattern Lens. Limitless, infinite Shadow is a terrible obstacle to finding anything. While a character with Advanced Pattern Imprint can look at anything, anywhere, it's still only possible to look at one place at a time.

For example, a character, searching for Corwin, might look into his rooms in Castle Amber, in the Library, the Dining Hall, the Practice Yard, and all around the grounds and gardens. After those few seconds of searching, it's up to the character to figure out where they'd like to look next. A complete and careful search of Castle Amber might take five minutes or more.

In another fifteen minutes, the whole City could be searched. Checking Forest Arden's common trails, camps and meadows, would take at least another half hour, but checking all of Forest Arden, looking behind every tree and in every glen, could take days.

Traces of Pattern. Searching out in Shadow is usually a lot easier. If the character knew about Corwin's long-time residence on Shadow Earth, and knew how to locate the place, then a search of that entire Shadow would take just a couple of minutes. Why so fast? Because the character wouldn't have to do any kind of place by place searching. Instead the entire Shadow would be observed, and any "real" things, stuff originating from Amber, would show up in contrast to the rest of the Shadow.

Traces of Chaos. Just as someone with Pattern is obvious in Shadow, so things connected with the Logrus seem to radiate a malevolent stench. Just a peek through the activated Pattern will reveal any nearby Logrus Master, or any activated Logrus artifact.

Traces of Trump. Searching for Trump with Pattern Lens is very difficult. The cards seem to have just a bit of Pattern within them (at least the Amber Trump have Pattern). That means they can be found with careful searching, but they don't radiate with any particular vigor.

Traces of Shadow. The breaks in Shadow, caused by those who have the Power to move from one Shadow to another, can also be detected through the Pattern Lens. The character looks for tiny disruptions in Shadow, attempting to find their source locations in the Shadow being studied, which other Shadows are involved, and what Power was used in the disruption. How long this takes depends on how old the trail and how major the disruption. If a single Hellrider passed through the Shadow just minutes before, the traces could be detected in just a minute or two. If the passage had taken place the day before, it might take an hour to find the traces. However, if the disturbance was major, as is the case if a vast army was led through, then the traces will be obvious and easy to find even days later.

Altering the Rules. By focusing the Pattern on a Shadow, or an area within a Shadow, it's possible for a character to affect the local rules. For example, some Shadows allow no magic, while in others technology won't work. With a bit of concentration, the Advanced Pattern Initiate can change these Shadows, allowing, or prohibiting, the use of different natural laws. In other words, the operation of things like gunpowder and electricity can be switched on and off at will.

Affecting someplace real, like Amber, is also possible, though it takes enormous energy. A character can bring the Pattern to mind, focus on a very limited area, say, the size of a gun, and impose the condition that gunpowder would work in that area. In this way, an Advanced Pattern Initiate could get a pistol to work in Castle Amber. Affecting things this close to the Pattern is very difficult, and tend to exhaust a character with an Amber Rank Endurance in just a few minutes. Once the character stops concentrating on the change, things will return to normal.

Creating Shadow Pockets. It is possible to create special "pockets" or rifts in shadow. How these work are up to each individual Game Master.

The basic idea is that all Shadow exists as reflections of Amber, diminished by the distortion of the distant Logrus. A Shadow Pocket would then be an artificial Shadow, with just a sliver of Amber's image, brought into existence by a character's minute warping of Pattern.

Shadow Pockets can be fixed to a particular location, or may serve as gateways from one Shadow to a distant other, or can be moved around by their creator. The environment inside the Shadow Pocket is also entirely up to its creator. One drawback to Shadow Pockets is their fragility. As artificial constructs they are easily destroyed by others with Advanced Pattern, or by the strong influence of Logrus.

Detecting & Disguising Traces of Pattern. The Blood of Amber is a substance that is very obvious to any Advanced Initiate of Pattern. If it is highly activated, as is the case with Advanced Pattern, then it's obvious from across a crowded room. Less activated blood can be detected at a touch, or when looking through the Pattern Lens.

The ability to "tune out" one's Pattern means that a character can, at will, appear to have no Pattern, or potential to walk the Pattern, or just a normal level of Pattern. This ability, depending on the Game Master, is something that is either gained with experimentation, or is something that must be purchased during Character Advancement for something in the range of five (5) points beyond Advanced Pattern Imprint.

Hellriding versus Advanced Pattern. Hellriding has several advantages over the use of Advanced Pattern, just as, at times, there are times when Advanced Pattern is far superior.

Seeking What is Lost in Shadow. When looking for something missing out in Shadow, when there is no clue as to where it might be, Hellriding works much better than Advanced Pattern. On a Hellride the character can simply picture the desired object, and Shift Shadow, zeroing in on the target. Advanced Pattern has no "intuitive" way of discovering a missing item, and a search through the Pattern Lens is always directed by the character, one Shadow at a time.

Seeking a Character's Shadow of Desire. Likewise, establishing a Shadow of one's desires is something that can only be done efficiently by walking or riding through Shadow. Advanced Pattern can be used to locate any place, including a character's Shadow of Desire, but only through a laborious searching process.

Moving Through Shadow. If, on the other hand, you already know where you are going, Advanced Pattern offer an instant travel ticket. This is especially useful for getting in or out of Amber, where normal Hellriding adds a day's worth of travel to any trip.

Affecting Probability. Both normal and Advanced Pattern Initiates can affect probability and the contents of Shadow. However, Advanced Pattern users don't need to move to shift things around, they can control the outcome of chance events while standing or sitting, with the power of their will alone.

Pattern Travel. Summon the Pattern to your mind. Now imagine walking it, traversing with your mind to its very center. Once you have done this (in a couple of minutes with total concentration, three times as long if you have distractions), you will be able to teleport yourself anywhere, instantly, just as if you had actually walked a Pattern. The only drawback is that your mental image of Pattern will be disrupted by the transfer, so Pattern will no longer be active in your mind.

Pattern Editing. The image of the Pattern in your mind can also serve as a tool to manipulate whole shadows and the pathways between them. With this method you can alter the structure of a shadow or even erase it altogether. Erasing shadows is not something you would do lightly as it seems to have a detrimental effect on the overall arrangement of Pattern and Shadow.

Erasing Shadow. Doing away with a Shadow, either on purpose, or by accident, will upset things in the larger system of Shadow. Shadow Storms, shifting of Shadows barriers, and the displacement of established pathways through Shadow can all result from a Shadow being wiped out. An even bigger problem is that elder Amberites and the Lords of Chaos will likely notice this kind of large-scale interference in the order of things.

Pattern Recognition. Any true initiate into the realms of Advanced Pattern will have a peculiar sensitivity to Pattern. Those of Pattern blood will be apparent by touch, and other Advanced Pattern masters will be obvious across a room, and sometimes even from a considerable distance, say, upon entering a common shadow. The energy of Primal Pattern, Amber Pattern, and the reflections of Amber, are all different, and you can sense which is which by sensing their Pattern energy.

LOGRUS MASTERY

45 Points.

NOTE: An absolute prerequisite for Logrus Mastery is Shape Shifting. Attempting to invoke the Logrus without the ability to Shape Shift to the demands of pure Chaos is subject to involuntary shifts in both form and persona.

Walking the Logrus is a dangerous task that inevitably leads to injury and madness. You accomplished your walk but spent a lengthy period in Chaos recuperating. Now that you have accepted the Logrus you may summon it to your mind with a few moments of concentration.

However, the Logrus is an ever-changing and never-repeating construct. Just keeping the slippery thing brought to mind requires a lot of concentration. For example, a character with Logrus brought to mind can't engage in a sword fight, except very defensively against vastly inferior opponents.

Logrus users tend to regard Pattern users as "flat-landers." That means they think of them as people who have access to a lot of power, but a power that is only effective in a limited area, in and around a pattern. Logrus, by contrast, works everywhere, even in the depths of Castle Amber (although it might be a bit risky to actually use Logrus in Castle Amber, as one of those nasty elder Amberites could take offense).

Attribute Tips for Logrus Mastery. Psyche drives Logrus more than any other Attribute. Amber level is really the minimum for effectively using Logrus. The higher your Psyche Rank, the more powerful will be your control over the Logrus. However, don't neglect Endurance. A character whose Endurance has dropped to Human level will not be able to safely manipulate the Logrus, and risks being absorbed by its dangerous instability.

Logrus Warnings: Madness & Vulnerability

The Madness of the Logrus.
In Zelazny's later series, we are told, by Merlin, that traversing the Logrus means going mad for at least a time. Here are two ways of imposing madness on player characters.

Blackouts. Characters who are not completely themselves may, for a time, simply lose control of their characters. What they did during their time "asleep" will be a mystery.

Delusions. Since you, the Game Master, are the eyes, ears and all the senses of the characters, you can inflict madness by betraying those senses. Perhaps the character will see monsters where friends are standing, or the eerie spectacle of monsters peering from around corners, just out the corner of the eye. Sure, the player may ignore it, but not safely. Delusions, in *Amber*, can turn to flesh so easily... In any case, the Game Master should mix this up enough so that the player can never be sure just what is real and what is not.

Vulnerability to Pattern & Trump.

Sensitivity to Pattern. Any time the Logrus, or any of its tendrils, comes into contact with something containing Pattern (like an Amberite), it sends a shockwave through the entire Logrus construct. The sensation will instantly travel along the tendril, and disrupt what the Logrus Master has brought to mind, as well as conveying a rather nasty headache. Any spells that have been hung on the Logrus can be destroyed by the touch of the Pattern.

The Hazard of the Logrus Touching the Pattern Itself. Using a Logrus tendril to touch a Pattern, can cause severe damage, or even death. Worse, should Logrus-charged blood come into direct contact with Pattern, it will instantly ignite. This means that characters who have Logrus, when wounded by a Pattern-based weapon, run the risk of their blood catching fire!

Sensitivity to Trump. While the touch of Trump is not so drastic as a touch of Pattern, it is dangerous in its own way. When a Logrus Tendril touches a Trump, or touches a character charged with Trump energy (anyone involved in a Trump contact is charged with Trump, as is any Trump Artist who wishes to exert the effort), it's like touching a pair of high-powered electric wires together. The Logrus Master will experience the shock, and will have to focus immediately on detaching the connection and on maintaining the Logrus.

Logrus Master Abilities

Logrus Summoning. You may use the arms of the Logrus to reach out into shadow and chaos for an item or person of your desire. Having found it you may then pull it towards you or pull yourself towards it. The reverse of this process, pulling yourself toward an object, is how Logrus Masters usually traverse Shadow.

Searching with Logrus Tendrils. Locating objects using the tendrils of the Logrus is always dependent on player character choices. Logrus Tendrils can be used to search selectively, at the direction of the Logrus Master, or automatically, relying on the chaotic movement of the Logrus itself. Here's how it works:

GM: The cupboard is empty. There's no sign of the Crystal Ball anywhere in the one-room house. What are you doing?

Beth: He got here before us! Let's go get him!

Kevin: We don't have time to go walking in Shadows. I'll just search for the thing with the Logrus.

GM: Okay. What are you doing?

Kevin: I reach out with my Logrus filaments...

GM: Wait a minute. You don't have Logrus brought up to mind.

Kevin: Oh, yeah. Okay, Roderick will concentrate on bringing the Logrus to mind.

GM: A few minutes later, you've got the Logrus up and vibrating in your mind's eye. Now what are you doing?

Kevin: I reach out with my Logrus filaments.

GM: The thin lines flail outward from Roderick. Where are you directing them?

Kevin: I'll picture the Crystal Ball, and reach the lines towards it.

GM: What are you picturing?

Kevin: A round, glass ball.

GM: No particular direction? Or are you just letting the Logrus find its own way?

Kevin: Whatever would get it quickly.

GM: That would be leaving it up to the Logrus. There's a sense of contact, your filaments lash over a sphere's cool, glassy surface.

Kevin: Great, I'll yank it back here.

GM: Into your hands?

Kevin: Hmmm. No, right back to the cupboard.

GM: No sweat, the crystal sphere appears, snugly fitting into the velvet depression.

Beth: I'll look into it, seeing if there's any impression of who stole it.

GM: Okay, Yvonne is looking into the crystal sphere.

Kevin: See, (to the rest of the group), wasn't that a lot easier than chasing around through Shadow?

GM: Beth, you see that the ball isn't really clear. It's filled with little specks of shiny metal. It's also got three holes drilled into the top part.

Beth: Holes? There aren't supposed to be any holes in the Crystal Ball. What do they look like?

GM: Well, it's clear enough so you can see that the holes aren't all that deep. One hole is bigger than the other two, and spaced farther apart.

Beth: Yvonne will pick the stupid thing up by the holes. Roderick, you idiot, you just retrieved a bowling ball!

Kevin: What?!?

GM: It does seem to be a bowling ball. Just made out of a glassy material. Kind of pretty.

Willy: Can we go search now? The longer we wait, the worse it's going to get.

Kevin: Hey, it wasn't my fault. I want to try again.

GM: Okay, you've still got the Logrus up to mind. What are you doing?

Kevin: This time I'll just stick to this Shadow. I send out the Logrus filaments again, searching outward, sensing for the crystal ball.

GM: You focus on the filaments, flickering them outward and away. Minutes pass as they spread miles and miles away.

Beth: Is this going to take long?

Kevin: Is it?

GM: No, you can probably search the whole world this way in ten or twenty minutes.

Kevin: Okay, I keep going.

GM: You make a contact! What are you doing?

Kevin: I haul it here.

GM: In your hands?

Kevin: Yes!

GM: Okay, you're now holding a crystal ball.

Kevin: The right one?

GM: I don't know, what are you doing to find out?

Kevin: I've got the Logrus up already, so I look at it through the Logrus.

GM: Not a trace of anything magical or mystical here. Seems to be a piece of ordinary crystal. Though just the right size and shape.

Kevin: I try again!

Willy: What's the chances of finding the wrong ball again?

GM: Well, most every Shadow is a whole world. I suppose most of them have at least a few crystal balls of the right size.

Kevin: Willy, that doesn't matter. My mistake was not describing the exact Crystal Ball. I need to add in the magical description, as well as just the way it looks. Then I can search for just the specific Crystal Ball we're looking for.

GM: Absolutely. You can either go Shadow by Shadow, or you can just extend your filaments outward, until you find exactly what you're looking for.

Willy: How long could that take?

GM: Depends. Shadow by Shadow, it could take forever because there are an infinite number of Shadows. Just reaching out with Logrus is kind of haphazard. It might be found in a few minutes, or a few days. Of course, if the ball is protected somehow, shielded from the Logrus, or in a Shadow barred to the Logrus, then it will never work.

Kevin: Hey, it's worth a shot. I'll concentrate on the exact object, and reach out, letting the Logrus find the ball.

Beth: Yvonne is leaving!

Logrus searching has advantages and disadvantages. The advantage of using Logrus is that you can find generic objects fast. Grabbing any defined object, sword, a plate of food, a lantern, anything defined in a general sense comes quickly as the Logrus finds the nearest occurrence in Shadow and snaps to it.

Unfortunately, the more specific you get, the longer it takes for the Logrus to find anything. In the example above, Kevin "pictured" the Crystal Ball and then said he wanted it "quickly," so he got something that looked like his description.

Logrus Combat and Manipulation. You may summon the Logrus and shape its force into useful shapes, extensions that are far more powerful, and more flexible, than normal Shadow matter. These extensions of Logrus operate with a Strength that is equivalent to the Psyche of the Logrus Master who controls them.

Logrus Sight. You may use the Logrus as a sensory lens, either "seeing" through its center, or by "touching" a distant place with a Logrus tendril. In this way you can see the aura of Pattern or Logrus, or feel the distinct sensation of Trump or Magic.

Using the Logrus as a lens, to examine anything in sight of the Logrus Master, will reveal several things. First off, anyone who is charged with either the Logrus or Pattern will be immediately obvious. Any item can also be seen to be either a thing of Shadow, of Amber, or of the Courts of Chaos. Magic, whether from spells, forces, or creatures, is obvious. A careful use of Logrus Sight, combined with touching of Logrus Tendrils can reveal the general level of a character's Psyche, and whether or not they are charged with any ambient magic (i.e. Power Words).

Logrus Spell Storage. You may use the Logrus as a "rack" for the storing of spells. However, it always takes at least ten minutes, plus the normal spell casting time, to "weave" the spell into the Logrus, and any spell will "decay" within days or weeks.

Logrus Defense. Bringing the Logrus to mind, you may then use it to defend yourself in one of two ways. One way is to fill your body with the force of the Logrus, making it resistant to the effect of Pattern, Magic, Psyche, or other forces, but providing no physical protection. Another way of using the Logrus is as a "shield," blocking in any one direction, which is effective against physical, energy and most Magical attacks, but does not protect against Pattern, Logrus, Trump or Psyche. You cannot use both techniques at the same time.

ADVANCED LOGRUS MASTERY

70 Points. Or, if you've already got Logrus Mastery, an additional 25 Points.

Attribute Tips for Advanced Logrus Mastery. As with standard Logrus Mastery, Psyche and Endurance are important.

Realms of Chaos. Unlike Amberites, who simply Hellride to their perfect Shadow retreat, those of the Courts of Chaos will find a Shadow, and then manipulate it to perfection. Since Shadows created with Advanced Logrus have a tendency to drift back to their original form, most Advanced Logrus Masters settle on a particular domain as home. It is then shaped and maintained with a combination of Logrus, Magic, and powerful Chaos creature servants. While the personal realms of Shadow will, with precautions in place, survive a long absence, very few Chaos Lords ever maintain more than one realm at a time.

Starting with a Primal Plane. Locating an unoccupied Primal Plane, conveniently located near the Courts of Chaos, is difficult. Some of the friction between Lords of Chaos comes from the frequent bickering over handy Primal Planes, and battles can often be fought over those abandoned or ill defended. However, once a Lord of Chaos obtains a Primal Plane, and uses it as the base Shadow for a personal realm, it has many advantages. First, it tends to have more of a base reality, meaning that it is more resistant to any attacks by forces or either Logrus or Pattern. Second, Primal Plane realms, if properly cultivated, create a sphere of secondary Shadows, making the Chaos Lord's domain all the greater.

Advanced Logrus Abilities

Manipulate Shadow. Using the Logrus, and the Chaos resident in any Shadow, it is possible to manipulate the structure of that Shadow, altering it to fit your wishes. In this way, for example, the physical laws of the Shadow, or the form of its inhabitants, or any simple detail, will be changed. It takes about as long as it takes Pattern initiates to Shift Shadow. Eventually, if not tended, the Shadow will drift back to its original form. The closer to Primal Chaos, the easier a Shadow is to manipulate, and Shadows close to Amber are nearly impossible to change.

Summon and Control Creatures of Chaos. Through observation and study, you can now recognize many of the Creatures of Chaos, and, more importantly, you have learned their weaknesses and the ways of binding and controlling them. Just as there are infinite Shadows surrounding Amber, so there are an infinite variety of powerful creatures adrift in the fragmented Shadows of the Logrus. A creature of Chaos is put together with the same rules as those for artifacts and creatures. Figure out the points, and then take an hour of searching for each point, plus another ten minutes of binding and controlling for each point.

Shape Logrus Servants. The Logrus itself, either in part, or in whole, can be brought to life and manipulated as a servant of the Advanced Logrus Master, but only so long as full concentration is devoted to the task. In addition, it is possible to separate a tendril of the Logrus, give it a simple instruction, and leave it, untended, to perform its duty. Logrus servants have Strength and Psyche each equal to Chaos Rank, and a Human level of Warfare. In place of Endurance they have a tireless Stamina, but can only heal or regenerate in a place where Logrus is strong. They have no "minds" as such, and will simply dissolve if attacked by a superior Psyche. The number of Logrus servants on "duty" at any given time depends on the Psyche of their creator.

Summon Primal Chaos. You may summon "Primal Chaos." This force, a direct connection with the untamed Chaos at the remote center of the Courts, will bring total destruction to whatever Shadow it occupies. Once summoned it must be dispelled quickly, or it will run amuck, eventually absorbing the entire Shadow back into nothingness. Summoning Primal Chaos too close to Amber or Pattern is very, very dangerous, as there is little control in those places, and the Logrus Master could easily be consumed. The merest touch of Primal Chaos will instantly destroy anything of Shadow.

Here's the perfect example of how, in *Amber*, the player characters have the ability to destroy whole universes. Primal Chaos destroys anything, and very little can stand against it (though someone with Advanced Pattern might like to try). Once released, unless halted before it gets out of hand, Primal Chaos will destroy an entire Shadow.

Summoning Primal Chaos into Amber. A place of true reality, like Amber, or any nearby place, are resistant to the effects of Primal Chaos. Unless a character continually "fed" the Logrus, and forced the Primal Chaos to remain in existence, it will simply fade away near Amber.

TRUMP ARTISTRY

40 Points.

Trump cards are an eternal link between image and reality. Trump can be used for communication, across limitless Shadow, or for transportation. The subject of a Trump may be a person, creature, place or thing. The cards of the family of Amber are only one aspect of this discipline. The royalty of Chaos (if they can be called such) also have decks of trump.

As a Trump Artist you are in a unique position. Trump cards are one of the very few items that can't be found in Shadow, can't be bought (except by a Trump Artist), and can't be Conjured. The only way to obtain a Trump is from a Trump Artist.

A Trump, once created, is a special artifact. Once filled with the power of the Trump, a card becomes invulnerable to any conventional forces. Likewise Trump are immune to Logrus Tendrils (attempting to Logrus "grab" a Trump can be painful), and the power of the Pattern. Trump sketches, on the other hand, have no particular innate power, and are only as durable as their paper medium.

Attribute Tips for Trump Artistry. Psyche is the most important Attribute for a Trump Artist. Not only is it essential to have a good Ranking to use the Trump, but the better your Psyche, the quicker you'll be able to absorb the essentials of any new Trump subject.

Trump Artist Abilities

Creating Trumps. This involves creating a card representing some particular person or place or object. When the user concentrates on the card there will be a psychic bond between the two. Creating a trump takes two days of concentrated work. Working from memory, without a live model, will double or triple the amount of time it takes to create a card. Cards made from descriptions, without the Trump Artist ever having observed the subject, have the large possibility of linking to a shadow, instead of the actual subject.

Trump Sketches. With simple drawing materials you can sketch a place from memory. This takes about 30-40 minutes. The resulting sketch has the same powers as a full trump but is not as strongly attuned to the subject. The image is not permanent, unlike permanent Trump, and, if the subject changes, the sketch will lose its power. Once a Trump Sketch is created it takes more concentration to use it than for a comparable Trump card. If the subject of a Trump Sketch changes form, or moves through Shadow, then the sketch will stop working.

Trump Identification. A Trump Artist can often tell the identity of a caller, without opening to the contact. Upon receiving a Trump Contact check through your Trump Deck. The card of whoever is making the call will be psychically active (though if another Trump conversation is already taking place, this will narrow down the possibilities). Obviously, this works only if a character happens to have the caller's Trump.

Trump Defense. Trump Artists, recognizing the power of the Trump, can use it for a defense. Evoking a Trump defense requires using a Trump card (your own card works best), and concentrating on the cool sensation. So long as the concentration is maintained, the Trump will protect the character against intrusions of Pattern, Logrus, Psyche and most Magic. Maintaining a Trump defense is dependent on the character's Endurance. Against major powers, maintaining a Trump defense is just as difficult as a mind to mind battle.

Sensing Trump. As a Trump Artist you can recognize the Trump quality in any object. Likewise, you'll be able to recognize any Trump object, telling the difference between a normal painting and one with Trump power.

Creating Trump Powered Artifacts. Trump Artists, when constructing items, specifically Trump Artifacts, can "build in" Trump Powers. See "Artifact & Creature Creation" for more details.

ADVANCED TRUMP ARTISTRY

60 Points. Or, if you've already got Trump Artistry, an additional 20 Points.

Attribute Tips for Advanced Trump Artistry. Psyche is all important.

Advanced Trump Artist Abilities

Trump Memory. In creating a card, either a new one, or redrawing an old one, you memorize the image, so it is permanently imprinted in your own mind. That means you can attempt to contact an image without an actual card being present.

Trump Spying. There is another level of concentration where you can simply observe the operation of Trump. This allows you to overhear conversations as the Trump is being used. However, you must actually touch the specific trump. If you have Trump for only one of the participants in a Trump Contact, you will overhear only what that one person says, and will be unable to "hear" the other end of the conversation.

Trump Gate. The ability to create an open doorway from any one place to any other. It requires the construction of a Trump Sketch (or card) for the destination point. It requires a tremendous amount of personal energy in order to be opened, and even more for the maintenance of the Trump Gate.

Trump Jamming. It's possible, with intense

concentration, to jam any one, or any number, of Trumps. It requires the actual card of the subject of the jam. This can work one of two ways. The first can be done on several targets, just concentrate on all the cards and none of them will be able to receive calls. Another possibility is to concentrate on a single card and the victim will be blocked from sending or receiving Trump calls. If the victims of this ploy simply give up, then there is no problem. However, should someone "push" the contact, the Trump Artist will be involved in a Psychic battle.

Trump Trap. The ability to capture others in trumps, or project them to other places against their wills. There are a wide range of possibilities, from Trump that work nearly automatically, so they are activated by the merest touch or glance, to Trump that, when activated, automatically transport the wielder to the place they depict. Trump Trap also allows the Trump Artist to create "disguised" Trump, where the image that appears on the card is not the true subject of the Trump.

SHAPE SHIFTING

35 Points.

This is the power to manipulate the tissue of your own body. There are certain inherent limitations upon this. For example, you may not add to or subtract from the total mass of your body using shape-shift.

Shape Shifting comes with built-in dangers. While the wielders of Pattern or Logrus run the risk of destroying the universe, Shape Shifters face the day-to-day challenge of being able to destroy themselves. For example, since the power involves self-change, there's always the danger of changing to a state where a character can't change back.

Fair Warning. There are a lot of nasty things that can happen to player characters with Shape Shifting. Most Shape Shifting is done at a character's leisure, carefully and without difficulty. However, under time pressure, or when the character has been over-doing it, things can get more difficult.

Time Needed for Shape Shifting. Changing a single feature, like toughening a character's skin into armor, or changing fingernails into claws, is very quick, taking only ten or fifteen seconds. The complete transition to a familiar form, one that the character knows well and uses regularly, can be attained in just a couple of minutes. The fastest complete change is when a character goes "home," back to one of their natural forms, and fastest of all if the character returns to Human form.

All of which assumes a character being in good shape, rested, well-fed, and uninjured. Lacking any one of these conditions, the character can take twice as long to perform a change. If in bad condition, starved, or exhausted, then Shape Shifting becomes arduous and time consuming and may become impossible altogether.

Endurance Limitations. The process of Shape Shifting is enormously draining. Each cell of the character's body is stressed, fatigued and drained of energy. Even one ordinary Shape Shifting will exhaust a character of Human Rank Endurance. Chaos Rank Endurance characters can Shape Shift a couple of times a day without great strain. Characters with Amber or better Endurance can Shape Shift at will, needing only adequate nourishment (which can be a heck of a lot for an Amberite!).

Learning New Forms. Shape Shifters can learn just about any new forms, given enough time and practice. For example, the first time a character attempts to imitate someone else, they'll need some kind of reference to the features they want to imitate, and a mirror to check the results. However, given a few days of dedicated practice, a Shape Shifter will have learned the form and will be able to take it on again with only a short investment in time. The change to one of these practiced forms isn't as quick as a change to a natural form, but most Shape Shifters can change to a well-known form in just a few minutes.

Attribute Tips for Chaos Shape Shifting. Psyche is

important for maintaining your original personality in Shape Shifting. However, even more important is Endurance. Anything less than Amber Endurance will drastically limit the number of Shape Shifts you can perform before becoming exhausted.

Shape Shifting Abilities

Shape Shift Wounds. Once wounded, you don't just heal quickly, you shape shift the body to repair itself. Open wounds can be closed quickly, even during combat. Repairing flesh, knitting bones, and, worst of all, regrowing nerves, takes more time. Shape Shifting drastically increases healing time, but at a cost of the body's energy. If you've actually lost something, a part of your body, like an eye, or an arm, then you have to go through the long, slow process of regeneration controlled by your Endurance Attribute. In the meantime, that feature will be missing, no matter what shape you take.

Shape Body Parts. The changes you make depend on the parts of the body. It's easier to manipulate soft tissue, then cartilage, then bone. Most difficult things to change are nerves. Holding a change requires full concentration, as your body always prefers one of its familiar forms.

Shape Facial Features. With a massive effort, and a mirror for reference, you can change your facial features, even to the extent of disguising yourself, or imitating someone else. However, this requires constant concentration. When concentrating on something else, like a Trump battle, you'll automatically start shifting back to your normal face.

Shape Natural Forms. Each character has three "natural" forms which will be much easier to attain than any others. The range of forms gives the character the capacity to cope with a wide range of environmental changes.

Human Form. The primary form for any player character will be human. In this form the character has no bonuses, no particular advantages, but there are no drawbacks either. While in Human Form the Lords of Chaos will be considerably more resistant to Pattern.

Chaos Form. The first form is designed for combat, and generally is demonic, with scales, fangs, claws, and other natural hardware. Even characters from Amber will have a Chaos form, although it's possible that they may not discover it for many years.

Often called the demon form, this is simply because of the character's appearance. Demon forms have the same basic personalities as Human. Chaos Dwellers recognize that this form is best for dealing with a wide range of environments, any of which can show up at any time in the Courts of Chaos. The character is naturally armored in this state and equipped with natural weapons like talons, fangs, horns, and even spine-tipped tails. In other words, Chaos form is the Shape Shifting equivalent of putting on a suit of armor, and wielding a range of weapons. While the form is well equipped for Combat, there are no Attribute advantages of any kind.

Avatar Form. The Avatar differs depending on whether the character is from Amber or the Courts of Chaos. If from Amber the Avatar form will be of some animal, usually of a size and weight comparable to the player character. Chaos dwellers have Avatar forms that are much more dramatic, causing the body to turn into living fire, stone, air or water. Elemental Avatar forms are suited for their particular environments, common near the Courts of Chaos, but are difficult to maintain in Shadows based on Amber's Pattern.

Avatars for Amber characters are simply animal shapes. However, for Chaos dwellers, the Avatar form is more exotic, designed to adapt the character for survival in incredibly hostile Chaotic Shadows. Elemental forms are usually based on either fire, earth (stone), air or water, and are best suited for those environments. Trying to maintain, for example, a water form, outside of a watery environment is extremely difficult.

Shape Shift Animal Forms. Visualizing the persona of an animal allows you to take that animal's form. This works best if you have first made a Psychic contact with one of the animals. Once the form is "learned" the character can accurately copy it at any time. You can Shape Shift claws, teeth, hooves and horns, and armor like scales or plate. You can Shape-Shift to a bird, or bat, capable of flight, but will be able to fly only if your character has sufficient Strength (Amber or better). After all, unless your character is really small, you're trying to fly with a pretty hefty load.

While it is pretty easy to change into animal forms, it can be difficult to learn to use their various abilities. The first time in bird form, for example, the character will know nothing at all about flying, and will have quite a difficult time of it.

Automatic Shape Shift. Responding to the sometimes rapid changes of environment in Chaos means developing the ability to "let go" and allowing the body to Shape Shift automatically in response to dangers or changes in the area. This also includes the ability to Shape Shift defensively, in response to various kinds of attacks, including those of Magic, Logrus, Pattern, and Trump.

While dangerous, there are times when there is no choice but to rely on an automatic shift. When a character is falling, drowning, strangling, being poisoned, bleeding to death, or otherwise threatened with immediate death, there's just no time to do a careful, well thought out Shape Shift. In these cases you just take your foot off the brake and hope for the best.

What happens is that the unconscious part of the brain takes over, linking directly into the body's Shape Shifting function, and changing to the form best suited for surviving the current threat. The character can try to resist the Automatic Shape Shift, and success is usually determined by the character's Psyche. This becomes very difficult if the body has been conditioned by a lot of previous use, so there's no time left for the character to try to resist.

Examples of Automatic Shape Shift. Here are some of the possible stresses that can trigger an Automatic Shape Shift, and how the body will likely respond. This is not an exhaustive list, and certainly there are other possibilities.

Falling. The response of the body's Shape Shift to falling is two fold. First, the body tries to catch as much wind as possible, in an effort to break the fall. Even in a short fall vestigial wings, or underarm skin flaps, will ripple out from the character (provided that the character's clothing or armor isn't too tough or too constrictive). Second, the hard parts of the body soften, while the body's internal organs develop layers of muscle. Having hit, the body will continue in the Automatic Shape Shift mode, proceeding with repairs and letting the changes slip back to normal.

Drowning. The first Automatic response is to cut down on the body's need for oxygen. Unless the player character overrides this impulse, the character will end up comatose, but well preserved. Survival in this form is indefinite. It's possible for the character to continue changing, slowly, even while in a coma, to Shape Shift into a form that can breath water.

Poisoning. On Automatic the character's body will go through incredible contortions and internal shifts, attempting to isolate and expel the foreign substance. The character will ooze from every opening in the body, including the sweat glands. Poisoning a Shape Shifter is extremely difficult, but not impossible. For members of the Courts of Chaos, poisons that interfere with a character's ability to Shape Shift, or force a Shape Shifter to take on a particular form, are pretty common knowledge.

Mortal Wound. The Shape Shifter's version of Shock is to shut down the body and devote all the Shape Shifting Power to fixing the character's vital organs. In this state it is possible for the character to stop all signs of life, so that no breathing, heartbeat, or even brainwaves are active.

Pursuit. With the panicky reaction of flight from danger the Shape Shifter can automatically start to shift to a body more suited for running. The legs lengthen, with an increase in muscle and tendons. Even more significant are the effects on the brain of the character. The character's personality will undergo a subtle change, giving the character a relentless desire to keep running.

Battle. There is a tendency for the body to automatically move into Chaos form when wounds and exhaustion start to take their toll. While common enough for those from the Courts, Amberite Shape Shifters sometimes discover their Chaos form in this way, but only when the stress is enormous.

Burning. The heat and sensations from burning throw the Automatic Shape Shift into compensating for the heat. The character's skin and exposed parts will grow a heat-resistant layer of armor, and the body will devote more of its energy to air conditioning, while at the same time acclimating to operating at a higher temperature.

Primal Form. The most dangerous, yet most durable form is that of the Primal Form. When a character takes this state they become a being totally wired in to their Shape Shifting ability, so that they instantly react and change to accommodate anything threatening in their environment. This means being able to always escape a predator, overcome prey, and survive nearly anything. Each character's Primal Form is different, a reflection of their deepest hopes, fears and drives. Some characters will become vicious and dangerous predators, attacking madly any vulnerable living thing. Others will be more flight oriented, fleeing from any and all possible threats. Any Primal Form can instantly adapt to changes in atmosphere, environment, temperature, or gravity. Adapting to breathing underwater or in any liquid medium is also automatic.

NOTE: No character is ever "conscious" while in Primal Form. The danger is that a Primal Form isn't really intelligent. The character's personality, while it may survive in some dormant form, is no longer in control of the character's body. For this reason a Primal Form is usually a measure of last resort.

Waking from Primal Form. Another problem is that Primal Forms may never sleep or rest. Driven by their pure urge for survival, they may never let down their guard long enough for the player character's personality to resume control. It's possible that the only way out is for the character's friends to capture the Primal Form and force it to change. Or the character may wake up in some incredibly distant Shadow, the only place where the Primal Form felt "safe" enough to nap. Another possibility is that the character's personality will only become conscious when the Primal Form is so threatened or trapped that it can't get out with just raw fury and Shape Shifting.

Shape Shifting Gone Wrong.

Eventually, if players push their Shape Shifting, they'll go over the line into some kind of personal horror story. The mistakes are almost never life-threatening, but they are agony for players. Still, getting burned by Shape Shifting is another great way to build up the player character's personality through adversity.

Losing Personality in Shape Shift. One of the most common problems facing a Shape Shifter is losing their own personality, their own sense of self. Usually this is a result of the character attempting to imitate another character, or trying to take on the form of some psychically powerful entity. When this happens a player may lose control of the character, making it necessary for the Game Master to take over. In the best of cases, the player character's persona returns to control whenever the body goes unconscious. However, it is possible that the "foreign" personality can take over for extended periods.

Losing the Power to Shape Shift. This is a particularly dangerous and debilitating problem that can take one of two forms.

Loss of Shape Shift Skill. The character's mind

can lose the ability to control the Shape Shifting mechanism of the body. The character will then have to relearn Shape Shifting through trial and error, a process that will probably take a couple of weeks.

Loss of Shape Shift Power. A more serious condition, the body's cells lose their special Shape Shifting property. The body becomes immutable and unchanging. Eventually, as normal healing and regeneration take place, the cells will recover. How long this will be depends on the character's Endurance. Those with Human Rank will need a year or more of restoration. Chaos Rank characters will be back in Shape Shift form in about a month. Having Amber Rank means being able to go back to Shape Shifting within a week. Higher Ranks are faster yet.

Loss of Both Skill and Power. The body's cells will have to recover completely before the character can start exercising the skill of Shape Shifting again. Add the recovery time and the relearning time together and that's how long it will be before the character is back to Shape Shifting normally.

Involuntary Shape Shifting. In some cases the player may end up conditioning their character's body into reacting in certain ways automatically. Most of the time this is good. For example, if a character falls from a great height, consciously thinking about making a change isn't nearly as fast as letting the body do the change automatically.

Infection with Primal Chaos. Player characters who push their Shape Shifting too far, inflicting incredible stress on their body, run the risk of their shape shifting going berserk. Every living thing must have a bit of Primal Chaos as part of their makeup. Without chaos there is no birth, growth or aging, only stasis. Shape Shifters are those who can control the element of chaos that dwells within every living thing.

Primal Chaos Cancer. Characters who have pushed themselves too far, through exhaustion, fatigue and starvation, can find that they lose control of some of their body's cells. This is almost always discovered too late for any preventative measures. Characters will find themselves constantly drained as both body and Shape Shift abilities attempt to contain the cancer. Both "tumors" and wild cells roving through the body, will Shape Shift with incredible speed, adapting to consume other cells, and multiplying every few hours. If left unchecked, Primal Chaos Cancer will eventually attack the character's vital organs, appear on the skin, and will generally start to eat the character alive.

Total Primal Chaos. If allowed to completely run its course, a Primal Chaos Infection will not necessarily kill the character. Instead every remaining cell of the character's body will become a separate tiny Shape Shifter. Effectively the character becomes a mass of constantly changing cells, and looking like an amorphous blob. No thought will remain, because the neural connections will be severed. However, even in this state it's possible for the character to be restored to normal form.

Curing Primal Chaos. Specific cures are up to each individual Game Master. One possibility is the use of Advanced Pattern, though this is incredibly destructive as the Primal Chaos literally explodes through the body. There are also "Cancer" specialists out in the Courts of Chaos trained in dealing with this kind of problem, though their methods are rumored to be barbaric and torturous.

Danger of Automatic Shape Shift. Letting go of your body's natural facility also means surrendering control of what form it may eventually take. That means that your brain may also change, along with the body. The new form may have urges, desires, or fears that are unnatural to your normal personality, but which you will be helpless to resist. In the worst case, you may lose your personality altogether, regaining consciousness hours, or even years after the Automatic Shape Shift.

ADVANCED SHAPE SHIFTING

65 Points. Or if you've already got Shape Shift, an additional 30 Points.

Attribute Tips for Advanced Shape Shifting. Equal in value are Psyche, for sensing the proper mind set of any new shape, and Endurance, for surviving the stresses that shifting puts on your body.

Advanced Shape Shifting Abilities

Shape Shift Aura. This shifts the mind's Mental Structure so it takes on a completely different Psychic aura. Note that it doesn't change the ability, the Psyche, or anything else. It just changes the appearance of the mind, so that a Psychic touch will not detect the difference. Some of the possible auras include that of a plant or animal, the lesser Psyche of a normal human, or the imitation of any person's Psyche known to your character.

Shape Shift Persona. Part of the problem with trying to imitate someone else, regardless of whether you look, sound, smell and feel like them, is that you can still give yourself away by not acting like them. Shape Shift Persona lets you change your personality so even slight, subtle mannerisms can be imitated.

Danger of Shape Shift Persona. If you attempt to Shape Shift into the personality of a particularly powerful mind, you run the risk of it overwhelming you. In essence, you actually become that person, and may totally lose control over your character. Usually this ends with sleep or unconsciousness, when your own natural Persona will come back into existence.

As the Game Master, you are constantly presenting to the

player their character's view of the surrounding world. This view can change in subtle ways, influencing showing how a character's personality may interpret different things. For a player character who has taken on a different persona, some of these subtleties may get a little confused:

GM: "You see a small shape dart in front of you, moving a little too quickly, some weapon clutched in its paw. It screams a threat and points at your head. What are you doing?"

GM: "You see a child running fast toward you, some toy weapon in its hand. Skidding to a stop, the youngster smiles, screams something you don't understand and points at you. What are you doing?"

The Game Master responses from this point can continue to be colored by the artificial persona. "You don't know what is being said, but it's obviously a threat," if the persona is suspicious or aggressive. Or, with a neutral tone, "You haven't the vaguest idea what the kid is talking about. It's just loud." Finally, if the persona is kindly, "The words don't mean anything, but the way the kid is talking you know it's part of a game and that he's telling you to surrender or else."

Of course, for all the player knows, from either description, the kid could be pointing a lethal weapon. The first description warps things so they appear more dangerous, but it could be that the situation really is dangerous. On the other hand, if a player has taken on a gentler, kinder persona it's possible that the second of the two descriptions will apply, even when the character is confronted with something threatening.

Example of Character Imitation with Advanced Shape Shift. Let's assume for the sake of this example, that Roderick, Kevin's character, has gone from Shape Shift to Advanced Shape Shift. He's facing a rough situation where his usual means of escape are cut off, and his Warfare (still Chaos Rank) is just not up to handling the situation.

GM: No good, the alley is a dead end.
Kevin: I look around, how far behind me are they?
GM: The crowd is gathering at the only opening to the alley. They seem uncertain. From previous experience you think they're waiting for the soldiers.
Kevin: I'll try bringing up Logrus again.
GM: Nothing. Even now, in your Chaos demon shape, something is blocking you from using Logrus. You hear a cheer from somewhere at the back of the crowd. What are you doing?
Kevin: Great, just great, the soldiers are probably here again. Any chance of fighting them off?
GM: Their pikes and halberds are really nasty, designed for armor a lot tougher than yours. Plus they've got too much reach, so they can keep you a good six or ten feet away. Yes, the crowd seems to be parting.
Kevin: I need Warfare. Where are Dorell and Harick when you really need them?
Cindy: Hey, it way your idea to go off alone, remember?
GM: You're also hearing some noise from somewhere high above, at rooftop level. What are you doing?
Kevin: Who do I know who has the greatest Warfare?
GM: Benedict, of course, but you've already lost your Trump deck.

Kevin: I'm going to change into Benedict. Right now.
GM: I've got to warn you Kevin, Benedict has a powerful personality, a sense of identity that's older than you can even imagine. You might never return to your own persona!
Cindy: Why don't you imitate Dorell? Wouldn't it be safer?
Alex: Yuch. No way would he want to turn into Dorell. Kevin, why don't you just go for Harick? You'll have Warfare to spare.
Kevin: Game Master, correct me if I'm wrong, but when I change into someone else, I don't get their full ability, do I?
GM: That's right. The copy is never quite as good as the original.
Kevin: Then I've got no choice. It's got to be Benedict. I'm doing it.
GM: Okay, you see the soldiers carefully assembling into their ranks in attack formation. As you watch your vision changes, it seems that you become acutely aware of their every mistake, as if you should really be over there giving them pointers. Also, your body is shifting to human form, but in a more awkward way than you've ever experienced before. Are you going to keep pushing it?
Kevin: Yes, I want the total Benedict package.
GM: You feel gradually shoved to the side, your perceptions taking over, your body becoming foreign and distant. And then, poof!
Kevin: What? What do I see?
GM: Nothing. I'll have to get back to you later.

Shape Shift Features. You can only accurately substitute yourself for another person if you've had close contact and the opportunity to practice. Partial shifting of features, like moving your facial proportions around so that you are no longer recognizable, is easy. However, doing this analytically, for example, trying to turn your hair red by picturing the color you want, doesn't work. Instead, you should try imagining some other red-headed character, someone you know well, concentrate on their hair, and then imitate the person. Substituting yourself for someone larger than you means you'll look like a skinny version of them. Changing into someone of a smaller form means looking like a fatter version of that character. However, it only takes about a week to fully adjust, losing or gaining weight as necessary, and to become an acceptable imitation.

Shape Shift Internal Structures. Your character can move around, change, duplicate, and augment the body's internal organs. Aside from the obvious benefits of generating two hearts, or a sub-brain, you can use this ability to regenerate lost body parts. Regeneration of lost parts takes about a week of total concentration, using Shape Internal Structures.

Shape Shift to Creatures of Power. Since some creatures have innate powers, it's now possible to take their form and use those powers. Certain shapes seem very interesting, such as the demon shapes and dragon shapes that promise tremendous power. Use the Item Creation Section as a guide for the possibilities.

Each Game Master's campaign can present different creatures of power. Dragons, for example, are mentioned in Zelazny's *Chronicles*, but are never really described. So the dragon form's powers will depend on the campaign. Likewise, in the minds of some Game Masters a dragon's personality could be very powerful, or very alien, and cause problems with either perception, or, as in the case of the Primal Form, could cause a loss of identity.

Shape Shift to Animal Abilities. Instead of having to shift into animal form to gain the animal's ability, you can simply change your own form to add the ability. For example, you can have your character sprout wings like a bird or a bat, grow gills, or develop claws or armored scales.

Shape Shift Others. Once a psychic contact has been made, it's possible for an Advanced Shape Shifter to impose Shape Shifting on someone else. If the subject resists, this is possible only after the Shape Shifter has completely dominated the other's mind with Psyche. On the other hand, if the subject of the Shape Shift is willing, the process can be quick and simple.

Creatures of Blood.

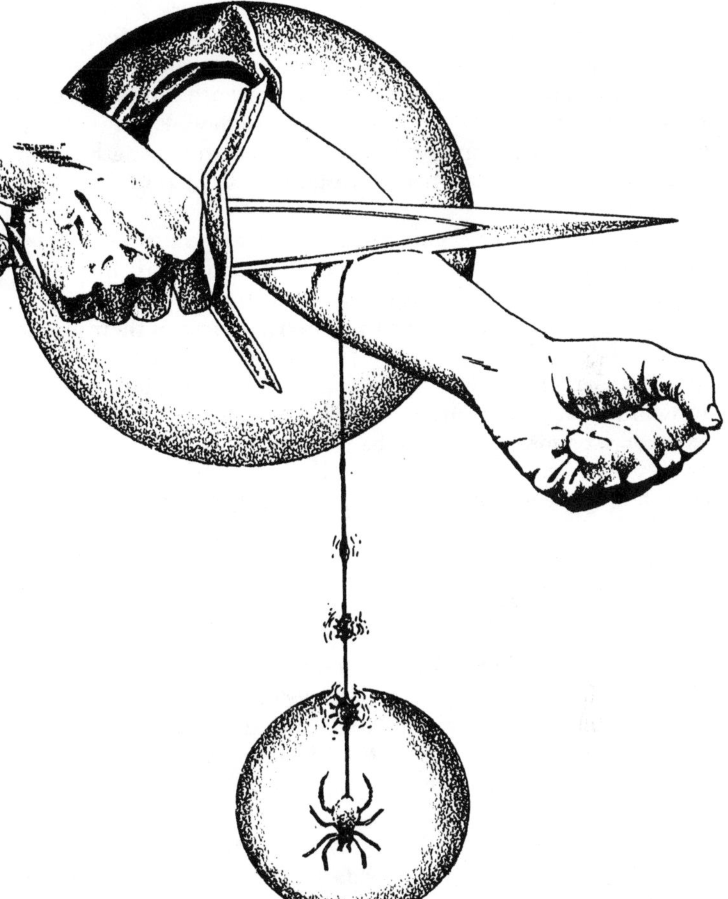

He took hold of my arm with his left hand and drew his dagger with his right.

I watched as he cut my arm, then resheathed his blade. The blood came forth, and he cupped his left hand and caught it. He released my arm, covered his left hand with his right and drew away from me. Raising his hands to his face, he blew his breath into them and drew them quickly apart.

A crested red bird the size of a raven, its feathers all the color of my blood, stood on his hand, moved to his wrist, looked at me. Even its eyes were red, and there was a look of familiarity as it cocked its head and regarded me.

'He is Corwin, the one you must follow,' he told the bird. 'Remember him.'

The Courts of Chaos

Shape Shift Blood. The blood of an Advanced Shape Shifter has very special properties. Cut yourself, bleed out a few drops, and you can Shape the blood into a creature of your desire. This creature will have a measure of your own powers. So, for example, if you have the Blood of Amber, your blood creature will have a bit of power over Pattern as well.

Corwin's *bloodbird* is a prime example of a creature made of Shape Shifted blood. In this case the bird has the power to fly through Shadow and has an innate "homing" connection to Corwin. It's also important to note that the bloodbird is a non-player character. It operated independently of Corwin, delivering Oberon's package, then leaving, then returning to rescue Corwin at that moment when his life was in grave danger.

MAGICAL POWERS

Magical Powers, whether in the form of Power Words, or Magic Spells, or Conjuration, are not primal powers. Instead, Magic is a way of manipulating the stuff of Shadow. Since all Shadow is built of the subtle interplay between Pattern and Logrus, Magic works with, and against, both those Powers.

In some Shadows, Magic is a real and potent force, easily manipulated by local Sorcerers. These masters of local magic are commonplace, extremely powerful, but are limited in that their spells only work in that single Shadow. Even in their home Shadow, changing the magical "mixture," something an Amberite or Logrus Master can do easily, renders them helpless.

Such Sorcerers become much more dangerous when they are exposed to other Shadows. Once they know that they must tinker with their spells, altering them to match the magical environment, then they have the potential to work their Magic anywhere.

There are also places where Magic, by definition, will not work. In such places only Power Words and magical items will function, and most spells will be disabled.

POWER WORDS

10 Points.

Power Words are a limited form of magic, used mostly in a defensive way. Power words are easy to cast, instantaneous in effect, but have no lasting effect. Because any Power Word involves channeling a burst of Psyche, they are always noticeable, never concealed.

Attribute Tip for Power Words. More important than anything in using Power Words is Psyche. The higher the Ranking in Psyche, the more powerful the impact that the Power Word will have on its target. This is especially important in battling Spell Casters, where an inadequate Psyche, weakly driving a Power Word, may not be enough to break their concentration. Endurance is also important, since each use of a Power Word is a drain on the character's life force.

Learning Power Words.
Well, they don't have to take the form of Power "words," they could as easily be gestures, signs, whistles, finger snaps, or whatever the character likes.

Unlike other forms of Magic, which use the power of Shadow, Power Words are based on the power of the character who speaks them. Power Words aren't just words, they're actually a linkage, a magical connection, built into the body of character who owns them.

Range, duration and the effects of Power Words are all restricted to the immediate time and place of the caster. The effects are somewhat uncertain since the raw power used does not take into account the variables of chaos and pattern influence in a particular place.

Each character is limited to *exactly* five Power Words. It is possible to increase this number later, at character advancement time, at a cost of one point per additional Power Word. Once the five are selected, they may not be exchanged, switched about, or otherwise substituted. That's the five you start with, and that's the only five you're going to get.

And, while we're on the subject, Power Words are the only "Power" in Amber for which there is no significant advancement. With any other power an advanced version confers greater potency and greater understanding. Putting more points into Power Words just gives the character more Power Words.

The power words listed below are the ones that are well known. Players can use the "word" listed, or invent their own.

It's also possible to invent new Power Words. Bear in mind that, like all other Power Words, they must have an *instantaneous* effect. The new ones can be substitutes for the first five, or, at a cost of a point each, can be added on later. New Power Words must be approved by the Game Master.

Counteracting Power Words.
A drawback to Power Words is that they are fairly simple and routine. If an opponent is prepared, it's possible to overcome them.

However, counteracting Power Words requires specific measures.

For example, the most common Power Word, Magic Negation, breaks up a spell in mid-casting, while the Sorcerer is invoking the Lynchpins of the spell. The Magic Negation Power Word wouldn't work against an instantaneous spell, one that is completely prepared, requiring no time-consuming lynchpins.

It's also possible for a Sorcerer to be prepared in advance for a Power Word, and to defend against it. The preparation would require that the Sorcerer would have had to come to the conflict prepared with a protective spell. Such a spell would only be good against the Power Words the Sorcerer knows. Since Power Words are unique to each character, a Sorcerer can't make any sort of general or "all purpose" magical defense against Power Words.

For this reason characters with Power Words tend to be rather secretive. They only use them in genuine emergencies. That way they greatly reduce the chance of being observed, and of having their Power Words countered by others.

Making up new Power Words.
So long as the restrictions are followed, players are allowed, or even encouraged, to make up their own Power Words. In general terms, Power Words have to be fast and easy. Fast, in that what they do should happen in a flash. Easy means they can't do anything complicated or that requires explaining or defining conditions. For example, most Sorcerer's spells are complex enough to require several parts or phases, something impossible with a Power Word.

Power Word Energy Drain.
Since Power Words use the energy of the character, rather than magical energies, they tend to be rather draining. For most Amberites, most of the time, this is no particular problem. However, any character with less than Amber Endurance (Chaos or Human) will be

tired by each use of a Power Word. A character of greater Endurance, while not usually troubled, should beware of using Power Words when engaged in some other difficult task, like using the Jewel of Judgement, Walking the Pattern, or engaging in a Psychic Duel.

TWENTY POWER WORDS

1. Power Word {NOGTZ!}: Magic Negation.

This spell allows the disruption of gathering magical energies. As a Sorcerer casts a spell, or releases a stored spell, the Power Word disrupts the elegant structure of a spell and renders it useless. If used fast enough it can also fizzle trap spells or wards before they are unleashed. The more powerful the Psyche of the invoker of the Power Word, relative to that of the source of the Magic, the more effective it will be. It can also be used to neutralize active magic, including shields and wards, and other things charged with magical energies.

2. Power Word {POLRZ!}: Chaos Negation.

Used only against yourself. It works internally, inside your own body, as self protection. Defends against destructive changes to your person that might be rendered by the Logrus, or by other damaging Chaos attacks.

3. Power Word {SHAGSK!}: Psychic Defense.

Creates a burst of internal Psyche, enough to disrupt any contact that you are involved in. Also works great against the pressure of others trying to break into your mind. Lasts only for a second but very useful against unexpected attacks during times of psychic contact, or, for example, during a Trump call.

4. Power Word {ASKIIR!}: Psychic Disrupt.

Directly affects the creature. Naming the target, especially with its "true name" ("Merlin Askiir!") makes it much more effective. It works to disrupt the concentration of the victim. So, someone with the Logrus or Pattern brought to mind, or someone in a tenuous Trump contact, would lose their focus and have to start over again. Likewise someone casting magic would be interrupted. Against a "named" target it will work no matter what the difference in Psyche. Used without a name, it is possible that the object of the Psychic Disrupt, if superior in Psyche, will be unaffected.

Repeated use of the Psychic Disrupt Power Word, against the same opponent, will become less and less effective. Having once felt its effect, the victim can brace for subsequent attacks.

5. Power Word {KROLAK!}: Neural Disrupt.

Directly affects a creature, throwing off their coordination, and causing them to twitch or flinch involuntarily. This is best used in a critical moment of battle, changing the Warfare situation for a moment. Out of combat it works if you need to grab something away from somebody, knock them over, or do something where a momentary break in their coordination comes in handy.

Naming the target, like "Oberon Krolak!" makes it effective against that opponent. If the target is unnamed, then the user of the Power Word must have a superior Psyche.

After the first time, it's possible for victims to steel their mind against the Power Word, and resist its effects.

6. Power Word {LIEVZ!}: Lifeforce.

This Power Word generates a burst of living energy in the target. Doesn't actually have enough time to do any healing, but does break destructive anti-life energies. Also functions well as a "wake up call" for those who are sleeping or unconscious. It's also useful in certain medical emergencies, where the heart, lungs or brain may have "seized" up momentarily. The target will feel "good" briefly, as if they were in the peak of health, regardless of their true condition.

7. Power Word {SCHANG!}: Resume True Form.

This is a command that forces a person or thing to go back to their true form. It is only effective on things that are not currently in their proper shape. It's really useful in dispelling magical alterations in just about anything. Shape Shifters, and things with the power of Shape Shifting, can resist the effects, but only with effort.

If used against an unnamed target the caster will need a Psyche advantage. However, even if the Power Word fails, it is likely that any Shape-Shifted subject will have reacted in some way. Repeated use against the same opponent will be gradually less and less effective.

8. Power Word {FORTZ!}: Defensive Luck.

Creates an instant of fortunate karma, sort of like a small dose of "Good Stuff." Results will be either fortunate events for the caster, or minor accidents befalling the opposition. It can only be used once in any given combat location. It is best used when some event seems to be dependent on chance. For example, it's timely to use it just as the enemy's foot is about to touch down on a patch of ice.

9. Power Word {LEGANT!}: Pattern Negation.

As with Chaos Negation, it's something you cast against yourself. Works internally as protection against the influence of Pattern.

10. Power Word {VOILE!}: Pain Attack.

Gives a "jolt" of pain to the person or creature at which it is directed. The momentary pain is a distraction, but does no real damage. Named victims are affected, regardless of Psyche. However, if the victim is not named, then resisting is possible for opponents with a superior Psyche. The main effect is one of surprise, so using it more than once against tough opponents is pretty useless. On the other hand, wimpy or cowardly victims can be driven into helpless terror by repeated uses.

11. Power Word {JASK!}: Trump Disrupt.

Can be cast against a Trump card, instantly blocking any active connection. Used either against a card in the caster's hand, or, by pointing, at a nearby card. If the caster is receiving a Trump contact, where the other party is holding the card, it can be cast through the contact. In this latter case, it will succeed only if the caster has Psychic superiority. Only works against a card that is currently in use, not a similar card in another deck.

12. Power Word {MAGIQUE!}: Process Surge.
In any process (fire, magic, pattern, etc.) this adds an extra *burst* or *push*. For example, casting it on a working auto engine would give it a quickie burst of extra power. A campfire would briefly flare up. The most it can do is double the energy exchange rate of the normal reaction.

13. Power Word {QUIMK!}: Process Snuff.
Momentarily dampens whatever is affected. Throwing it at a car's engine would cause it to slow slightly, or even stall out. A campfire would dim, but a flickering candle flame would likely go out altogether. The effect is to dampen the reaction rate down momentarily.

14. Power Word {OMBRE!}: Shade.
Adds to the darkness of all shadows in an area. The effects are spooky, frightening, scary and forbidding, but with no real effect. Basically used to disorient and confuse.

15. Power Word {LUUM!}: Light Strobe.
Releases a bright "strobe" or burst of white light. Can emit from the character speaking the Power Word, or from an object that is being touched. Useful in temporarily blinding an opponent (at least the first time), but pretty useless for illumination.

16. Power Word {AFLAK!}: Spark.
Creates a single spark, like that generated when a match is first struck. Only lasts a moment, but it's enough to start a fire, ignite something, or, against a living creature, cause a bit of pain. Spark must be generated at a fingertip, or somewhere on the caster's body.

17. Power Word {HURG!}: Burst of Magic.
Generates a "pulse" of magical energy, either inside the caster's body, or at some point outside of it (the fingertip, for example). Sometimes useful for activating magical items. Definitely visible as a "flare" of magic for anyone observing with mystic Logrus sight or by other means of detection.

18. Power Word {MARSK!}: Weaken Structure.
Temporarily weakens the molecular structure of an object. Works best on rigid objects made of metal, stone, glass, and plastic. It's ineffective against organic material, including flesh, bone, wood, cotton or wool cloth, and most rope. Won't work against magic or mystic items. Timing is important when using this Power Word, as it is best used at the exact moment of impact. For example, so an enemy's weapon or armor might shatter when struck.

19. Power Word {HAGGK!}: Thunder.
The Power Word, as spoken, is transformed into the sound of thunder bolt, just as if a stroke of lightning banged right from the caster's mouth. Loud, with a sound that carries, and startling. About the worst damage possible is temporary deafness if shouted directly into the ear of a victim.

20. Power Word {ZZAAQ!}: Burst of Psyche.
The caster's Psyche is boosted, just for an instant. This may be enough to assist at a critical moment of an attempted Trump call, or during a Psychic battle. It also serves to brighten the character's Psyche, so that an observer will notice the character.

SORCERY

15 Points.

Attribute Tips for Sorcery. Although Psyche is crucial for powering Spells, don't neglect the Attribute of Endurance. Casting spells is time consuming and physically draining. A less-than-Amber Endurance will limit the number of spells you can create before dropping from fatigue.

Spell-Hanging and House-Cleaning

A lot of elder Amberites have full knowledge and power of Sorcery. Most of them rarely use it. Spells have to be studied, pre-prepared, maintained, refreshed, and constantly studied. All in all, a lot of trouble. Frankly, it's just too much bother.

A careful, well-prepared sorcerer should have a dozen or so spells on hand at any given time. Which means spending about twenty hours a week on nit-picky maintenance. How would you feel about spending twenty hours on the same chores, week after week? How would you feel a year later? A century later?

More important, what kind of person would spend centuries on this kind of busy work?

Immortals, like those with the Blood of Amber, aren't thrilled with that kind of busy-work. Especially when every pleasure imaginable is out there in Shadow for the taking. Even megalomaniacs, plotting to conquer everything and everybody can spend their time more productively by researching *real* power like Pattern, Logrus and the like.

How Spells Work

Sorcery operates differently from Power Words (characters who plan on engaging in magical combat would be well advised to get both). The big difference between Power Words and Sorcery is time. It takes anywhere from ten minutes to several days to cast a spell. The way around this is to prepare spells in advance.

Which leaves two main problems.

First you can only "memorize" one spell at a time. Even then you've got to concentrate on the spell constantly. If you take a nap, get hit in the head, or work on other magic, the memorized spell is going to be lost.

Which means the only practical way of storing prepared spells is by putting them in some kind of container. Storing, or "hanging," spells isn't something you can do just anywhere. Spells are basically a weaving of power and information, and they only last in an appropriate receptacle. The magical container must be either a living thing, capable of remembering the spell, or a magical artifact containing power that can be forced into the form and structure of the spell.

Second, the other problem with prepared spells is that they are very specific. A completed Invisibility Spell designed to cloak someone on Shadow Earth, will do nothing at all if cast in another Shadow. A Lightning Bolt Spell readied for casting in the Courts of Chaos will be a sparkly dud in Castle Amber. A spell designed to bind a human of Shadow Earth is harmless against a Lord of Chaos. A magical barrier cast to keep out Pattern energy is transparent to the tendrils of the Logrus.

No problem, if you happen to know exactly what, where, why, when and how your spell is going to be used. Not likely, eh?

The way around this is to prepare spells that aren't quite complete. Instead of preparing the entire spell, you leave out a few key instructions, which can be filled in when the spell is finally released. We call these instructions *lynchpins*.

Lynchpins and Spell-Casting

I lined up the spoken signatures and edited them into a spell. Suhuy would probably have gotten it down even shorter, but there is a point of diminishing returns on these things, and I had mine figured to where it should work if my main guesses were correct. So I collated it and assembled it. It was fairly long—too long to rattle off in its entirety if I were in the hurry I probably would be. Studying it, I saw that three linchpins would probably hold it, though four would be better.

I summoned the Logrus and extended my tongue into its moving pattern. Then I spoke the spell, slowly and clearly, leaving out the four key words I had chosen to omit. The woods grew absolutely still about me as the words rang out. The spell hung before me like a crippled butterfly of sound and color, trapped within the synesthetic web of my personal vision of the Logrus, to come again when I summoned it, to be released when I uttered the four omitted words.

Blood of Amber

Lynchpins are the words you use when triggering a hung or memorized spell, defining last-minute conditions.

The most common lynchpin is *Magic of Shadow*. This defines the type of Magic present in a given Shadow. Without a specific description of the Magic of the environment where a spell is cast, no spell can work.

Other common lynchpins define the target of the spell, the duration, and the conditions for the turning the spell off.

Although lynchpins make stored spells flexible, they have disadvantages.

Lynchpins take time. Each lynchpin adds time to the Base Casting Time. Lynchpins also add time to the unleashing of a spell. The more lynchpins there are, the more opportunity there is for others to interfere in the spell.

Psyche and Spell Casting. Your character's Psyche isn't crucial with many Magic Spells. However, the "Invasive" spells, those that must enter an opponent's body, require a contest of Psyche.

Spell Building. Aside from the "Basic" spells listed below, you can create and name your own spells. See the following section on Magic Creation for more details. Any spells that you overhear, you may also attempt to memorize and imitate, though this can be dangerous.

Mind links and Magic spells. If the Sorcerer has established a mind to mind Psyche link, either through the willing or unwilling-but-helpless cooperation of another, it's possible to release any of the spells through the linked mind. The spell can be cast on the subject of the link, or, using that person's mind as a conduit, as if the Sorcerer were present on the scene. While a simple matter, most Amberites are highly suspicious of having spells cast through their own Psyche, first because the spell could turn out to be a direct attack, against which they'd have no defense. Secondly, because magic can be rather unpredictable in the hands of inexperienced users.

Sorcery and the Logrus. Logrus Masters and Advanced Logrus Masters have a huge advantage over other Sorcerers. That's because the Logrus itself can be used for "hanging" spells. In addition, spells can be cast out along a Logrus Tendril. In other words, the Logrus Master can send out the Tendrils, seeking a distant object or Shadow, and, when contact is made, spells can be cast out of the end of the tendril.

Personal "Style" and Sorcery. Each character brings their own unique signature to spell-casting. On one hand this means that Sorcerers can come to recognize distinctive spells and discover spells created by Sorcerers they know well.

The other aspect of personal style is that characters can create their own special effects. This can be built into spells, or used as another lynchpin, defining the character's trademark. Sparks, smoke, lights, glitter, radiance, or glow can be complemented by a unique color, scent, sound, or whatever. Each Sorcerer should pick a unique mark, such as a green flame, and stick with it. Personal special effects are harmless, but can serve as a good warning to others.

BASIC MAGIC SPELLS

The following spells are those taught to a beginning Sorcerer. Any basic spell can be cast the long, safe way. Spells that are to be stored for later use must be either memorized by the caster, or "hung" in some way.

INVASIVE MAGIC SPELLS

Spells where the magic is brought forth directly inside of the target or victim. Typical invasive spells are attack oriented, and attempt to alter or change the victim. Note that the "victim" of any Invasive spell can choose to open themselves to it, negating the need for superior Psyche.

1. Basic Spell: Mind Touch. Opens a mind to mind link between the caster and the victim. This link performs the same way as touch or a Trump contact, allowing the two minds to touch. Requires overcoming the Psyche of the victim, so an advantage in Psyche is required. The exact location of the victim, specifying which Shadow, and where in that Shadow, is needed for the spell to succeed. Which is obviously not a problem if the victim is in sight of the caster. The duration is indefinite, lasting as long as the Psyche contact persists.

Base Casting Time: Thirty Minutes

Lynchpins: Additional five minutes casting time each. Name/Description of Subject, Name of Caster's Current Shadow location, Name of Subject's location in Shadow, Duration (Optional), Dispel Word (Optional).

2. Basic Spell: Quell. By influencing and altering the nervous system of the victim, this spell puts them to sleep. It requires overcoming the Psyche of the victim. Duration depends on whether or not anything disturbs the victim (a person who is put to sleep and falls into a batch of stinging nettles isn't likely to stay "Quelled" for very long.).

Base Casting Time: One Hour

Lynchpins: Additional ten minutes casting time each. Name/Description of Subject, Magic of Shadow, Duration (Optional), Dispel Word (Optional).

Variations: Can be modified to affect any natural bodily function of the victim, either speeding up or slowing natural processes.

3. Basic Spell: Cardiac Arrest. Stops the heart of the victim. Effect is instantaneous. Those with Human Endurance will simply collapse and die. A Chaos ranked Endurance character will suffer the equivalent of a heart attack, requiring a week of bed rest to recover. For characters of Amber or better Endurance, the effects will be alarming, roughly equal to a sudden blackout, but harmless in the long run.

Base Casting Time: One Hour

Lynchpins: Additional ten minutes casting time each. Name/Description of Subject, Magic of Shadow, Duration (Optional), Dispel Word (Optional).

4. Basic Spell: Stone Binding. Changes the time rate inside the victim, so they are slowed down into immobility, relative to their environment. The resulting statue remains this way indefinitely, in any one Shadow. Requires a Psyche advantage.

Getting rid of the spell is fairly simple. A touch of Pattern or Logrus power will easily dispel the binding. Likewise, moving the "statue" around in Shadow tends to decay the magic, so that the spell will wear off in a few hours.

Base Casting Time: One and a half Hours

Lynchpins: Additional fifteen minutes casting time each. Name/Description of Subject, Magic of Shadow, Duration (Optional), Dispel Word (Optional).

5. Basic Spell: Invisibility. Turns the subject invisible, or, if desired, partly transparent. One big drawback is that with complete invisibility all light passes through the body, so there's no light for the eye to pick up, and the subject is effectively blind. Therefore it's best to become only partly invisible, so there's still enough light to see, though dimly. During bright daylight 95% invisibility would leave enough light to see. On a dark night 50% invisibility would make the character impossible to see, but would leave at least some light to see by. Invisible objects make normal amounts of noise,

smell, taste, and are just as noticeable with Psychic or mystical powers as anyone else.

Base Casting Time: One Hour.

Lynchpins: Additional ten minutes casting time each. Name/Description of subject, Magic of Shadow, Degree of Invisibility, Duration (Optional), Dispel Word (Optional).

Variations: Turning insubstantial instead of invisible is one possibility. However, with a total insubstantial form there are two drawbacks, first, one would be deaf, and second, unless somehow prevented, the subject would start falling down through the ground.

6. Basic Spell: Strength Drain. Reduces the Attribute of Strength to Human level. Dispelled by a touch of Pattern or Logrus, or by a movement through Shadow. Lasts indefinitely in any one Shadow.

Base Casting Time: Two Hours

Lynchpins: Additional twenty minutes casting time each. Name of Subject, Magic of Shadow, Duration (Optional), Dispel Word (Optional).

DEFENSIVE MAGIC SPELLS

All are designed to protect the caster from a variety of threats. Among others, there are those that "block" attacks using barriers, those that "absorb" threats, and those that fill the body with defensive energies.

7. Basic Spell: Bodily Defense. Fills the body with magical energy devoted to blocking out external influences. Although the spell can be tailored for any specific threat, the basic spell is designed to work against Magic, Trump, Logrus, and/or Pattern. Lasts up to one hour or until dispelled by the caster.

Casting Time: One Hour.

Lynchpins: Additional ten minutes casting time each. Name/Description of Subject, Magic of Shadow, Duration (Optional), Dispel Word (Optional), Protection Range (Naming Magic, Pattern, Logrus, Trump, or any combination).

8. Basic Spell: Defensive Shield. A magical shield, fixed in place, is cast as a barrier. The basic spell works against Physical Attacks, plus Fire, Heat, Lightning, Wind, Rain, and harsh Weather. A determined individual, of Amber Strength or better, or any largish moving object (a bulldozer, or a dragon), will be able to penetrate the barrier. Will last for a maximum of two hours in Amber.

Casting Time: One Hour

Lynchpins: Additional ten minutes casting time each. Magic of Shadow, Placement of Shield, Dimensions of Shield, Shape of Shield (Optional, will be square unless otherwise defined), Limits of Effectiveness (name the forces it blocks - Optional), Duration (Optional), Dispel Word (Optional).

9. Basic Spell: Magical Drain. Creates a hole through which all loose magic is drained away. Effectively limits the magic in an area, up to two miles in diameter, where spells will be drained as they are cast, or as lynchpins are inserted, or, for those successfully cast, any magical energies they generate. Duration, if undisturbed, is thirty minutes in Amber. However, the drain can be made self-sustaining, so the

more magic the spell drains, the longer it lasts. If enough magic enters before any thirty minute time segment expires, it can continue indefinitely (unless, of course, a duration is defined). Although it can be dispelled by other energies, dispelling a Magic Drain with magic is near impossible, since the magic just goes away.

Casting Time: One Hour

Lynchpins: Additional ten minutes casting time each. Magic of Shadow, Placement of Hole, Drain Exit (can be simply into the ground, or dispersed, or off to another Shadow location), Area of Drain Effect (Optional), Duration (Optional), Dispel Word (Optional).

10. Basic Spell: Defensive Psyche Ward.

Placed on a doorway, on walls, or simply as a circle drawn on the ground, it creates a Psyche barrier to any who attempt to pass. Creatures of less than Amber Psyche will be unable to pass the barrier. If left untended, the Ward has the equivalent Psyche of an Amberite, and can be forced in about as long as it would take the intruder to make a Trump contact. However, it is possible to link the Ward to the mind of the caster, or one who the caster chooses, in which case anyone attempting to force the Ward would end up in a direct mind to mind link. Lasts up to twenty-four hours in Amber, or, if linked, until dispelled.

Casting Time: One Hour

Lynchpins: Additional ten minutes casting time each. Magic of Shadow, Line of Ward (Touch, the caster must actually move a finger or pointer, tracing the line or boundary of the Ward's limits), Unattended (Optional), Linked (Optional), Name of Link (Optional - only if Linked), Duration (Optional), Dispel Word (Optional).

11. Basic Spell: Defensive Material.

Gives any object the qualities of a shield that is Invulnerable against Conventional Weapons. Basically, the object then becomes invulnerable to missile fire, blades, and normal energy like fire and electricity. Can only be cast on a single object, made of a single type of material (a cotton shirt, a wooden shield, a rock wall). Lasts indefinitely in any one Shadow. Has no resistance to being dispelled with Psyche, magic or in any other way.

Casting Time: One Hour

Lynchpins: Additional ten minutes casting time each. Magic of Shadow, Name of Object, Duration (Optional), Dispel Word (Optional).

SUMMONING MAGIC SPELLS

Although summoning spells can be used for a variety of things, they are most often used offensively, to draw objects and forces out of Shadow as weapons.

12. Basic Spell: Lightning Bolt.

The magic actually opens up a conductive line between some highly charged place in Shadow, and a target area. Duration is instantaneous, but the conductive "channel" must be defined before the Shadow hole is opened, other wise the caster will be the first target. Unfortunately this "lag time" is enough for those with Psyche sensitivity to dodge out of the way, but it works great against stationary targets and massed soldiers. Chances of hitting a target depend on the character's Warfare.

Casting Time: One Hour

Lynchpins: Additional ten minutes casting time each. Direction of Conductive Channel, Distance of Channel (Optional), Magic of Shadow, Trigger.

13. Basic Spell: Pressurized Lava.

A quick opening, from the current location, to a place where lava is under high pressure, allows for a chunk of the super-heated rock to blast out from the caster. The size of the lava chunk is defined by the caster, but anything over 200 pounds is risky because of the intense heat. Hitting, as with all projectiles, depends on the Warfare of the caster. The range is limited by the size, so the biggest chunks travel only a few feet, where a tiny blob can arc out over a couple of hundred feet.

Casting Time: Thirty Minutes

Lynchpins: Additional five minutes casting time each. Magic of Shadow, Hand Motions (gestures are used to specify the location of the opening, and the tilt of the hands show the angle through which the lava will exit), Size of Lava Chunk, Trigger.

Variations: Any fluid material under pressure can be substituted for the lava. For example, sea water or fresh water taken from the depths, rain or wind from a Shadow with great air pressure, even liquid sulphur, mercury or white-hot iron.

14. Basic Spell: Replicate Shadow Object.

Forms a duplicate out of Shadow of any object. Making a replicate of a living thing could result in a living duplicate, or if the Sorcerer prefers, a dead duplicate (a *corpse*). No duplicated item, living or otherwise, will have the powers or magic of the original. Duplicates last until dispelled, or until moved from the Shadow where they were created.

Casting Time: Thirty Minutes

Lynchpins: Additional five minutes casting time each. Name of Object, Magic of Shadow, Living or Dead (Optional), Duration (Optional), Dispel Word (Optional).

15. Basic Spell: Self-Teleport.

Transports the caster to a specified location. The location can be built into the casting, or can be defined as a series of lynchpins. Takes place instantly. Comes in two varieties. One, where if you are bound, or chained, or otherwise held, the spell does not release you, but may transport you along with whatever is attached to you. The other, where you are teleported free of anything bound or attached to your body, but also leaving behind all your other possessions, clothing, weapons, jewelry, and even objects held in your hand.

Casting Time: Thirty Minutes

Lynchpins: Additional five minutes casting time each. Magic of Shadow, Name of Destination (Optional), Transport with Possessions (Optional, the caster will bring along possessions, and anything loose that is being grasped, up to about a ton, but it won't work if the caster is chained or attached to something really large), Transport without Possessions (Optional, the caster will arrive naked, leaving all possession behind, but escapes bindings, chains, or the grasp of any hostile).

BUILDING CUSTOM SPELLS

Creating spells is just that, a *creative* process. Each spell is assembled from the four component spells listed below. No matter what you may wish to do, generating earthquakes or insects, requires assembly from the same set of "microspells."

The amount of time it takes to cast a spell depends on the components used, combined with the time it takes to build in each lynchpin. Each spell takes as much time to cast as the total of all the micro-spells involved (30 minutes each), plus five minutes for each of the lynchpins built into the spell.

Inventing new spells also takes time. A variation on a single micro-spell, like a Shadow Opening that releases an insect swarm, takes a week's worth of experimentation. Combining different types of micro-spells effectively acts to multiply the amount of time needed for research. For example, a spell to drain a victim's body of fluids would need two components, a Mind Touch and a Shadow Opening. Each Micro-Spell takes a week, adding up to two weeks, which is then multiplied by the number of different kinds (two), leaving a total of four weeks needed for experimentation. A triple micro-spell would take three times three, or nine weeks to perfect.

MAGIC RULE #1:
No spell can invoke a power, unless the caster of the spell actually possesses that power. Powers include, but are not limited to, Pattern, Logrus, Trump, and Shape Shift. For example, a sorcerer without the Blood of Amber can't create a spell that calls upon the power of the Pattern.

MAGIC RULE #2:
Spells can only be combined with a power if the caster has an advanced version of the power. Powers include, but are not limited to, Pattern, Logrus, Trump, and Shape Shift. For example, a spell that combines its actions with a Logrus Tendril could only be created by an Advanced Logrus Master.

1. Micro-Spell: Shadow Opening.
The first basic spell of the Sorcerer is that of creating openings, or "gates" from one Shadow to another, or simply from one point to another within a Shadow. These gates are usually temporary, maintaining themselves just with the Psychic energy of the caster, for only an instant. The most common Shadow Opening spell is Teleportation, used by the caster to travel through Shadow. However, with the addition of magical energies, either from a Magical Energy spell, or from other tapped energies, the opening can be kept open longer. Given sufficient energy, a Shadow Opening can be maintained indefinitely.

Sometimes, as with the Pressurized Lava spell, a Shadow Opening spell is useful as a weapon. Similar gateways can be used to channel in water, wind, or other natural elements.

The Lightning Bolt spell uses a Shadow Opening to gate in the lightning, but the area must first be prepared with a Shadow Manipulation; otherwise the bolt would pass through the nearest object, namely the caster.

Base Casting Time: Thirty Minutes

Lynchpins: Additional five minutes casting time each. Name of Current Shadow, Name of Destination Shadow, Name of Destination Shadow Location, Duration (Optional), Dispel Word (Optional).

2. Micro-Spell: Shadow Manipulation.
A spell to alter or change the stuff of Shadow can be used by itself. Shadow Manipulation can be used to change the form of a piece of Shadow stuff, like changing a steel door into one made of clear glass. In combination with a mind touch spell, Shadow Manipulation can be used against a person, inducing sleep, pain, or some kind of change in a victim. Shadow Manipulation can be prepared for the use of other magics.

Here are the possibilities:

Shape Shift Shadow Items. Objects can be changed in shape, size, mass or qualities. This is also what one would use to change a creature's body.

Prepare Item or Area for Magical Energy. Used to prepare objects that are to be powered by magic. Also used to prepare an item that can be used for hanging spells.

Define Channel for Magic Energy. Used for magic wards, circles, barriers, and other vectors.

Each change requires a separate Micro-Spell: Shadow Manipulation. For example, casting a single spell that turns a rock into an amulet (Shape Shift Shadow Item), and then preparing that amulet for storing magical energy (Prepare Item for Magical Energy), would require two Micro-Spells: Shadow Manipulation.

Base Casting Time: Thirty Minutes

Lynchpins: Additional five minutes casting time each. Name of Current Shadow, Duration (Optional), Dispel Word (Optional).

3. Micro-Spell: Magical Energy.
This spell creates raw magical power. Limits of the energy that the spell can generate is more a matter of how much spells and objects can contain, rather than any innate shortage of magic. If a spell is to be maintained, it must have a "pool" of magic to work with. Objects are usually limited to an hour's worth of magic, and fields, such as magical walls, wards and so on, are usually limited to about twelve hours of power.

Base Casting Time: Thirty Minutes

Lynchpins: Additional five minutes casting time each. Name of Current Shadow, Duration (Optional), Dispel Word (Optional).

4. Micro-Spell: Mind Touch.
Opens a mind to mind link between the caster and a subject. This link works the same way as touch or a Trump contact, allowing the two minds to touch. Taken by itself, this spell simply creates a link. It is also the basic building block of any spell that can directly affect another creature.

Overcoming the Psyche of the victim is always a consideration (unless the object of the spell opens their mind and becomes a willing subject). This means a Psyche superiority, where the caster dominates the victim, is usually needed.

Base Casting Time: Thirty Minutes

Lynchpins: Additional five minutes casting time each. Name/Description of Subject(s), Name of Subject's Current Shadow location, Name of Subject's location in Shadow, Duration (Optional), Dispel Word (Optional).

CONJURATION

20 Points.

Conjuration is the ability to "conjure" up items or creatures. It can also be used to "empower" or "ensorcel" people, places or things. It is really using Magic to shape Shadow, and to give that bit of Shadow power.

Since many of the details of Conjuration relate to putting together items, you'll want to refer to the Item Creation Section for a lot of the specific details.

Conjuration, in addition to allowing for the creation of artifacts and creatures of power, and for the investment of existing items with magical powers, also allows for the combination of magic with other powers.

Conjuration doesn't require Sorcery. However, in order to use Magical Spells with conjuring, the character must have Sorcery.

Attribute Tips for Conjuration. The duration and number of Conjured items depends more on Psyche than any other Attribute. It is also good to have sufficient Endurance to engage in the longer Conjurings.

1. Basic Conjuration. Using Magic to shape Shadow, and to give that bit of Shadow power. For example, a character could "conjure" a creature (like a horse), or an item (like a rope), seemingly out of thin air. Many of the details of Conjuration are found in the Artifact Creation Section.

2. Conjure Shadow Shape. Using the "stuff" of a particular Shadow, the Conjurer can "build" any artifact or creature (see Worksheet). The process is fairly quick, needing only about an hour, but the conjured shadow exists only in the Shadow of its creation, and quickly fades away, depending on how much Pattern exists in the environment (six hours in Amber, much longer in Shadows where there is strong magic, years in the Courts of Chaos).

3. Empowerment. The conjurer temporarily implants some magical power into some mundane object or creature. The basic preparation of the subject takes about thirty minutes, which defines the target and opens it up for the change. Referring to the Creature and Artifact Creation Worksheet, empowering each point of a quality extends the total time needed by another ten minutes, and each point of a power takes another hour. The multipliers, for quantity and form, apply to the total spell casting time. Extending Empowerment to a more permanent effect takes far longer, up to ten times as long.

4. Complex Conjuration. The ability to conjure magical items or living things. Use the Creature and Artifact Creation Worksheet. The total point cost for each item represents the amount of time, in hours, needed for conjuration of the desired item. Such items will lose their power if moved from the Shadow of their creation.

Combining Conjuration with Shape Shifting and Advanced Shape Shifting

One of the most powerful and potent combinations available is putting Shape Shifting together with Conjuration. This gives the character the ability to conjure items, both creatures and artifacts, out of his or her own blood. These items can be created with a measure of the character's powers.

Advanced Shape Shifting takes this one step farther, allowing the character to conjure items from the blood of others.

Combining Conjuration with Power Words

If a character has *both* Conjuration and Power Words, it's possible to conjure items containing Power Words. The only Power Words available will be those known by the character.

A drawback is that the items will use identical Power Words. This means that a Sorcerer witnessing the use of a Power Word by a Conjurer's creation would subsequently be able to construct a defense against that Power Word.

There is a difference between how creatures (anything living or intelligent) and artifacts (other stuff) use Power Words. Creatures with Power Words can use them independently. Artifacts with Power Words are used by their wielders.

In other words, an inanimate ring with a Power Word is designed so that the wearer of the ring can invoke the Power Word. If that ring had a psyche or intelligence, then the ring itself could invoke the Power Word, not the wearer of the ring.

Combining Conjuration with Spell Casting.

Without the spell casting ability of a Sorcerer, a character can only perform Conjuration in a methodical, step-by-step process. In other words, if a thing is to take four hours to Conjure, there's no way the Conjurer can do it in less than four hours.

However, if a character has both Conjuration and Sorcery, it's possible to create *Spells of Conjuration*. These spells don't do anything differently than any other kind of Conjuration, the difference is that they can be prepared, stored, and released as other spells.

For example, the four hour Conjuration described above could be stored in a spell. It would still take four hours to cast, but could be defined with various lynchpins. Then the spell could be cast at the caster's whim.

Creating Items with Magical Powers. The other cross-over effect of having both Conjuration and Sorcery is that it's possible to Conjure items capable of storing and activating spells. Any of the magical powers defined in the "Artifacts & Creatures" section (starting on the next page) can be invoked into an item by a character with Conjuration and Sorcery. Items created with magic can be made part of character's permanent inventory by adding advancement points.

Artifacts or creatures may be set up to contain a Sorcerer's magic spells. As with any Conjurer's creations, or with items created using points, these powers are installed in items. Note that such spells cannot be conferred upon a user or wielder.

CONSTRUCTING PERSONALIZED ARTIFACTS AND CREATURES

After a time, I stopped at a hollow tree that had to be there. I reached inside and drew forth my silvered blade and strapped it to my wist. I mattered not that it had been somewhere in Amber. It was here now, for the wood that I walked was in Shadow.

The Guns of Avalon

Spending points on artifacts or creatures means that the character has made them part of his own personal "reality." Greyswandir can be pulled out of shadow by Corwin at any time, even after it's been lost, or even, presumably, destroyed. If something is really destroyed, then those points return to the character (although the character may not be aware of it, it's up to the GM to figure out how to use the points).

Item Building in Five Easy Steps

You can *Build Personal Items*, when you first create your character, or when you upgrade your character. Building Personal Items means you spend points on the item. While expensive, items bought with points are always reliable, nearly always return to the character, and become part of the character's personal legend.

Step One - Start with a normal item.

Any normal artifact or creature will do. Typical artifacts are swords, rings, suits of armor, or amulets. Creatures are often dogs or horses, but can be anything at all, from humans, to vultures, to lizards, to elephants. The basic artifact or creature is free, no matter how weird or exotic.

Step Two - Add in Qualities.

Qualities are the item equivalents of character Attributes. Qualities give the item things like strength, speed, stamina, improved armor and weapons, plus the item's intelligence and psychic potency. Inanimate objects can be brought to life with qualities, while creatures can be augmented physically and mentally.

Note that both Intelligence and Psychic Sensitivity, for otherwise inert objects, makes such things vulnerable. A sword, for example, would usually be immune from Psychic manipulation. Give that sword the ability to Speak, or any form of Psychic Sensitivity, and it becomes vulnerable to Psychic attack and manipulation.

Step Three - Add in Powers.

Powers give items extraordinary command over the universe. Control over Shadow, or over the substance of the item itself, are examples of item Power. Additional item powers, such as those of Trump and Magic, can be added only if the character has the related Power.

Step Four - Transferals.

There are items that work best if they can confer, or transfer, their abilities unto whoever wields them. Each quality, and each power, must be separately transferred.

Step Five - Quantity Multiplier.

Take the total number needed to make the item, including Physical Qualities, Psychic Qualities, Powers and Transferals. Then, depending on how many of the item you'd like, figure out the multiplier. If you just want the one, multiply by one.

BUILDING ARTIFACTS AND CREATURES WITHOUT SPENDING POINTS

No matter how you put something together with points, there are many ways to get an item in Amber.

Instead of spending points, you can *Find Items in Shadow*. In this case, the points can be used to figure out how long it will take to find the thing. This mean you pay no points, but there is no particular connection between your character and the item. It's also a rather time consuming way to get an item.

Hellride to the Item is using Pattern to get something. This means you just go out into Shadow and keep moving until you find exactly the item you seek. The more complex the item, represented by the number of points required, the longer it will take to find it. This usually works out to a day of Hellriding for every point of Artifact.

Logrus Tendril Search is the method for Lords of Chaos. A character with Logrus can simply extend Logrus Tendrils through Shadow, "feeling" for the desired item. Again, the more points it takes to build the item, the longer it will take for the tendrils to locate it. Since Artifacts and Creatures are pretty specific, it takes about four hours for every point.

Creating Items with Conjuration can also be used to put together artifacts and creatures. For Conjured items, the points determine how long it takes to conjure the item. While this is the cheapest way to get items, in terms of either points or time, Conjured items are fragile, and are more likely to be disrupted or destroyed by your enemies.

Conjure Shadow Shape Item. Limited to a particular shadow, the Conjurer can "build" items taking just a minute of Conjuration for each construction point. Shadow Shape Items will not survive very far from the Shadow where they were Conjured. Only takes an hour to Conjure, but the resulting item's qualities and powers will fade away six hours later in Amber, or in non-magic places.

Empowered Item. Starting with a natural artifact or animal, qualities and powers, and/or a quantity multiplier, are added by the Conjurer. After a period of basic preparation, about thirty minutes, each Quality Construction Point takes ten minutes of Conjuration, and each Power Construction Point takes an hour. So long as the Empowered Item isn't moved through Shadow, it will retain all the Conjured changes. If the Conjurer spends the extra time, ten times as long, then the change becomes permanent regardless of where the item is taken in Shadow.

Complex Conjuration. Starting with a mundane creature or artifact, the conjurer adds in the various qualities and powers. The total point cost for each item desired represents the amount of time, in hours, needed for conjuration of the desired item. Once complex conjured, the item will remain until somehow it loses its magic (through disenchantment spells or by changes in its Shadow environment), or until it is lost or destroyed.

Artifacts of Power

It is also possible to create items, both artifacts and creatures, that contain powers such as Pattern, Logrus, Trump and so forth. Right off, any character with Trump Artistry can create Trump-based items. However, creating Pattern and Logrus items require powers beyond the powers of starting player characters. It may be possible, in the course of a campaign, to acquire additional powers, or to pay points for items that a character comes across, turning them into personal items.

QUALITIES

Vitality. Not just physical strength, but also the measure of life in any item. Note that someone using an item that transfers vitality will be able to perform the equivalent feats of strength, but will not have the durability that the Strength Attribute gives.

Animal Vitality. Used to bring inanimate objects to life. Animal Vitality gives objects the equivalent strength of a normal animal. Costs 1 Point.

Double Vitality. Ordinary animals, being fairly strong anyway, will be capable of jumping, leaping, or carrying double their usual limits. Raises the strength of a human to a Chaos level. Costs 2 Points.

Immense Vitality. Animals become far stronger, able to jump or leap phenomenal distances, and capable of exerting force many times their own weight. Humans with Immense Vitality have their strength raised to Amber level. Costs 4 Points.

Movement. Used for either animating things that are otherwise immobile, or for speeding up the natural movement of artifacts and creatures. Note that movement improvements are not particularly useful in combat, since they usually refer to long range movement of running, rather than the quick, jerky movements of combat.

Mobility. For those items which are usually immobile. The movement speed depends on the item's physical apparatus. If there are wheels, legs or other means of locomotion, terrain permitting, the speed is roughly equal to a walk. Items without movement equipment (like Merlin's strangle cord) have to slither or creep along. Even then, if the item lacks Animal Vitality it will move slowly and will be too feeble to climb, push, or pull. Costs 1 Point.

Double Speed. Creatures who already have the power of movement, like deer, dogs and horses, have double their usual speed. This usually works out to something like 30 miles per hour. Objects that start without movement, but with legs or wheels, will manage similar speeds. Items without means of locomotion will slither or creep quickly, at about the pace of a run. Costs 2 Points.

Engine Speed.
For those creatures and items with good running speed, like tigers and bicycles, this accelerates them enough to keep up with powered vehicles (cars, motorcycles). Dogs and wolves can get up to 60 miles per hour, and horses can manage 75 miles per hour or more. Items that start out without means of locomotion will wiggle around at speeds of up to 20 miles per hour. Costs 4 Points.

Stamina.
Item Endurance should not be confused with Amber Endurance. Although Stamina gives duration in exertion, it does not give the healing and regeneration abilities that come with the Attribute of Endurance. Inanimate objects that are given vitality or movement start out with only feeble stamina, and tire quickly.

Double Normal Stamina.
Creatures will be able to run, fight, or perform difficult tasks for twice as long as their normal counterparts. Gives the equivalent of Chaos Endurance to humans or items. Costs 1 Point.

Amber Stamina.
Creatures are able to run or fight at full exertion for hours without getting tired. Humans, and inanimate objects, get a stamina equivalent to the Amber level Endurance Attribute. Costs 2 Points.

Tireless, Supernatural Stamina.
The creature or artifact will simply never get tired. Costs 4 Points.

Aggression.
Unlike a character's Warfare Rank, an item's Aggression refers only to its basic reflexes and combat skills, not to general weapon, tactical and strategic training. So, for example, a dog with Combat Reflexes could fight at an Amber level with teeth and leaps and so forth, but would know nothing about swords, guns, or, for that matter, anything about chess or military strategy.

Combat Training.
When conferred on a creature, or a number of creatures, this gives them the training to work together as a team, with intelligence enough to use practical combat tactics. In addition, the creatures will be expert in their weapons (natural or artificial), and quicker than any normal human. Items with Combat Training have a specific expertise of up to a Chaos level of Warfare. Costs 1 Point.

Combat Reflexes.
Creatures with Combat Reflexes are trained to work intelligently, but also with extraordinary speed, making them faster and more adept than opponents on a Warfare level of Chaos. An item, or a human, will be improved to the point where they are quick enough to compete with someone of Chaos rank in Warfare. If the item is devoted to a specific type of combat, then the skill will improve to equal Amber rank in Warfare. Costs 2 Points.

Combat Mastery.
Living things with Combat Mastery have an intuitive grasp of any tactical situation, are aware of ambush and other bad combat situations, and will quickly adapt to new weapons and other threats. Their speed is comparable to someone with Amber normal Warfare. Items specializing in a particular combat will surpass the reaction time of an Amberite. Costs 4 Points.

Resistance to Damage.
This is the Amber equivalent of an Armor Rating. In both creatures and artifact there can be natural armor against various threats.

Resistant to Normal Weapons.
Normal swords, arrows, spear points, and most weapons will be deflected by the armored surface. Even weapons that penetrate will have a reduced effect. Costs 1 Point.

Resistant to Firearms.
In addition to protection from pointed and edged weapons, the artifact or creature is also immune to gunfire, explosions, and fire. Costs 2 Points.

Invulnerable to all Conventional Weapons.
Cannot be damaged by any normal means. This includes edged and pointed weapons, firearms, and energy based attacks. Costs 4 Points.

Weapon Damage.
These are the damaging effects installed in an item, especially weapons and items with edged or pointed surfaces, or in the natural weapons in a creature's biological arsenal such as claws or teeth.

Extra Hard.
For an item, this gives the striking surface the capacity to impact with great force without being damaged. Extra Hard Weapons, applied to animals, means that the natural "hard points" of the creature, whether fangs, or claws, or horns, or hooves, won't be damaged when they try attacking something solid or impenetrable. For example, normal claws raking a concrete building might become damaged, and would be unlikely to do much damage. If the claws were Extra Hard, they wouldn't be damaged and they would make some kind of impression on the concrete. Extra Hard weapons have an easier time penetrating armor. Costs 1 Point.

Double Damage.
As with Extra Hard, these weapons give both durability and an extra measure of target penetration. Double Damage also increases the amount of damage done, so, where a sword would usually inflict a light wound, a sword of Doubling Damage would inflict a serious wound. Double Damage Weapons easily penetrate armor "Resistant to Normal Weapons," and, with time and skill, can break through armor "Resistant to Firearms." Costs 2 Points.

Deadly Damage.
The teeth or claws or weapon, in addition to having extraordinary durability, can do severe damage to a target. Wounds inflicted by Deadly Damage Weapons are always serious and often fatal. Penetrates any type of armor, including armor "Invulnerable to all

Conventional Weapons." Costs 4 Points.

Intelligence and Communication Skills.
Most items, including both artifacts and creatures, can be made with a range of intelligence.

Able to Speak. The item can speak in the language of the user, usually Thari, but with a limited vocabulary. If applied to an inanimate object, this means the item can understand and obey commands. In either case, the item will act as the equivalent of a very clever animal, such as a well-trained dog. The item has loyalty to its owner, and will operate independently if commanded to do so. Items with the Ability to Speak, unless commanded otherwise, will always attempt to rejoin or assist their owners. Cost 1 Point.

Able to Speak and Sing. Beyond simple speech, the item can speak with a normal voice, sing, hum, whistle, and otherwise express itself. Clever enough to understand complex orders, to make plans, and to operate independently toward a goal. It has a good memory, can handle numbers, logic problems, and literacy, but only if instructed in these arts. Costs 2 Points.

Able to Speak in Tongues and Voices. The item has intelligence the equal to a normal human, capable of feats of mathematics, literacy, and philosophy. Capable of learning a wide range of languages, and the ability to "mimic" the speech of others. Costs 4 Points.

Psychic Quality. Confers a limited version of the
Psyche Attribute, but only that portion dealing with sensing Psychic phenomena. For example, a human, normally void of psychic sensitivity, given Psychic Sensitivity, would be able to sense the "coolness" of a Trump, but would have no particular resistance to a Trump contact.

Psychic Sensitivity. Equal to Chaos rank. The item can communicate with anyone it touches, by a mind to mind contact. It would have only the basic intelligence of a normal version of the creature, or, if given some form of speech (as above), that level of intelligence. Inanimate artifacts, with no speech modifications, become receptive to commands and can obey specific orders. Cost 1 Point.

Sensitivity to Danger. Gives the ability to sense danger, but not in any specific way. The item can pick up on the general distance and direction of hostile thoughts and emotions. This does not mean the item can detect inanimate traps or threats. Also includes the same abilities and limitations as Psychic Sensitivity. Costs 2 Points.

Extraordinary Psychic Sense. Flashes of insight into what is coming means the item can sense potential future events. Regardless of distance, the item will always "sense" when its owner is in danger. In addition to being able to establish mental contact with anyone in the immediate area, the item can also reach through Shadow to attempt to contact with any known mind. This takes a lot longer than a Trump contact, and is nowhere near as reliable, and does *not* allow for any physical transferral. Items with Extraordinary Psychic Sense will, if lost, usually attempt to contact their owner. Intelligence is no more than the basic creature, or, if an artifact, that of something barely capable of communication. Has no Psychic "power" that can be used in Psyche battles. Costs 4 Points.

Psychic Defense. Although artifacts and creatures
with Psychic Defense operate against Psyche contacts and attacks, they do so without any offensive Psychic power, or any power to create Psychic links.

Psychic Resistance. The item is resistant to most Psychic attack or tampering. Its defenses can be overcome, but doing so would be the equivalent of Psychically overwhelming a character of Chaos Psyche. Cost 1 Point.

Psychic Neutral. Has the ability to "cloak" itself, so it's Psyche will not be detected. It also covers the Psychic emanations of intelligence (speech), mental powers (Psychic Sensitivity) or any other powers the artifact may possess. Of course, if a Psyche attack succeeds, or if a mental contact is established, the disguise will be penetrated. Actual defense is equal to someone of an Amber level of Psyche. Cost 2 Points.

Psychic Barrier. The item will be resistant to ALL Psyche attacks. Note that this is incompatible with any Psychic sensitivity or communication. In other words, while the item cannot be affected by psychic attack, neither can it use psychic communication (including Trump), nor can it be psychically sensitive. It is possible to build an artifact or creature with a Psychic Barrier that can be turned on and off, but there must either be a mechanism (a switch), or enough intelligence to obey a spoken or tactile command. Anyone protected by an active Psychic Barrier artifact will also be blocked from using Psyche, including Trump, and will also be unable to "scan" with Psyche. An artifact, or creature, with a Psychic Barrier, is quite obvious to those with high Psychic sensitivity. Costs 4 Points.

POWERS

Item Movement Through Shadow. These powers allow for a limited amount of movement through Shadow. In most cases the artifact or creature uses an innate form of magic to perform the feat, rather than Pattern, Logrus or Trump.

Shadow Trail. The item can follow the path left by someone else moving through Shadow. Costs 1 Point.

Shadow Path. The item can find a path to any Shadow it knows well. If the item is just backtracking, that is, going back along a route it has taken before, the speed is the equivalent of a Hellride. However, if the item is stranded somewhere unfamiliar, and doesn't know the specific way back, it could take a very long time, bumbling around, to arrive at the Shadow destination. Costs 2 Points.

Shadow Seek. The item can control a movement through Shadow toward a person or item in Shadow. Note that this is *not* the same as being able to move freely through Shadow, because the item can only move in the "direction" of the desired subject. The item also tends to hit the worst of "Hellriding" routes, and may require a lot of hit-and-miss tries crossing difficult Shadows. Costs 4 Points.

Item Control of Shadow. Power over various

aspects of Shadow. Each level includes the powers of the lesser forms.

Able to "mold" Shadow Stuff. The item can change

simple Shadow items and features. For example, coins can be shifted to match the current environment. Costs 1 Point.

Able to "mold" Shadow Creatures. The item can

affect creatures and inhabitants of Shadow. For example, a dog can be friendly rather than hostile, or a person could be informative instead of mercenary. Costs 2 Points.

Able to "mold" Shadow Reality. The item can

change some of the features, including the "odds" of the current Shadow. For example, the magical environment of the Shadow could change slightly, making it difficult for the resident Sorcerers. Costs 4 Points.

Item Healing. If applied to an inanimate object, any

type of healing gives the item a self-repairing ability. Even if totally destroyed, the item will eventually pull itself back together. For creatures, healing powers accelerate their normal process of mending and restoration.

Self Healing. For an inanimate object, this allows for a

slow, gradual, self-repair, at about the same rate that normal creatures heal from comparable wounds. Creatures with Self Healing recover from wounds at twice their normal rate. Costs 1 Point.

Rapid Healing. Any item, artifact or creature, will heal

like someone with an Amberite's Endurance, recovering from slight wounds within a day, and healing even the most severe damage over a period of weeks. Missing parts will gradually be regenerated, even though it may take years. Costs 2 Points.

Regeneration. Quickly, in a matter of minutes, the item

can repair any damage or injury it has suffered. Even regrowing lost parts is quick, requiring less than an hour. Costs 4 Points.

Item Shape Shifting. For any shape an item takes, it

will have all the normal attributes and instincts of that shape. So, for example, if shifted into a flying creature, the item will be able to fly normally. Also, any qualities and powers of the item will be carried over into all its shapes. So a sword with Deadly Damage, changed into a cat, will have Deadly Damage claws and teeth.

Exotic Shape Shifts allow for items to gain seemingly magical properties. For example, an item could shift to transparent material, rendering it invisible. Or an item could shift to a very hard substance, giving it resistance to some weapons and attacks.

Alternate Form/Shape. The item has an alternate

form that it can change into. The alternate shape must be described in detail when the item is first created. Costs 1 Point.

Alternate Named and Numbered Forms/Shapes. The item has a number of forms,

with an absolute maximum of twelve, all determined when the item is first created. Costs 2 Points.

Limited Shape Shift. The item can take the shape of

any animal or item of comparable size and mass. Initially this is limited to those creatures or things that the item has already practiced, but the item has the capacity to learn new shapes. Shape Shifting powers or persona is impossible. Limited Shape Shift is useless for healing or repairing damage. Costs 4 Points.

Item Trump Images. Artifacts and Creatures can

be created with "built-in" Trump images. However, *only characters with the Power of Trump Artistry can create such items!* Trump artifacts and creatures are **never** "found" in Shadow or Conjured, but can only be created by Trump Artists.

Contains a Trump Image. The item doubles as a

Trump, containing a unique image drawn by the Trump Artist. The item also has the innate durability of Trump, and a Trump's resistance to the manipulations of Pattern, Logrus or Magic. The Trump Image can be painted on a card in the usual way, or embroidered on a piece of cloth, or inscribed on metal. This power can<u>not</u> be installed in a creature. Costs 1 Point.

Personal Trump Deck. As with other point artifacts, a

personal trump deck becomes innately connected to the character. This means that, if lost, the deck can always be sought in Shadow. The cards of the Personal Trump Deck are limited to those created by the Trump Artist. Once the personal deck is created, the Trump Artist can add in any cards they create without additional cost (however, this apples only to cards created by the Trump Artist). Multiple copies of personal decks, bought with "Named and Numbered" (total cost times two), can be shared with other characters. Costs 2 Points.

Powered by Trump. The item, created by the Trump

Artist, is "Trump-Like," containing the power of Trump. While conferring a Trump image is possible, the main point here is that the item works by the power of Trump. Costs 4 Points.

Power Words in Artifacts & Creatures.

Each Power Word contained in an item can be cast numerous times. Some Power Words depend on the Psyche of the caster. Power Words in items depend on the item's Psychic Quality. If the item does not have either Psychic Sensitivity (Chaos Rank), Sensitivity to Danger (also Chaos Rank), or Extraordinary Psychic Sense (Amber Rank), then the item operates Power Words at Human Rank. Item Intelligence or Psychic Defense do not affect an item's operation of Power Words.

A character wielding an item with a Power Word normally casts the Power Word from the item. So an artifact speaking a Psychic Defense will protect the Psyche of the item, not the wielder. Unless, of course, the item is built with "Confer Power."

Contains a Power Word. The item has a single Power Word, which must be identical to one of the Conjurer's Power Words. Costs 1 Point.

Contains Named and Numbered Power Words. The item contains a number of Power Words. The maximum number is equal to the total number of Power Words possessed by the Conjurer. Only Power Words known by the Conjurer can be used. Costs 2 Points.

Spell Storage in Artifacts & Creatures.

"Racking" or "Hanging" spells in items requires a special power. Once a spell in stored in a Creature or Artifact it can be cast using the required lynchpins. Spells must be "loaded" by someone with the Power of Sorcery. Each time a spell is used, it is considered gone from the item and can must be constructed all over again.

NOTE: Available only to characters with both Sorcery *and* Conjuration.

Capable of Racking a Spell. The item is capable of having a spell stored within it. Note that the spell is incomplete, and will require the services of a magic wielder to invoke it and cast it forth. Costs 1 Point.

Can Rack Named & Numbered Spells. The item can rack up to a dozen spells. Note that the spells are incomplete, and the services of a magic wielder are needed to invoke them and cast them forth. Costs 2 Points.

Rack & Use Named & Numbered Spells. Aside from being the mere container for spells, the item is itself capable of casting the stored spells. The maximum number of spells the item can rack is twelve. Creatures, or anything "alive" (with intelligence or psyche) will cast the spells independently. For inanimate, unintelligent items, this power means that the wielder of the artifact can command the triggering of the spells without taking the effort of casting. Costs 4 Points.

TRANSFERAL

Sometimes items are created to lend qualities or powers to their owner. A player character, having sold down Strength to the Human level, may want an item to make up for the loss. Creating a "Belt of Strength" with Equals Amber Strength, four Points, and Confer Quality, five Points, would end up costing nine Points. Wearing the belt would give the wearer the equivalent of Amber strength.

It's important to note that the effects of transferal are not additive. In other words, if the character has Amber Strength, and is using an item with Confer Amber Strength, the character gets no improvement in Strength. In fact, nothing really happens. Conferring a quality substitutes for the character's abilities, and doesn't add to what the character already has. So, for example, using a sword with "Combat Reflexes" would give the wielder that level of Warfare, even if the character were Ranked more highly.

Confer Quality on Wearer/Owner/User. For example, a sword with the quality, "Armored, Resistant to Firearms," which costs two points, combined with "Confer Quality on User," another five points, will cost a total of seven points, but will mean that whoever holds the sword will also be protected by the sword's armor quality. Costs 5 Points per Quality.

Confer Power on Wearer/Owner/User. As with qualities, each power must have a separate conferral. Costs 10 Points per Power.

QUANTITY MULTIPLIER

Any item, no matter what form or shape, must also be defined by quantity. The total points spent on the item, including qualities, powers, and conferrals, is subject to a multiplier that varies according to the desired quantity of the item. There are six variations on Quantity Multipliers.

1. Unique. There is but one of the artifact or creature. No additional cost, since the multiplier is one.

2. Named & Numbered. If you want more than one of a particular item, without paying for each separately, then use a simple multiplier of two times the original cost. Named and Numbered means giving either an individual name to each of the duplicates, or identifying each one by a number. The number depends on what makes sense for the item and the character. For example, a character might get four duplicate daggers, one for each sleeve, one for a boot, and one to wear on the belt. Or seven dwarves. Or eight tiny reindeer (Comet, Donner, Blitzen...). No matter what explanation, Named and Numbered is limited to a dozen or so. Costs twice (*2) the total points for the artifact.

3. Horde. A horde is a large number of the item. They usually are found around a particular location, in a particular Shadow. So, for example, Julian's Hellhounds are a horde that infest Forest Arden. At short notice, in the right place, the character will only be able to gather together fifty or more members of a horde. Given time, many more can be gathered. Double the initial numbers in the first hour, double again in the next eight hours, double yet again after a full day's summoning. After that, gathering more of the horde slows down, because of the limitations of distance and numbers. A horde costs three times as much as a unique sample of the item.

4. Shadow Wide.
Within a particular Shadow, the item is everywhere. Regardless of where you go in that Shadow, you'll find the item in horde quantities. Costs four times the basic cost of the artifact or creature.

5. Cross-Shadow Environmental.
Found in every one of the infinite Shadows that contain the particular environment. For example, if warm-water swamp creature is Cross-Shadow Environmental, then the creature will be found in every warm-water swamp thoughout Shadow. A limitation is that the item must have a natural appearance. So alligator creatures will have to blend in with the natural alligators of each Shadow's swamps. Another example might be a sword that will be found in any middling sized armory, regardless of where it might be in Shadow. There will be a Named and Numbered quantity in any one environment, though probably just one of the item in any given place. Costs five times the basic cost of the basic artifact or creature.

6. Ubiquitous in Shadow.
They're everywhere, they're everywhere! Anywhere, throughout all of Shadow, the item is always waiting to be found. The only exception to their ubiquitous appearance is in the two poles of reality, the Amber and the Courts of Chaos. Outside of those places, one of the item can always be found in just a few minutes. As with Cross-Shadow Environmental, the items must be things that appear normal, and, moreover, they should appear to be really commonplace. Costs six times the basic cost of the basic artifact or creature.

Quantity Multiplier Example.
For example, let's look at a three point sword, having Double Damage (2 points), and Psychic Sensitivity (1 point). This three point sword is just three points if Unique (3 times 1 equals 3). The swords, all identical, in the small quantity of Named & Numbered, would cost six points (3 times 2 equals 6). A great number of the swords, such as might be used to equip a body of soldiers, is said to be a Horde, and would cost nine points (3 times 3 equals 9). The same swords, available and common throughout a single Shadow, costs twelve points (3 times 4 equals 12). Swords, all of the same type, scattered such that there is at least one in every decent-sized armory, or museum weapon collection, in all of Shadow, are Cross-Shadow Environmental, and the total cost is fifteen points (3 times 5 equals 15). Finally, if the character is able to find one of the swords anywhere at all, anywhere in Shadow, then it is Ubiquitous, and costs eighteen points (3 times 6 equals 18).

SHADOWS

'And Lorraine?'

'The country? A good job, I thought. I worked the proper shadow. It grew in strength by my very presence, as any will if one of us stays around for long—as with you in Avalon, and later that other place. And I saw that I had a long while there by exercising my will upon its time-stream.'

'I did not know that could be done.'

'You grow in strength slowly, beginning with your initiation into the Pattern. There are many things you have yet to learn. Yes, I strengthened Lorraine, and made it especially vulnerable to the growing force of the black road. I saw that it would lie in your path, no matter where you went. After your escape, all roads led to Lorraine.'

'Why?'

'It was a trap I had set for you, and maybe a test...'

Roger Zelazny,
The Courts of Chaos

Just about the most fun thing you can do in Amber is create a world. For a single measly point, you can be the god-like being that decides how the show will be run for a whole world full of happy slaves. Or uppity natives. Or, for the loners among you, that barren and bare place where you can really get away from the hustle and bustle.

First off, just as with artifacts, remember that you don't have to pay points to get a Shadow. Once you start playing the game, you can just as easily wander away and find whatever Shadow tickles your fancy. It's just that later on, coming to the Shadow of your dreams can involve a certain amount of interaction with the Game Master. And we all know how those power hungry, plot-driving GM maniacs can sometimes put a crimp on your style. Mainly because they start thinking in terms of who might be watching, what kind of Shadow trap you might be walking into, blah, blah, blah...

How does spending points on a Shadow differ from simply heading out and finding a Shadow of one's desire?

A Shadow that is paid for, like an artifact paid for, becomes an innate part of the character's portfolio, a built-in part of that character's personality. Just as Corwin contains a piece of Avalon, a Shadow he insists is destroyed, and which he never really visited during his adventures in The Chronicles of Amber. *Therefore, buying a Shadow makes it part of the character, part of the campaign, and part of the game master's grand design.*

To make absolutely sure that you get what you want, just spend the points at the beginning of the game. You don't have to spend a lot, just one point is all it takes to put the definition of a Shadow under your control.

Any Shadow can be cheap, just so long as it's static, relatively unchanging and vulnerable to any power that happens to sweep over it. What makes a Shadow more expensive is the degree of control you have over it.

Basic Shadow Type

Personal Shadow. Gets the character an entire Shadow, or universe, set to personal tastes. Any inhabitants, any combination of technology and magic, any society. For example, you could "buy" a version of Shadow Earth, but one that happens to still be experiencing the Eighteenth Century. Or, if you prefer, a fairie Shadow inhabited by elves, dwarves, dragons and magi. Or, just as easily, a high-tech Shadow featuring a galactic empire and ships travelling faster than the speed of light. Costs 1 Point.

Shadow of the Realm. A personal Shadow, but located in close proximity to Amber. Such a Shadow will be close enough to the center of things so it's only a relatively short trip. It's possible that the denizens of the Shadow will be capable of moving through the Golden Circle. On the down side, a Shadow of the Realm is too close to Amber to allow for really free shifting of Shadow. Another drawback is that the royal family of Amber will know of the Shadow, and will likely have an interest in what happens there. If you prefer, a Shadow of the Realm can be placed near the Courts of Chaos, in the "Black Zone," having similar advantages and disadvantages. Costs 2 Points.

Primal Plane. As with a personal Shadow, the player can shape it in any way. However, this Shadow contains a bit of reality, something left over from the creation of Pattern, and possibly even pre-dating the Logrus. 4 Points.

Shadow Barriers

A Shadow becomes less of a sanctuary if anybody can enter any time they please. Restrictions on admission keep your hidey hole your own. More expensive barriers include the features of the cheaper ones.

Communication Barrier. The Shadow is barred against methods of reaching from one Shadow into another. Pick any or all from Pattern, Logrus, Trump, Magic and Psyche. The barrier can restrict incoming calls (from outside the Shadow), outgoing calls (from inside), or both. Communication barriers are all-inclusive, which means the player can't make a barrier that blocks all Trump "except mine." Costs 1 Point.

Restricted Access. Shadow can only be entered through a limited number of means or access points. The entry points can be limited to a certain geographical feature, say across a great mountain range, or through a particular type of feature, like any rug merchant's shop. Or a character would have to hop on one foot and hum a particular tune in order to cross into the Shadow. Costs 2 Points.

Guarded. Some guardian, or class of guardians, will be set to intercept any and all who attempt to pass into the Shadow. The instructions given to the guardians are, of course, up to the character creating the Shadow. Costs 4 Points.

Character's Degree of Control over Shadow

Any Shadow that is "owned" by a character, can be changed by the character's relation with the place. Select one, and only one of the possibilities below. More expensive types of control include all the features of the cheaper ones.

Control of Contents. The character can manipulate the Shadow's material, creatures, society, history, reality, and physical laws. For example, the character can decide that magic will cease to function within the Shadow. Or, as another example, the royal family of a kingdom will not have existed, and another form of government will take its place. Costs 1 Point.

Control of Time Flow. Character is able to influence the Shadow's "clock," either speeding it up, or slowing it down, relative to the universe's only absolute; Amber. For example, if a character, fleeing some threat, needs a few hours to heal and recover it's possible to speed up the Shadow. That way the hours of healing can take place while only a few minutes pass in Amber. Costs 2 Points.

Control of Shadow's Destiny. A Shadow can exert its influence over those who walk through infinite Shadow, putting itself in the way of a person or class of creatures. For example, it's possible for a Shadow to become part of Martin's destiny, so the next time he Hellrides, he'll find himself entering the Shadow. 4 Points.

YEAH, I KNOW IT BLOWS HOLES IN THINGS BACK HOME, BUT WILL IT WORK OVER HERE?

Pick up a .38 revolver on Earth, or a Wand of Ignition over in Gazreal, and they'll work just fine. Try firing the gun at a charging dragon on Gazreal, or triggering the wand's spell at an on-coming truck on Shadow Earth, and, in either case, you might as well be holding sticks. Nothing is going to happen.

Why?

In both cases, the technological gun, and the magical wand, the items were built according to natural law. In the case of the revolver, the gun won't work unless both the primer and the powder ignite and burn at the proper temperatures, a matter of chemical laws. Also, unless the physical tolerances of metal are exact, the triggering and advancement mechanism of the gun won't work.

It's no different with the wand. Magical artifacts must make use of built-in spells. Spells must be defined by the magical laws of their location. Take the wand to a place where the magical laws are even a tiny bit different, and it won't work.

The more sophisticated the technology, or magic, of an item, the more sensitive it will be to changes in natural law. So a one-shot pistol needs less precision than an automatic rifle. Electronics, depending on the exact numbers of Maxwell's Equations, are even more sensitive. And the smaller the electronic parts, from transistor to microchip and down into super-scientific miniaturization, the more finicky they get.

When you combine magic and science, taking items from those Shadows where the two exist side by side, it's even worse. Then both sets of conditions, physical laws and magical properties, have to match the item's tolerances.

Of course, the technicians and magicians who built these items in the first place dealt with these problems. That's why they became techs or magi, because they had the patience to calculate the various finicky details that make things work. They just didn't take into account the possibility that some slob would be hauling their creation over to where molecular bonding, or entropic decay, would be so different.

In short, take any item dependent on magic, or technology, away from its home Shadow, and it will stop working.

Except.

Remember that two of the powers, the Pattern and the Logrus, have the ability to manipulate Shadow. Items like flashlights and magic lanterns are made out of Shadow. Therefore characters with Pattern or Logrus can alter technological and magical items.

A character with power can make something work when it shouldn't.

There are, of course, limits to what can be done with this.

The limit is usually *where* you're trying to do the manipulating. In general, it's easy to find a place where things can be made to work. It's places where things can't be changed that give Amberites headaches. One place where changes don't work is Amber. The closer one gets to Amber, the more difficult it becomes to Shift Shadow (yeah, for Logrus Masters too). Since it's rough shifting things close to the Pattern, very little actually works in Amber. No advanced technology, and very limited magic.

SHADOWS FOR THE SHADOW LAME

Moving to Shadows is pretty obvious if you're an initiate of the Pattern or a Logrus Master. Trump Artists can always make Trump, or Trump sketches, of their Shadows. Moving through Shadow is pretty easy for those with the right magical spell. Having an artifact or creature with the power of Shadow Movement is another way to get around. And, finally, characters with Advanced Shape Shifting can just take a form that's capable of moving around between Shadows.

Still, it's possible that a character might have none of these things available. Even powerless characters can walk through Shadow. It's all a matter of following a marked pathway. Like a ship's navigator following an invisible course, a knowledgeable character can follow a charted road through Shadow.

The big limit here is that the character must start somewhere along the marked path. Starting in an unknown Shadow, or in a Shadow that doesn't lay along the marked path, makes it impossible to get anywhere.

When defining the Shadow for the first time the character, whether capable of independent Shadow movement or not, can define any number of "pathways" leading from their personal Shadow to any other known point in the universe.

For example, there could be a path from Amber to the

Shadow, or from the Shadow to Ygg, or from the Shadow to Shadow Earth. Each of these pathways can be two way, allowing for travel back and forth, or if desired, any of these pathways can be designated as "one way." A one way Shadow Path can only be followed either to, or from, the Shadow to another point in Shadow.

As with any other Shadow definition, fixed Shadow paths can also be routed through the Shadow barriers, and will be subject to the same restrictions, costs and limitations...

A good example of fixed Shadow Pathways are the routes to the "Golden Circle" shadows that border Amber. In other words, from any of the Golden Circle shadows it's possible to manipulate Shadow (albeit with difficulty), and vice versa. However, since it's hard affecting Shadow so near to Amber, it's easier to just use the established trade routes, those known to ship captains and merchants, travelling to and from Amber without manipulating Shadow. Things are comparable out in the "Black Zone" around the Courts of Chaos, in that they each border directly on the Courts, and are also jump-off points for the more remote (from the point of view of Chaos) Shadows.

So, if you've got a problem getting around in Shadow, or if you anticipate having followers who may need to get around unassisted, be sure to add in a set of pathways to your Shadow's definition.

HOW TO PLAY A CHARACTER IN *AMBER*

Playing a character? For gamers it's the easiest thing in the world. What can be so hard about it? Why should there be a whole chapter on something so simple?

It is simple. As simple as "just pretend." As simple as shrugging off your worldly cares, and stepping into your alter ego, your *Amber* character.

Simple doesn't mean easy.

Some players handle their characters better than others. Every player is capable of improvement. Role-playing may take a minute to learn, but it takes a life-time to master.

This chapter shows you how you can role-play better. Not just in *Amber*, but in any role-playing situation.

Each player brings a different style, a different perspective, to their character. Nothing wrong with that. The idea here is not to make everybody play the same, but to make everybody play *better*.

Love Your Character.

The first rule of role-playing in *Amber* is to love your character.

Loving your character is really the main point of *Amber*, both the books and the role-playing. Ask most folks why they enjoyed the Zelazny books and they'll tell you, "I loved the characters!" Ask most gamers why they enjoy playing *Amber* and you'll hear, "I love my character!"

Loving your character can be as simple as trying to stay alive. As simple as enjoying the details of the character's life. As simple as building a personality strong enough to make the character come completely alive.

Making a character come alive is an act of faith.

Walking down Baker Street in London with a thirteen-year-old, we noticed a memorial placard. Like thousands of others scattered all over the historical city, it announced the famous former resident of the building. "Sherlock Holmes, Private Investigator," it said.

"Sherlock Holmes isn't real!" said my thirteen-year-old friend, "is he?"

"He's real to me," I said. "I believe he's real."

We had quite an argument, my friend and I.

You see, I love Sherlock Holmes. Yes, I know that his exploits fill the pages of fiction, and are not in the history books. In his time you had to read about Holmes under the by-line of Doyle, and not in the front page news.

That doesn't mean he doesn't exist somewhere, somehow, in some reality we haven't yet found. He'll always be real to me.

In the same way my *Amber* character Theazipha Jak will always be real to me. Maybe we haven't actually met, perhaps my character has never been to this version of Shadow Earth. But I love her, and I believe she's real.

Why is this important?

It's important because the more you love your character, the more involved, the more intense your play in *Amber* can

become.

It's also important because playing *Amber* is not always a pleasant experience. Sometimes it's uncomfortable, sometimes it's painful, and sometimes it's sheer mental torture.

Pain is a universal element in character development. If the main character of a book doesn't struggle against the odds, doesn't overcome nasty situations, and doesn't have to make bitter choices, then there won't be much of a story.

Corwin, the main character in the *Chronicles of Amber*, goes through a lot of pain. Imagine you were the player, and Corwin was your character. How would you feel about being a helpless captive of your worst enemies? Would you still keep playing after Corwin's eyes (*your eyes!*) were burned out of his head? Would you come back, session after session, to endure Corwin's stay of years in a cramped dungeon cell, blind and with no hope?

You might. If you loved your character.

Play in Character.

When you play in character, you try to look at everything from your character's point of view. That means your character's outlook, feelings, and everything your character does, is based strictly on what you've got from inside *Amber*.

It's the Game Master's job to describe the world to you. The GM tells you what your character sees, hears, smells, tastes and feels. It's up to the Game Master to supply you with your character's memories and background.

When playing in character, ignore anything you hear that your character has not heard. Yes, this is sometimes hard. Sitting in the room with the Game Master and the other players, you'll often hear things that are of vital interest to your character.

If you've heard there's a trap, or a sniper, right around the corner, that doesn't mean your character knows anything about it. Your job, playing in character, is to ignore what you've heard, and react naturally to what the character senses.

Even more difficult, when you're playing in character, you've got to ignore everything *you* know. Something you, the player, may know, is separate from what your character knows. If you know how to drive a car, shoot a gun, or cook an omelet, don't assume that your character shares this knowledge.

Live your Character.

Don't be afraid of your character's emotions. Characters who laugh and cry, who feel confused or angry, are interesting characters.

You probably wouldn't enjoy reading about a character who never cares about anything. Characters in books are interesting because they feel.

There's an attitude among some people that role-playing games have to be cold and calculating. Those folks would say that players shouldn't get emotionally involved in their characters.

Yet the very best kind of role-playing is where you do get involved. Where you can laugh, and cry, and feel very, very confused, and have the very best of times.

Why should your character cry? Well, sometimes it's a matter of life and death. When Corwin loses people he loves, it hurts him. Why shouldn't it?

Sometimes things get mixed up, and you say something to hurt somebody, or someone says something that hurts you. Real life is like that. So is life in *Amber*.

And yes, sometimes the character's emotions spill out onto the player. That's okay too. Why shouldn't you feel as excited, or sad, or elated about a role-playing game as about a football match?

Keeping Secrets.

Learn to keep your secrets.

You start with a basic set of four secrets, your Attributes. No matter what they are, unless you are ranked first, there's no reason why anyone, except for you and the Game Master, should ever know what they are.

Other secrets are the Powers, Shadows, Items, and Allies that you develop for yourself. Again, there's no reason to share these things, unless you can share them in character.

In character, in your character's voice, should you reveal secrets?

Not usually. Not in *Amber*. First, you don't know who to trust, who to believe. Anyone, including the character of the innocent looking player next to you, can be another character, Shape-Shifted.

Then there's the problem of who might be listening. Powerful, unknown, off-scene, non-player characters have abilities known only to the Game Master. Who's to say that somebody isn't using some exalted version of Pattern, or Trump, or Logrus, or Magic, to spy on your character at any given moment? Can you ever be sure your character is really speaking in private?

In *Amber* secrets are the most valuable commodity. An elder Amberite will laugh at an offer of gold, or jewels, or any mere trinket. What they want is knowledge. Don't give it away.

Keeping your secrets isn't just right, it's also fair. It doesn't matter that if player sitting next to you is your best friend, your spouse, or a loved one. That person is also trying to play a character, in character, and revealing your secrets to that player hurts them by making it harder for them to play.

Find your character's voice.

A role-playing game is mostly verbal. The Game Master describes everything, and the players respond, almost always, with speech.

For players there are two kinds of speech, two voices. The first voice is your own, the one you use all the time. You use that voice to talk with the Game Master and the other players. "What does my character see?" is in the player's voice, just as "when are we going to play again?"

The other voice is the voice a player uses when the character says something. "Fellows, there's nothing here," says the player, in the character's voice, "what do you say we start searching for something important, like lunch?" Or, "Caine, I don't trust you. I never have, and I never will. I don't trust you, but I'll obey you."

When a player speaks in character, it is in exactly the character's words. Spoken only when the character would speak. The character's voice comes out mostly when speaking to other characters. It also comes out when characters mumble to themselves, or yell out against fate.

Finding the right voice is tricky. It can be the hardest thing about playing *Amber*. For some players it can take years. You'll never have a perfect character voice, but you'll know you're on the right track when you don't have to explain the difference between your two voices.

Don't get funny. A player character's voice shouldn't be a "funny" voice. We can all make up squeaky voices, or deep ones, or talk like we're drunk, or pretend to stutter. Funny noises don't make for good character voices. For one thing, they get pretty boring. For another, it's hard to really believe in a character with such a limited range.

The best way of developing a character voice is role-playing a lot. Practice makes perfect. Aside from speaking, it also helps to write in the character voice. And handy opportunities for writing comes with the character quiz and the personal diary.

The Art of the Quiz. Your Game Master designs your *Amber* Campaign with your character in mind. Each game is designed to give your character, and you, a fun and challenging experience. If you want the Game Master to be able to do the job right, you need to put some thought and effort into the quiz questions you are given. The quiz sheets are also a way to get you thinking about your player character. Try answering some of the questions in character, in your character's own voice.

Don't be afraid to change your answers as time goes by. Characters change, and *Amber* characters sometimes change a lot. Stress, and the occasional save-the-world responsibility will do that to a person.

Dear Diary. Not only is a diary an easy way to get points, it also really helps make for a better player character. It doesn't have to be long-winded, or professional. Sure, some players put a lot of time and energy into a character diary. All you need to do is jot down your character's thoughts and feelings while you're playing the game.

Writing the diary after the game helps get all the details right. Sometimes, and it happens to everybody, you find important details you'd otherwise miss.

The most important thing about character diaries is that they get the player in touch with the character. You learn more about the character's "voice" and feelings. Any novelist can tell you about characters that seem to take over a book, somehow coming to life as they are written. This really works with Amber diaries, where the act of writing helps to firm up the character in your own mind.

Another useful aspect of character diaries is the way they let you introduce your own details into *Amber*. When you describe events you can also add in a lot of detail that can be useful later on. Examples can be bits and pieces of the stuff you carry around, non-player characters who could be useful later on, and even old memories that fill in your character's background. If your diary from last session was okayed by the Game Master (and they almost always are), then some detail you introduced, a servant, a secret room, a favorite meal, can come in handy later on.

Player to Player Interactions.

Amber characters are pretty interesting! Usually they're interesting enough to amuse themselves without a Game Master around.

The idea is that players talk to each other, making up details of background and setting as necessary, while staying in character. This can be pretty challenging, but it's an important aspect of the game, since groups are often scattered, and the Game Master can only handle one group of players at a time.

Amber gives you some really high quality, intense, role-playing experiences. You are allowed to operate independently or in very small groups, have lengthy conversations with non-player characters, and get the Game Master's undivided attention.

In return, you've got to wait when another player, or player group, is monopolizing the Game Master's time. Trading your own quality time for "down-time" while waiting around.

Don't waste that time. The very best Amber players don't bitch. Instead they recognize the time as an opportunity, and they use it creatively.

Use the other players. So long as you work with at least one other player, you can keep role-playing with each other. Working out your tactics and strategies, but also having a lot of fun exercising your role-playing talents. After all, in most journeys, you spend a lot of time just chit-chatting with strangers and fellow travellers. Take advantage of your down-time by perfecting your character's voice.

ROLE-PLAYING TACTICS

Aside from playing in character, there are certain tactics that work well to get your character ahead in *Amber*.

Finding your Character's Focus.

Each player character has different strengths and weaknesses. By focusing on your strengths, your best Attributes and Powers, you'll find your way of imposing your character's will on the universe.

Don't assume that there is any one best way of using any of your Attributes or Powers. There are an infinite range of possibilities. In play-testing no two players have ever used Shape Shifting in exactly the same way. In fact, when players from different campaigns meet each other, they're often astonished at what others can accomplish with exactly the same powers.

Finding your Focus also means to find a philosophy, some kind of view of the universe, that explains your power, how it works, and how you can expand on it.

The Art of Asking Questions.

One of the problems in Amber, or any role-playing system, is that the player doesn't know as much about things as the player character. Your *Amber* character may have been around for hundreds of years, lived on scores of Shadows, and lived through a dozen careers. Getting a handle on all this can be rough for mere human players. It may seem obvious to some, but you've got to ask about things from the player's point of view.

For example, in some kind of Warfare situation, a lot of players find themselves lost, unable to visualize the complexities of fencing or hand to hand martial arts. No problem, just ask the Game Master what the character thinks. "What," you might ask, "does my character know to be the best options in this situation?"

See the examples of combat. Notice how characters ask the Game Master for their options, and their character's evaluation of situations. A good Game Master won't tell the player what to do, but will narrow the possibilities down to what the character would find acceptable.

A personal example has to do with riddles. I hate riddles. Can't stand them. So, if it looks like riddles are going to be regular fare from some Game Master, then I have the character solve the problem. I have the character become an avid riddler, reading books and exchanging riddles with passing strangers. I just try to throw all the solutions back, informing the Game Master that surely my character would know the answer. Then, when my character gets hit with a riddle, I say, "Quick as lightning, my character spits back the right answer."

Why argue when you can ask? Arguing with the Game Master is always a bad idea. If you have a problem, try asking about it. Instead of backing the Game Master into a corner with yes-or-no questions, keep your questions open-ended. Rather than questioning your Game Master's actions, instead ask questions from your character's point of view. Not only does this reduce aggravation, but you're liable to gain even more information.

Using Elder Amberites.

Remember when you were just a little tyke, and all the adults were giants? It seemed like Mommy and Daddy were like gods, all-knowing, all-seeing, and all-powerful.

Your character's relationship with the elder Amberites should be something like that.

Only maybe it's more like this. Your character is a high school senior. A bright kid, maybe good at some things. Maybe a star athlete, or chess player, or math wiz.

Imagine your high school character is suddenly sent to summer school at the most prestigious university in the world. One where four star generals teach history, and presidents and prime ministers teach political science. Gold-star olympic athletes coach the sports teams, Nobel prize winning scientists are laboratory instructors, and you learn English from the best writers in the world.

Do you think you'd be a bit, shall we say, intimidated?

The difference between your character, and an elder Amberite is even greater.

The elders don't have many weaknesses. There are three you can count on. First, and this is not to your advantage, they aren't particularly wise, and, powerful as they are, they can make bone-headed decisions like anybody else. Pointing out the error of their ways usually invokes their second great weakness. They are egotistical and touchy as hell! Being god-like for a few centuries does that to an Amberite.

Elder Amberites also have one important soft spot. They are suckers for young, helpless, and vulnerable relatives. They know their siblings, all the other sons and daughters of Oberon, are cynical, jaded and pushy. They don't trust their siblings.

Each elder Amberite thinks that other elder Amberites are likely to take advantage of the youngsters, the player characters. They feel protective, plus, they like foiling the plans of their equals.

This is something that each player can use.

A couple of years ago, in the first Amber campaign, a group of players finally reunited with their long-lost father, Corwin. It didn't seem to faze them that the Game Master had described their father, Corwin, as worn, tired and battle weary.

Each player character had a million questions, and even more problems. They asked questions, brought out old grievances and quarrels for "dad" to arbitrate, asked for power and assistance, and generally complained about the state of the universe.

Then the last of their group arrived. The player's character threw himself at his father, hugged him, and started crying! I missed you so much, he said, I worried that I'd never get another chance to tell you how much I love you.

It was a pretty dirty tactic. Corwin, worn and frazzled and tired, was a sucker for this kind of attention. The late-comer had brought out the "father" side of Corwin, by being more of a son than an Amberite. And Corwin, naturally, looked upon the rest of his children coldly, seeing them to be acting suspiciously like the brothers he mistrusted.

Building on to *Amber*.

Amber is pretty open-ended. No Game Master, not even the best in the world, can fill in all the holes in *Amber* and infinite Shadow. Which leaves the players with a marvelous opportunity. Filling in the gaps with your own creations, things, places, and people.

"Look at the Board" is a phrase some New Jersey players use to describe a tactic in playing *Amber*. It comes from a case where one player character was playing chess with a non-player character. The player, a decent chess player, was trying to actually play the game, visualizing all the pieces and all the moves.

However, Amber doesn't work like that. Few mere humans from Shadow Earth could hold their own in a chess game with an Amberite. After all, chess, like any game of pure strategy, is covered by the Warfare Attribute.

No, the idea is not to actually play the game. The idea is to make your moves *sound* interesting and realistic. Don't worry about the exact layout of the board, just say something like, "I'll open with a cautious, armored strategy, and wait for white's first attack," "I'll open things up for white to put me in check, I want those pieces moved around," "I'll move my bishop into jeopardy, setting things up to endanger my enemy's king and queen," and "I'll start a series of exchanges that will surely lead to a draw."

Then, when confronted with similar lines from the Game Master, ask about the results of various options. After all, chess, like other forms of combat, has a range of responses, from purely defensive to purely offensive.

Since then, the New Jersey group has broadened their definition of "Looking at the Board" to include the whole process of players introducing pieces of the shared fantasy, the Amber universe, into the game.

It's really a matter of visualizing where you are. If you need a weapon, and you happen to be in Castle Amber, doesn't it seem reasonable that weapons would be on the walls? Or that there are busts of Oberon decorating most tabletops? Wouldn't it seem reasonable to find throw rugs on the stone floor and tapestries hanging from the walls?

Sure. Each piece of the shared reality, the universe presented by the Game Master, but reflected in the player's imagination, is an opportunity. If you use it.

"Creating non-player characters" is usually the province of the Game Master, but also an area of potential for players. Start out simple, asking non-player characters questions in order to fill in the blanks in their background and personality. Talk to the non-player characters. The more you interact with a non-player character, the more real it becomes.

Then, sooner or later, players run into a Game Master's limitations. After all, the Game Master can't prepare in advance all that many characters. This is the point where you, as a player, can be really helpful.

Offer to fill things in for the Game Master. "Hey, I've got some ideas," you might say, "how about if I just fill in the names and details of our encounter in my diary?"

Once you've the go-ahead, make the characters as real as you can. Give them names, families, backgrounds, motivations, and everything it takes to bring them to life. Once you've got a hold of non-player characters, don't forget to use them. Bring them up again and again. Rely on them, build up their characters and personalities. Give them motives and conflicts and make them a little more real every time you play them.

The more time and effort you lavish on the non-player characters, the more real they become. Which makes it much more difficult for the Game Master to just kill them off.

If they are killed, any or all of them, then mourn the hell out of 'em. Build memorials, visit grieving relatives, inflict revenge on their enemies. Keep them alive in the game forever. If necessary tell lies about 'em.

"Walking worlds into existence" is as simple as using Pattern to travel through Shadow. Remember, the player defines the worlds. The more real and detailed the description, the more likely it is that it will become a continuing part of your *Amber* campaign.

COMBAT

Running Combat: The Short Course.

The following pages cover the *Amber* Combat system in great detail.

Most of which you don't need most of the time.

Most Combat is resolved quite simply, using the following two steps.

Step 1. Compare the Attribute Ranks of the participants in any Combat.

If there are weapons involved, or troops, or strategy, then use the characters' *Warfare Attributes*. If the combatants are in a hand-to-hand fight, actually grabbing each other, then use their *Strength Attributes*. If the battle is one of the mind alone, through Trump, or some other means, then use *Psyche Attributes*.

Step 2. The character with the larger Attribute Rank wins.

That's it. Everything else is just a matter of adding details, figuring things out when it's a close call, and making things seem realistic.

Combat: Following Zelazny's Example.

'It was a very foul blow, about four inches below the belt buckle, I'd say, and it left him on his knees.'

- from the third page of
Nine Princes in Amber

That's the very first example of combat in Zelazny's *Chronicles of Amber*. No heroic duel, no picture-perfect battle, just a quick, unexpected fist in another guy's groin.

Corwin fights like this all the way to the end. In fact, his last true battle costs him dearly, mostly because, as usual, he fights dirty.

It's also unconventional combat. Even in the midst of set duels, where you'd expect the combatants to act in some kind of "normal" way, Zelazny always has the characters throwing curves, doing the unexpected.

When you read Zelazny his combat seems to go quickly. He skips over the boring parts, and snaps your attention to the eye-openers, to the critical decisions, and to the drama.

Zelazny writes of Bleys fighting up Mount Kolvir, and of his slaying of hundreds, perhaps thousands, in just a few paragraphs. Just one line, *"...half an hour I watched him, and they died and they died..."* is all it takes for Zelazny to get across an episode of combat that would take hundreds of pages to describe in detail.

When a battle is important, Zelazny isn't coy about supplying details. When Corwin, the main character, faces his ancient rival, Eric, we read, *"I tried a head-cut, which he parried; and I parried his riposte to my heart and cut at his*

wrist." The battle goes on with this kind of detail for pages.

The point is, Zelazny moves the story along. When Combat will help make it more exciting, he concentrates on the details. When it looks like Combat will slow down the story, Zelazny cuts to the chase.

As a Game Master, that's what you want to do. Present your players with Combat that is always exciting. Not boring with details, no matter how "accurate" or "true to life" they may be. You want them to always experience Combat as something tense, challenging and exhilarating.

USING ATTRIBUTES TO RESOLVE CONFLICT

When two or more characters come into direct conflict, duking it out, things are usually resolved by comparing the two Attribute ranks. Then, the one with the higher rank usually wins.

Not always, just *usually*.

When Corwin fought Benedict in *Guns of Avalon*, Corwin knew that Benedict was the better swordsman. No ifs, ands, or buts about it, Benedict could beat Corwin in any fair fight.

Corwin won.

Why?

Well, Corwin fights dirty.

In this case Corwin knew about something, about a patch of grass that served as a trap for anyone stepping on it. Benedict didn't know about it.

Corwin cheated. And Corwin won.

In *Amber* the character with the higher Attribute rank should win against anyone with a lower rank.

Compare this with chess.

Chess is a game of pure thought. No luck, no chance. Just pure strategy.

Every one of the billions of people in the world can be ranked in the hierarchy of chess players. You, you the player, are probably ranked rather low in the scheme of chess masters. Maybe you don't know how to play at all, in which case you'd be so much meat for anyone who knows the rules. Or maybe, like many people, you've just played for fun with friends and relatives. From the point of view of serious chess players, somebody who just plays for fun, who doesn't participate in the tournament circuit, is "unranked."

Once you get in the chess world, things get more formal. Moving up the ranks, the higher the number, the better the player. Players ranked close together, say a couple of guys ranked 1150 and 1170, are so close it's hard to tell which will win any particular game. But take any two guys with a spread of a couple of hundred points, and you can bet hard money the guy with the higher ranking is going to win. Masters, with over 2,000 points, can cream everybody else, Grand Masters are even better, and they can do tricks like play forty people, all at the same time, while wearing a blindfold.

So, what if you, you the player, had to play a Grand Master? Let's say, for some odd reason, you had to play this guy. Not for fun, but for something really, really serious, like the fate of the world, and, as a side bet, your own life.

What do you think, should you play fair?

We'll call your opponent Kurmenkov, one of the Russian's top chess guns. Flat out, the best there is.

Here's what you might do.

First, you make a few arrangements. Like learning the

game, but also getting a team of the finest chess coaches money can buy, and a little microphone so you can hear their advice.

Then, you set up the room where the game is going to take place. Figure out everything Kurmenkov hates, heat, loud colors, rock music, whatever. Give him the works, but only at critical moments, on and off.

And if that isn't enough?

Then you go for some real equalizers. Have his wife call, mid-game, threatening divorce, madness, and/or suicide. Then a high-ranking Russian officer, informing Kurmenkov that if he has been accused of anti-state activities, and his victory in the game will be sufficient proof to warrant his arrest and/or execution.

Still not enough?

When his back is turned, steal pieces from the board. Drug him, and move twice for every move he makes. At least once, during some critical move, just as Kurmenkov reached for a piece, have someone put a gun to his head and offer to blow his brains out if he makes that particular move.

Now could you win?

Depends. Frankly, if you're a total novice at chess, even all these advantages may not assure you of victory. On the other hand, if you're a master rank player yourself, you ought to be able to take him out. And, if you're the second ranked player in the world you should have no problem whatsoever.

Silly? Yeah, but it's the way things work in real combat. Armies spend a lot of time and money on military intelligence to pull exactly the same things that you were trying on Kurmenkov. Psychological warfare, dirty tricks, surprise attacks, and quick flanking maneuvers is what it takes to score victories out of proportion to the strengths of the forces involved. To use one example from recent history, when you own the air, and the other guy doesn't, you bomb the spit out of him. Not exactly "fair," sort of on the level of stealing the other guy's chess pieces, but pretty effective.

Combat as Story Telling.

With Zelazny in the *Chronicles*, each fight is an important piece of the bigger story. Each Combat is a way of advancing the story.

Advancing the story means several things.

It can mean finding out information about the characters, whether they are participants, manipulators or observers. Each time we see how well someone fights, as when Bleys kills hundreds, or how poorly, as when Julian is over-powered by Corwin, it says something about those characters.

In role-playing this is even better since the players can actually experience being kicked around by somebody, or have a chance to beat up on a lesser character. Each time something like that happens, it speaks a lot more clearly than the Game Master just describing the abilities of a character.

Example of Story-Telling Combat. Here's a typical combat, where a player character battles a non-player character with swords. Notice that the Combat tells the characters several things about their opponent. It also reveals an opportunity for Dorell to make a moral decision. Should she slay her opponent, who is concentrating elsewhere, or should she risk losing him?

G/B	PSY	STR	END	WAR
0/0 Cindy - DORELL	A	4th [9]	A	2nd [45]
0/1 Willy - GARVIN	1st [52]	C	A	A
0/0 The Stranger	[10]	C	C	[30]

Note: In the summary the first number is a column is the character's rank, and the number in brackets "[ ]" is the number of points spent on that Attribute. "A" means Amber Rank, "C" means Chaos Rank, and "H" means Human Rank.

CINDY: (looking around a tavern) I saw him run in here. Where is he?

GM: All the drinkers look the same to you.

CINDY: Garvin? What do you think?

WILLY (to Game Master): What do I sense about this crowd?

GM: There is one man who seems to have a strong presence.

WILLY: I'll point to that guy, the one with the high Psyche.

GM: As Garvin points a man leaps backward, reaching for his weapons, his left hand going for a big sword on his back, the right hand snatching a small dagger from his belt. What are you doing?

CINDY: How fast does he look?

GM: It looked really fast, but it also looked like sort of a practiced move. That's all you're doing, looking?

CINDY: No! I back up a step and draw my own sword.

GM: And just in the nick of time, as he furiously lunges forward, thrusting with his big sword. How are you reacting?

CINDY: So far, just defensively. Let's see if he's good enough to make me sweat.

GM: It isn't easy, but you parry all his blows. He's still coming, still varying his attacks.

CINDY: Do I hit him?

GM: Since you opted for a purely defensive tactic, you wouldn't hit him unless he stumbled into your blade. He's unhurt and continuing to press the attack.

CINDY: I don't feel particularly scared. I think I'll shift to a stance where I can try to hit him.

GM: An all-out attack?

CINDY: No, nothing that crazy. I don't quite believe this guy. I think he's just testing me. I just want to draw him in, see if I can open him up with some fancy defensive maneuvers.

WILLY: Good going! Caution is probably a good idea.

CINDY: Garvin! Butt out! I don't need your opinion.

GM: He keeps up his attack, but doesn't show much in the way of weakness. What are you doing?

CINDY: I'm not that impatient. I'll keep it up. After all, I can probably last longer...

GM: Fine. You continue being defensive, but keeping your options open. After a dozen or so engagements of the blade, you finally see an opening. Are you going for it?

CINDY: You bet.

WILLY: It could be a fake! Have you considered that he might just be trying to sucker you?

CINDY: I ignore Garvin...

GM: Just as well, by the time he says anything you've already moved. Your blade slips around his guard, and slices the skin on his forearm. He curses with some unknown word, and follows it up by yelling, "Immutable Progeny of Amber!"

CINDY: I press him!

GM: All-out attack?

CINDY: Yes. Wait! No!

GM: Well? Which is it?

CINDY: I just want to move aggressively, not put myself at risk.

GM: He's forced back by the fury of your attack. His wound seems to be closing up, almost fast enough for you to see it happen.

WILLY: Oh great! He must be a Shape Shifter.

CINDY: Garvin, for the last time, Dorell did not ask your opinion.

GM: He's moved to a more defensive posture.

CINDY: Dweller of Chaos, we have you cornered! Surrender!

GM: He says, "Never, not to anyone whose fathers consort with beasts!" Meanwhile, he's not attacking, just looking at you warily, and... Hmmm... Perhaps he's concentrating on something.

WILLY: Crud! Is he bringing up the Logrus?

GM: He's doing something eerie. It feels creepy.

WILLY: Dorell! He's bringing up the Logrus or something.

CINDY: I ignore Garvin and keep pressing my attack.

GM: Cindy, he's definitely distracted. As if he were getting a Trump call. Dorell could kill him right now if you attack all-out. What are you doing?

Choosing the Detail of Battle.

Amber combat has to be free ranging. It can't focus on each individual sword stroke. It can't always describe each individual hit, block, sidestep and deflection. But sometimes a single blow, a single dodge, a single quick advance or retreat, becomes a critical part of the story.

What's the point of wasting time on describing the details of slaughter? Better to do as Zelazny does with Bleys, skip ahead to when Bleys is tired, worn, and on the verge of making a mistake.

Even when a Combat is really important, and should be covered in detail, certain boring parts have to be skipped. For example, if two top-notch fighters are battling, and are so close as to be tied or nearly tied in rank, there will be long stretches where the two are just clashing swords looking for weakness in the other. In this case the Game Master can skip over these lengthy "probes," and concentrate on when the characters make decisions, take risks or change tactics.

Here are some points to consider when deciding whether a Combat be conducted in detail, or skipped over lightly.

When Old Enemies Come Face-to-Face. Kind of obvious, but anytime traditional foes get together, that's when we want to follow things in detail. Sherlock Holmes fighting Moriarty. Robin Hood against the Sheriff of Nottingham. In *Amber* players may not even realize they're facing familiar enemies, because of Shape Shifting and magical disguises.

When the Combatants are Equally Matched. Anytime two combatants are equal enough so that the outcome is in doubt, or equal enough so they are unsure about each other, that's a good time to focus on the Combat.

When there is a Moral Choice Involved. Making a choice of life or death, good or evil, is a time when the Combat should be slowed to the detail level. The classic moral choice in

Amber is whether or not to kill a helpless foe, when Combat really turns into murder. For example, "she stumbles, drops her blade, and, for the moment, is a clear and ready target. What are you doing?" or "Frozen by the assault on his mind, your enemy is suddenly motionless, helpless to defend himself. What are you doing?"

When the Unknown May be Revealed. There are mysteries aplenty in *Amber*. Sometimes the only time they can be penetrated is in the heat of Combat. This can be a case of a literal exposure, as when someone who looks ordinary turns out to have machinery under their skin, or the flaming blood of a Chaos dweller, or is built from magically animated sticks and straw. Other times it is far from obvious. An opponent, easy to kill, may possess some secret, a secret that will be lost in their death.

Ambush, Back-Stabbing, and Other Surprises. Any time players are confronted with particularly nasty shocks, as when enemies show up unexpectedly, or friends turn out not to be friendly, it's time to focus on the details again. The reverse, when the player characters pull something on others, is also a time to expand the details.

When a Player Character is in Danger. Player characters, especially when totally out-matched and in dire danger, are owed the courtesy of a full explanation of their defeat. While this provides a means of escape, there will be the occasional final call for a player character.

Example of Detail Combat. Some Game Masters are comfortable with a great deal of detail in the combat, while others just want to get it over with. Here are four examples of the same Combat, with the same player, Peggy, squaring off against the same beast. The first example is non-Combat oriented, the second uses a bit of Combat for realism, the third uses the Combat in detail, and the fourth uses a great deal of imagination to supplement the combat. See what you think.

G/B	PSY	STR	END	WAR
0/0 Peggy - Iresa	4th [20]	1st [24]	1.5 [16]	C
0/0 The Beast	H	[5]	C	A

IRESA VS. THE BEAST: NO DETAIL VERSION

We start by examining what would happen in a simple case, where the encounter is fairly unimportant.

GM: There's a big *thump*, and it sounds like it came from behind you.

PEGGY: I whirl around, what do I see?

GM: There's something moving right at you, something hard to see.

PEGGY: I back away quickly!

GM: Yes, it seems to be some kind of ape, it's fur is a mottled brown and green, making it hard to see against the backdrop of jungle. It's on it's hind legs and attacking with swinging arms.

PEGGY: I'll go on the defensive.

GM: With swinging arms it's better than you.

PEGGY: I'll go for Strength. I want to grab it.

GM: You move in, getting scratched by it's claws, but you grab it. You seem stronger. What are you doing?
PEGGY: I want to break its neck!
GM: Okay, it dies.

IRESA VS. THE BEAST: SIMPLE VERSION

Here the Game Master is looking to get across more of a mood.

GM: There's a big *thump*, and it sounds like it came from behind you.
PEGGY: I whirl around, what do I see?
GM: There's something moving right at you, something hard to see.
PEGGY: I back away quickly!
GM: Yes, it seems to be some kind of ape, it's fur is a mottled brown and green, making it hard to see against the backdrop of jungle. It's on it's hind legs and attacking with swinging arms.
PEGGY: I'll go on the defensive.
GM: Totally? Just defensive.
PEGGY: No, I want to hit if I get an opening.
GM: Okay, you're being opportunistic. Let's see... You don't get hit, but you can tell it's faster than you are.
PEGGY: Ooh. How strong is it?
GM: You don't know yet. Are you staying opportunistic?
PEGGY: No. I'm going for Strength. I want to grab it.
GM: You move in, taking a rake from it's claws, but you manage to grab it. Now what?
PEGGY: I want to break its neck!
GM: It's struggling. You've got the advantage. The two of you squeeze, and you hear a loud snap!
PEGGY: What? What broke?
GM: It did. It's dead.

IRESA VS. THE BEAST: MEDIUM VERSION

In this case the player introduces a bit more detail, describing her character's actions specifically. In response, the Game Master makes the combat more descriptive. The final result is the same, but Peggy gets a better idea of the Beast's Warfare.

GM: There's a big *thump*, and it sounds like it came from behind you.
PEGGY: I whirl around, what do I see?
GM: There's something moving right at you, something hard to see.
PEGGY: I back away quickly!
GM: Yes, it seems to be some kind of ape, it's fur is a mottled brown and green, making it hard to see against the backdrop of jungle. It's on it's hind legs and coming at you, swinging its arm, trying to hit you with its claws.
PEGGY: I'll try to parry it with my arm, and then punch it right in the face.
GM: You barely escape being raked with the claws, and it nimbly avoids your punch. You think it's a lot better than you.
PEGGY: Tell me about it! Any idea how strong it is?
GM: How can you tell at this point?

PEGGY: Never mind, I'm going to try to grapple with the sucker.
GM: If you go in close right now, it'll have a shot at you.
PEGGY: Yeah, I'll try to dodge the blow, but I want to get in there.
GM: You get caught by its swing, taking a light wound.
PEGGY: Ouch. Do I get a hold on it?
GM: Yes, you've got it, and it's responding by pulling you in a bear hug. What are you doing?
PEGGY: I want to break its neck!
GM: The two of you squeeze, and you hear a loud snap!
PEGGY: What? Who broke?
GM: You've broken it's back, it's dead.

IRESA VS. THE BEAST: DETAILED VERSION

We start with the same *thump*, same swing, same miss, same try at grabbing. Now that Peggy's character has engaged the beast, the Game Master decides to do things a little differently.

GM: There's a big *thump*, and it sounds like it came from behind you.
PEGGY: I whirl around, what do I see?
GM: There's something moving right at you, something hard to see.
PEGGY: I back away quickly!
GM: Yes, it seems to be some kind of ape, it's fur is a mottled brown and green, making it hard to see against the backdrop of jungle. It's on it's hind legs and coming at you, swinging its arm, trying to hit you with its claws.
PEGGY: I'll try to parry it with my arm, and then punch it right in the face.
GM: You barely escape being raked with the claws, and it nimbly avoids your punch. You think it's a lot better than you.
PEGGY: Tell me about it! Any idea how strong it is?
GM: How can you tell at this point?
PEGGY: Never mind, I'm going to try to grapple with the sucker.
GM: If you go in close right now, it'll have a shot at you.
PEGGY: Yeah, I'll try to dodge the blow, but I want to get in there.
GM: Okay, it's been swinging first from the left, then with it's right arm, aimed a bit lower. What are you going to do?
PEGGY: Hmmm. I'll wait for the right arm, then dodge the left, ducking underneath.
GM: Okay, you almost make it, but you get caught by its left swing, and it rakes your left shoulder.
PEGGY: Ouch. Do I get hold of it?
GM: Well, you're close and inside, your face inches from its left shoulder, what are you going to grab?
PEGGY: Is it a lot bigger than me?
GM: No, bigger, but not a lot bigger. Just under six foot.
PEGGY: Good, I'll get a solid grip on the left upper arm.
GM: Done! It's pulling the left arm in, and also moving to hug you with the right arm.
PEGGY: No problem, has my right hand got a firm grip?
GM: Like iron, but the ape's right arm is squeezing you pretty tight.
PEGGY: I want to get my arm around its back, so I can pull its

head back with my left, and keep pulling with my right.

GM: Okay, it's trying to restrict you, but you seem to be stronger. You've got your other arm around its neck. What now?

PEGGY: Squeeze!

GM: It struggles, it's grunts getting more and more shrill. It's going limp, are you keeping it up?

PEGGY: You bet!

GM: Snap! It breaks.

How to Role-Play Multiple Combat.

There are two tricks to Combat with a lot of opponents.

First, only so many attackers can gang up on any one defender. In hand-to-hand, where characters are grappling with each other, the maximum number that can actually have any effect is four. Backing up a little, with swords, a character can be surrounded, but the most that can attack without getting in each other's way is six. Of course, if a character is foolishly standing out in the open, and the target of ranged weapons, then there's no limit to the number of attackers.

Second, any good and talented fighter is going to move around quickly enough so that only one, two or three of the enemy can attack at any given time.

G/B	PSY	STR	END	WAR
0/0 Cindy - DORELL	A	4th [9]	A	2nd [45]
1/0 Mick - FARLEY	5.5 [11]	3rd [12]	3rd [9]	4th [12]
1/0 Kevin-RODERICK	2nd [51]	A	A	C
0/0 Armed Rabble	H	C	H	C

Example of Multiple Opponent Combat. Three characters, Dorell, Farley and Roderick, are faced with five strangers. Here's how the combat might come out, based on the following Attributes.

GM: There are five of them, and only three of you. How are you going to handle the situation? Mick?

MICK: What do you think Cindy? Should we go defensive, back to back, or spread out and take our chances?

CINDY: Let's keep this in character. (She changes her voice to a more haughty tone) Farley, this rabble means nothing to me, take your ease and I will dispatch them all!

MICK: (to GM) I give a slight bow in Dorell's direction. (to Cindy), Mistress Dorell, I hope you aren't biting off more than you can chew. Try your luck, and I'll jump in when and if you need me.

GM: Okay, Dorell, during your little chat, the five have spread out into a semi-circle, leaving you with three in front, and one each on your right and left flank.

CINDY: How does Dorell evaluate the situation?

GM: Experience tells you that your best bet is to either force your way through their line, and then turn them around. Or you can go for one on the edges, and position him between yourself and the others.

CINDY: Hmm...

KEVIN: Hey! What tactics does Roderick see here?

GM: Roderick thinks the best bet is to run like hell, and find a couple of dozen big guys in Shadow who can kill these suckers.

MICK: Not much of a Warfare rank, eh Kevin?

GM: Back to the fight. Cindy?

CINDY: I'll charge straight through the line, attempting to take out one quickly, using my combined strength and skill to bust through any defense.

GM: Fine... Hmmm. You sure did a job on that guy. He tried to attack in kind, and collected a bad sword stroke for his trouble. Now you're on the other side of the line, and two of the remaining four are turning toward you. The two guys who were flanking are looking nervously back at Dorell, but keep facing Farley and Roderick.

MICK: Does that mean they're in range?

GM: Yes. One is moving toward Farley, and another toward Roderick. Which brings us back to Dorell. Cindy, what now?

CINDY: Only two? Dorell laughs, salutes with her blade, and waits for them to attack.

GM: One attacks immediately, the other hanging back and looking for an opening. Mick, Kevin, you're both being approached, what are you doing?

MICK: I'll engage this guy, attacking cautiously.

KEVIN: Roderick is going to back away, using his sword defensively.

GM: Fine. Cindy?

CINDY: He's attacking me, remember?

GM: Yup, just wanted to see if you were doing anything different. Dorell slays him easily, before the other one even moves in range. What are doing now?

CINDY: Dorell commands the other to disarm.

GM: He doesn't seem to understand your language.

CINDY: I'll gesture for him to drop his sword.

GM: Fine, he throws it down and holds up his hands, babbling something you don't understand. Meanwhile, Farley seems to be having an easy time of it, since he's already wounded his opponent in the arm. Mick, what are you going to do?

MICK: I think I'll try to use a bit of Strength and disarm this guy.

GM: No problem. He's looking at his empty hand, and then back up at you. Kevin, you're being attacked, are you still falling back?

KEVIN: You bet I am. I'm waiting for Farley or Dorell to give me a hand over here.

GM: Okay, no problem. You seem to be able to hold this guy off indefinitely.

Judging Combat.

Judging means deciding the outcome, based on the relevant factors. As a Game Master, your job is to moderate the Combat. You've got to describe what's happening, give the player options, figure out what the non-player characters are doing, and act as judge, deciding which actions will succeed and which will fail.

The most important factor is the Attributes.

If they're far apart, the whole thing is easy. Close together, and that's where your job as Game Master starts to get challenging.

Here is the range of possibilities.

1. The Player is Far Superior to the Opposition. Most of the time, the Game Master merely inquires of the player how they might like to proceed.

For example, on hearing that the player character intends to slay each opponent, with a neat thrust to the heart, the Game Master can then respond, "Eight bodies lay about you, bleeding but slightly, and you scold yourself for having only slightly pricked the heart of one man, since he holds on to life for longer than you had intended. The others are running away. What are you doing?"

Or, if the opposition is of Chaos rank or better, the result might be, "Your blade reaches out once, twice, three times, each time finding it's place deep in an enemy body. Now all of them are badly wounded, and only two still stand and face you. What are you doing?"

Against a single opponent, of Amber or better rank, the Game Master could work things out so, "You break through her defenses with ease, and put your blade to her shoulder, ripping through her jacket and cutting her badly. In her eyes you see fear, and defiance. What are you doing?"

Note that in each case, the player is shown their advantage, and is then given the chance to respond with fury or with mercy. In each case the enemy could be spared (though only one remains alive among the merely Human Ranked opponents).

2. The Player is Clearly Better than the Opposition.
The difference between being far superior and clearly better is the amount of time it takes to dispatch the opposition. Instead of being able to instantly kill or disable the enemy, the player character will have to wear them down. It's possible, if the character is overly aggressive, and if the defender is cautious enough, that the player character may also pick up a scratch or two.

3. The Player is Very Close to the Opposition's Rank.
There are three possibilities here. The player can be "Just a Bit Better than the Opposition," or "Equal to the Opposition," or "Just a Bit Worse than the Opposition."

In the case where both opponents are opportunistic or aggressive, wounds can go either way, but eventually, if the Combat keeps going, whoever has the best Attribute will start inflicting more wounds than they receive. When that happens, when it becomes noticeable, that character will start inflicting greater and greater injury.

Subtle moves, like feints and revealed openings are places where the exact advantages are more plain.

So long as the player is conservative, remains defensive, neither gives an opening, nor takes any opening that is offered, then this combat can last a long, long time. Eventually other factors, like Endurance, will come into play.

Example of Very Closely Ranked Combat. This is taken from a session back in March of 1988. **Godfrey**, played by John Speck, dropped into a tavern in one of the seedier parts of City Amber. Notice how any wounds are based on whether or not tiny deceptions work or fail.

GM: A woman, long of leg, approaches Godfrey. She smiles and holds out a remarkable hand. The hand features well-trimmed nails and fencing calluses. She wears a sword at her side, and it looks pretty efficient. What are you doing?

John: I'll shake her hand, smile and say hello.

GM: She also says hello, and she seems to have a very firm handshake. Are you going to test her strength?

John: Maybe later.

GM: She asks if you are familiar with the swordsmen of the castle. It seems to be a reference to your uncles in Castle Amber.

John: Yeah, I know them.

GM: And, she asks, are you any good with that thing? She glances at your sword.

John: I hold my own.

GM: And Benedict? she asks, the smile dropping a bit, how is your skill relative to Benedict?

John: Hmmm... I think Godfey will just say that I'm in his league.

GM: Hardly, she says, I think you exaggerate your own skill. And I think I could cut your heart out. What are you doing?

John: I'll push back, drawing my sword and Main-Gauche.

GM: You barely avoid the drink she threw in the direction of your face. She whirls, pulling her blade simultaneously. She's advancing, her blade pricking at you. What are you doing?

John: I defend, holding my ground. But I'm not going to show my full ability yet.

GM: Okay, you hold back, but slowing down means you have to retreat several paces and barely escape getting wounded. Are you still faking?

John: No way.

GM: All right, you fight the best you can, and you're just holding your ground, parrying her lightning thrusts, and having your own attacks blocked. This goes on for a full minute, and you finally see an opening in her defense. Are you going to attack through the opening?

John: Is it real, or is she faking me out?
GM: You can't tell. It looks real.
John: I'll use the opening, but only for a feint. If she goes to parry the feint, then I'll attack.

GM: She doesn't parry the feint, but instead she uses it to attack. You feel the shock of her blade banging against your sword-guard. A close thing.
John: I'm going to be a bit more defensive for now, to see if she tires at all.

GM: No sign of that. What are you doing?

John: This time I'll launch a series of feints. Let me see if I can lead her into ignoring them.

GM: Okay, let me know when you want to do something. She parries the first one. The second one. The third one...

John: Okay, now I'll convert this one into a thrust, hoping that she'll be in the habit of reacting one way.

GM: Not only does she avoid your thrust, she also gets past your defense. You take a long rake along the outside of your extended forearm. Doing anything different?

John: No, just fighting.

GM: A moment later she puts up her blade. She asks if you wish to withdraw.

John: No way! I attack, all the way!

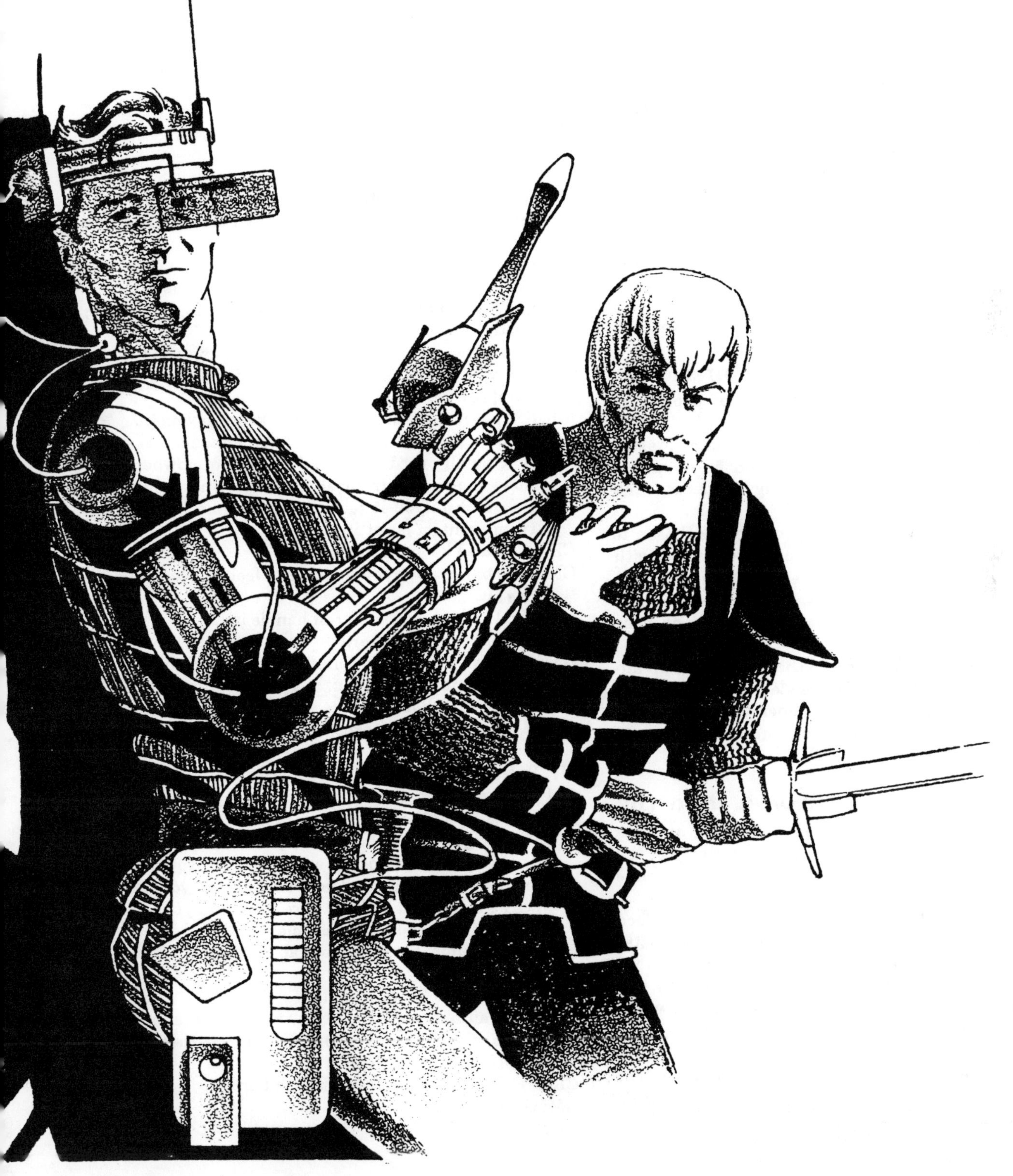

GM: She barely defends in time, and you force her to pull her foot back half a step.

John: Am I closer now?

GM: A bit. You had been fighting at the tips of your swords. You're closer now.

John: I'll try for a bind.

GM: She pushes back.

John: Using my greater Strength I'll force her blade out of line, opening her up for an attack.

GM: With greater Strength than yours, perhaps because of that wound on your arm, she circles your bind around, pushing your blade out of her line of attack.

John: Rats! I'll pull my sword back...

GM: That opens you up...

John: I'll use my Main-Gauche to parry!

GM: Good work! You succeed in blocking her thrust before it entered your armpit. You're relatively safe again.

John: If she extended that far, she might be vulnerable. I'll circle her blade and go for an attack on her hand.

GM: You manage to prick her thumb. A small line of blood appears.

John: I'll pull back a bit, and see how she reacts.

GM: Again, she offers you the opportunity to withdraw. "Seeing as we are both wounded," she says.

John: I accept!

GM: She smiles wide, embraces you, and seems really pleased!

4. The Player is Significantly Inferior to the Opposition.

The results of the Combat depend on what stance the player has taken.

If the stance was fairly defensive, then the Game Master might say, "You stop a flurry of attacks, and find yourself pushed to the limits of your skill, and bleeding from a cut to the back of your hand. What are you doing?"

An opportunistic choice, where the player has chosen to attack with caution, will have, "Your thrusts are stopped with precision, while your enemy's attacks seem to be pushing past your blocks all too easily. There are two cuts on your sword-arm, and you're lucky that thrust to your eye merely cut your cheek. What are you doing?"

Against a superior, the player who charges in with an all-out attack is going to be reeling, "Your blade does no harm, and is extended too far to defend you as the enemy cuts open your shoulder, pulls back to slash your chest, and then thrusts to your mid-section. What are you doing?"

5. The Player is Far Worse than the Opposition.

If the player is of Human rank, and has attacked blindly, their survival depends on the mercy of the enemy. Otherwise there will be nothing left but to make funeral plans.

If the player has at least a Chaos rank, or a bit of Good Stuff, or took a really defensive stand, then the Game Master might announce, "With blinding speed the enemy's sword flashes through your feeble defenses, and enters deeply into your body. You don't even feel pain yet, but you know you won't live through another thrust like that. What are you doing?"

Obviously, there's no point in continuing this fight. The player should seize any means of escape, or surrender, or do whatever it takes to avoid getting killed.

Characters with at Amber Rank or better, no matter how fine the enemy, will have at least two serious wounds before their lives are threatened.

Example of a Player Out-Matched. Here's an example, starting with Dorell and Farley in Trump contact. "Red" is really Bleys in disguise, a fighter superior to Farley in every way.

GM: A large man, clad in full armor, his head encased in a battle helm, his chest plate covered with a loose cloth of red.

GM: The guy in red says, "I am, for the moment, guardian of this place. I will not allow you to pass."

MICK: I tell him to move aside.

GM: Let me hear the words.

CINDY: Yeah, Mick, give us the voice of Farley.

MICK: Hey bozo, move it or lose it! You're risking irritating a Princeling of Amber!

CINDY: Yea!

GM: He shakes his helmeted head and says, "The power of Amber is weak in this place, your Pattern Blood will not help you to pass."

MICK: I pull my sword.

GM: As does he. What are you doing, combat-wise?

MICK: I'll go in offensively, not gonzo all-out, but strong and steady.

GM: You move in and, BOP!, while parrying your sword, the guy whacks you in the head with his left fist. Ow, you see stars! Are you still pressing your attack?

MICK: That makes me mad! Yeah, I keep moving in.

GM: You're a little dizzy, and your opponent seems to be dancing all over the place. From out of nowhere, you feel blows land on your cheek, forehead and shoulder. Every time he hits you it feels like a hammer. (to Cindy) The Trump contact is getting weak...

CINDY: Can I keep it going anyway?

GM: Yes, barely, but you're going to feel it, every time Farley gets hit.

CINDY: That's okay, Dorell can handle it.

MICK: Am I thinking clearly?

GM: Oh sure, the world's spinning a bit, there are a trio of red knights pounding on you, but you're thinking fine.

MICK: Three? I thought there was only one?

GM: Well, it looks like three, everything is kind of blurry. Are you still attacking?

MICK: No! I back off until my head clears.

GM: Back on defense?

MICK: Yes. I want to keep this guy off me.

GM: Okay, you move back a bit, and keep the big guy at sword's point distance. He doesn't push you, and in a few moments your head clears enough for you to feel your body's aches and pains. There's blood in your mouth, and it feels like he loosened a couple of teeth.

MICK: Next time, I'm going to grab him and teach him a thing or two.

GM: You want to try to grapple?

CINDY: Wait a minute! Do I know what he's planning?

GM: Sure, you've got a strong Trump contact, his surface thoughts are very clear.

CINDY: Then Dorell will speak with Farley.

GM: Go for it.

CINDY: Farley, beware this scoundrel! Remember, if he's

stronger than you, once you've grappled, you may never get free of him.

MIKE: Yeah, yeah.

CINDY: Test him. See if you are stronger before you commit yourself.

MICK: Okay. Is this guy stronger than I am?

GM: Hard to say. Unless he's got a roll of coins hidden in his fist, he's got a punch that's definitely better than human, maybe Amber level, maybe better.

MICK: I need to know for sure. I'm going to try to engage his blade, pushing him, see if he's a match for me.

GM: Here we go... You bind his blade, and he moves forward to the clinch. Are you going to use both hands?

MICK: Yes.

GM: He does the same, and the two of you are pushing, tensing, trying each other's strength. Which lasts just an instant, before he shoves you back reeling.

CINDY: Farley, it's best you knew.

MICK: Dorell, back off! I'm busy. (to GM) Am I falling down?

GM: No, not unless you want to...

MICK: No! I want to back up, and recover my balance, hopefully without getting stabbed or grabbed.

GM: The man in red seems honorable enough. He could have taken advantage of you, but he's giving you room to recover.

CINDY: I like this not, Farley. (to GM) Does Dorell think she can take this red knight?

GM: In Strength, you can't be sure, but you doubt it. On the other hand, the knight hasn't shown himself to be much better than Farley when it comes to the sword.

CINDY: Farley, let me come through, I am sure I can best this man.

MICK: No. I don't want us stranded here. If you come through, we'll have no easy contact back to Amber. I'll handle red here.

THE THREE FRONTS

In playing *Amber*, there are basically three fronts where you're likely to come into conflict with someone else: on the field of battle, where you use Warfare; hand-to-hand combat, where Strength is the deciding factor; and when minds collide, where Psyche is all-important.

You already know that the higher rank in each of these Attributes will determine the winner. That's no mystery.

The mystery is figuring out where you stand. Are you better? Worse? Or pretty close?

No matter which arena, you've got certain choices, each with their own risks.

Attack furiously, and, if you're better, even just a bit better, you may succeed quickly. Better yet, you may convince an opponent, even one who is slightly better, that you will win. On the other hand, if it turns out you're out of your league, an all-out attack could prove fatal, charging into a clinch you can't escape.

An *opportunistic stand*, lets you attack cautiously, taking advantage of openings, while cautiously defending yourself. This is safe, but also slow. If you've met your better, you haven't made a serious error. On the other hand, it may take a lot of precious time to figure it out, if you happen to be facing someone weak.

Going *Defensive* is good if you're worried, for it reduces to the absolute minimum the chance of taking injury. On the other hand, it gives your enemy a clear shot at escaping, or the room to take other actions.

Those three choices, the furious attack, acting as the opportunist, and staying defensive, are available no matter whether you fight with swords, bodies, or minds.

CHEATING ON ANY FRONT

No matter what kind of combat you get into, there are certain handy things that can be used to equalize the odds.

First, if you've got an appropriate Power Word, it can be used instantly, anytime during combat.

It's also possible to use something like Logrus tendrils or Advanced Pattern. These things can be used instantly *so long as they are prepared before the combat begins*. Either invoking the Logrus, or walking the Pattern in your mind, is time-consuming job that needs your full attention. So, unless you've got something ready to go before the combat starts, forget it!

Magic spells are also fairly quick, but only those that are memorized or "hung." Problem is, if there are any Lynchpins involved, they're going to work pretty slow. A skilled swordsperson, in range, can pierce your body many times in the time it takes to utter a single Lynchpin.

SWORD BATTLES & WARFARE

Why swords?

One reason is that it beats having to tear things open with your bare hands.

Still, given that high tech is available, why don't Amberites carry guns? Assault rifles, grenade launchers, laser-guided mini-missile pistols. They even could carry blaster rifles if they wanted.

Yet they still lug those blades around.

The main reason is that swords are reliable.

It doesn't matter what Shadow you happen to stumble into, a big piece of steel with a sharp point and a sharp edge is going to work. Shadows of high tech, low tech, or magic, you name it, a pig-sticker is dependable.

Even if the Shadow is so screwy that the edge is dull and the point is blunt, you've still got a handy, heavy thingie useful for bashing people.

A sword can be used to stab, chop, or cut, block or parry. Better yet, it keeps the enemy a good couple of steps away. Just in case something oozes acid, or arcs electric charges, or has other unpleasant personal habits, a sword leaves you a bit of breathing space.

Anyway, in Amber swords are traditional. And Amber tradition is based on a lot of hard-fought battles.

So Amberites use swords. Also quarterstaffs, axes, hammers, spears, and anything else that comes to hand, but the combat system works pretty much the same way for all hand to hand weapons. The following combat choices also apply to how one handles everything from troops on the field of battle, to strategy in a chess game.

SWORD MOVES & CHOICES

So now you're hauling sword.

What can you do with the thing?

In any combat you've got choices. Let's break 'em down into two major kinds of choices, defending or attacking. Each move can then be modified with deception, or using Strength in combination with Warfare. First off, what everybody learns first, defensive moves.

Main Combat Choices - Defensive.

Any move where your major concern is saving your own neck is considered defensive. Usually the opposition has to be clearly superior to you in Warfare before they can hurt you in a defensive position.

Pure Defensive. The heck with trying to hurt the other guy! Saving your own skin is your number one priority. If you've got even a suspicion that the other guy is better than you, go pure defensive. It's also a handy choice if you've got to deal with some kind of distraction, like a Trump contact.

Pure Defensive minimizes your chances of getting hurt.

Basically you stand your ground, moving back when pushed, and wait for the opponent to come to you. Since you leave it to the other guy to make the first move, it gives you an advantage.

Up against two or more opponents? Again, pure defensive means you keep moving, moving so that they get in each other's way. Better yet, move so that only one can attack you at a time. And, if you end in a situation where you can't seem to keep the enemy from circling around behind you, you know you're being out-matched.

What if you do get hurt when you're in pure defensive? Run!

Anyone who can score on you while you're staging this kind of fortress maneuver is too good. Way too good.

A wound when you're in a Pure Defensive position is the universe's way of telling you something is seriously wrong.

On the other hand, if you manage to hurt the other guy when you're in Pure Defensive, you can be pretty sure you've got the upper hand.

Measured Retreat. The idea is that you back off, hoping that your opponent will come after you, pushing a bit too far, and creating an opening. Works two ways, because you're protecting yourself, and, if the enemy doesn't follow you, then you've got a clean chance to escape.

Hasty Armor Defense. When in doubt, pick up something large to use as a shield, or squeeze yourself into a protected position, and make it into a makeshift fortress.

It might work, it might not, but it beats being skewered. Not only is it a low risk option, but the first blows will probably be absorbed by the furniture in use.

Main Combat Choices - Offensive.

Attacking always carries a certain risk. If you really want to make a measure of the opposition, preferably with your sword as a dip stick in the other guy's chest, then you've got to come out of your shell.

Fight! Most fighters take this stance. It just means you fight in reaction to your opponent, taking advantage of openings, defending against attacks, and holding your ground. This is also an option that gives you a chance to respond to opportunities where you can test the opposition's Strength.

When both combatants choose Fight, it means the battle will be determined purely on Warfare. Pretty quickly too.

The Feint. The idea is that you pretend to strike in one direction, then, when the opponent goes to defend, you strike in another direction.

If you're a lot worse than the enemy, this is going to be super obvious. Expect to take damage.

If you're a lot better than the enemy, they'll either fall for your ruse, which lets you hurt them, or they'll chicken out and retreat.

Finally, if the two of you are pretty even in Warfare, this is going to depend on a judgement call on the part of your opponent. The enemy guesses wrong, and you'll score a wound. If the enemy guesses right, it might be you taking the wound.

The Revealed Opening. Deliberately open yourself up, make something that looks like a mistake, or otherwise try to sucker your opponent into coming in a little too hard and fast.

If your opponent is equally matched, or has less skill, then it will look exactly like a real vulnerable opening. If they take the bait it's an opportunity to do damage.

On the other hand, if your opponent is a lot better than you, then your trap will be obvious. At which point it may be you who is in danger.

Advance. Steady, relentless pressure. You move forward, pushing for all the openings, shoving your opponent back.

Psychologically, a great move. Show your confidence! Intimidate the enemy!

All of which will work great, and you should damage the enemy, if you're really better.

If you're evenly matched, or out-matched, then this is a move that'll set you up for getting hurt.

All-out Attack. You launch yourself on the enemy. Forget about defending yourself, you're willing to take a scratch if necessary. Kill! A cocky, foolhardy thing to do.

Hope you're really a lot better than your opponent. In which case this option will definitely turn things your way, and you'll score the most damage possible.

On the other hand, if you're relatively equal, or worse, this is the choice that lets you collect the most serious wound.

Combat Modifiers: Faking It.

Aside from just bashing away, you can also try your luck at deception in swordplay. In *Amber* it's often not how good you are, but how good somebody thinks you are. These moves are also handy for maintaining a disguise, when you'd rather not have anyone know your true identity, or true abilities.

Feigned Inferiority. Since you are deliberately moving more slowly, and reacting with less skill, you may take slightly more damage than you might otherwise. On the other hand, there's a lot to be said for getting the enemy to incorrectly evaluate your skills.

If you are considerably better than the opposition then Feigned Inferiority should be easy. You'll take no more damage than you choose to take (sometimes it's very convincing to take a couple of small wounds). You can also look as klutzy as you like, from just below your natural rank, all the way down to Human rank.

If only marginally better, or equally matched to the opponent, then you're likely to get hurt. Certainly you'll take damage faster, and more seriously, than if you fought at your peak.

In the worst case, if you are significantly worse than the enemy, Feigned Inferiority can get you killed. By trying such a complex tactic, you're making yourself a lot more vulnerable (on the other hand, if you want to appear dumb and helpless, and you're sure the other guy won't kill you, this might be a good option).

Feigned Superiority. The idea is to pretend to be much better than you actually are. It's helpful to have a model to imitate. For example, if the character has had the opportunity to fence with, or better yet, take lessons from, Benedict of Amber, it might be possible to imitate his superior style and flair. At least for a little while. Unfortunately, not having the true ability means you're opening yourself up for any really serious threats.

If you're a lot better than your opponent, you've got a good chance of pulling this off. And no increased chance of taking any damage.

For bouts where you are fairly equally matched, things are a little riskier. Imitating someone else won't work quite as well as fencing in your own style. You're somewhat more likely to get hurt.

When your opponent is significantly better than you, this is a good way to get killed. The risk is high, and, unless the ruse succeeds right away, the style you imitate won't make up for your vast disadvantage.

Combat Modifiers: Combining Strength and Warfare.

Strength can be a natural complement to Warfare in a sword battle. At a critical moment, a character with superior Strength can vastly improve the Warfare equation. On the other hand, Strength isn't something you can use all the time, since most swordplay requires no particular muscle.

Beat or Bind. Beat is the fencing term for bashing the other guy's blade with your own. You hammer on the other guy's blade, adding your Strength to your blow. Bind is similar, except that you attempt to catch the opponent's blade in a tight little circle, forcing it away. The Bind requires more skill, and more of an advantage in Warfare, than the Beat.

If you have the better rank in Strength then how a Beat or Bind turns out depends on how you compare in Warfare.

If you are roughly equal in Warfare, but have a superior Rank in Strength, this can create a significant advantage. By bashing or moving the other blade out of the way, you gain a shot at a quick thrust or cut.

If the other fencer is clearly better than you, and you have a superior Strength, then this might work the first time, but only the first time. After that, knowing your advantage, the enemy will avoid future contests of Strength.

If you're the better swordsman, and better in Strength, then you can use the Beat or Bind to clear away the opponent's sword. Another choice with the Bind is to disarm the enemy, flicking the weapon out and away. A Beat, in this situation, could also be used to break an enemy's weapon.

On the other hand, if both parties have roughly equal Strength, then a Beat won't do much of anything other than waste a valuable opportunity. Remember that this particular Strength battle depends on just a couple of tiny muscles in the thumb and forefinger, which have to broadcast their power down a long lever (the blade). There has to be a really large difference in rank for there to be any advantage on either side.

Worst case, where the other guy has a much better rank in Strength, can be really bad. Failure will leave you open for a cut or slash, even from someone whose Warfare is a bit worse than yours. Then, after the Beat, your enemy will know your weakness, and attempt to exploit it.

Rough Housing. In the midst of a fight, sometimes there's the opportunity to get in a punch, a kick, or a good shoulder bash. Which is fine if you've got a superior Strength, but not too smart if you're weaker. Your chance of success depends on your Warfare.

If you're a lot better in Warfare, then you've got a good chance of getting in your blow, and it's unlikely you'll take any damage.

If both you and your foe are equal in Warfare, then the results of Rough Housing will depend on your respective Strength. Also, when you are both roughly equal in Warfare, and you initiate Rough Housing, it gives your enemy a free shot at grabbing you, and changing the battle to Strength.

If you are definitely inferior in Warfare, a Rough Housing move can be pretty risky. The enemy can choose to either avoid the physical blow, or take it and simultaneously wound you with the weapon at hand. Worse, if you are badly outmatched in Warfare, the foe can avoid the Rough Housing attack *and* get in a good wound.

Bait and Switch. Getting the combat away from Warfare and into Strength. Walk right in to the opponent's range, and grab whatever is handy, hopefully switching the combat from a Warfare to a Strength contest.

If you're a lot better in Warfare, this is easy and carries no risk. In fact, it's pretty common for Amberites, facing a bunch of normal Humans, to just grab and toss anyone in their way.

If you're equal, or slightly better, in Warfare, this move is sure to involve some risk. Chances are, you'll have to take a wound in order to get your hands in place. Since wounds get worse as combat progresses, it's a good idea to do this kind of thing early.

Switching to Strength is particularly bad if you are outmatched in Warfare. Then the enemy can cut you up pretty good as you move in. If you are a lot worse in Warfare, you may never even get your hands on the antagonist, and still take damage.

STRENGTH COMBAT

'...I can kill you, Corwin. Do not even be certain that your blade will protect you, if I can get my hands on you but once...'

Gérard,
Sign of the Unicorn

STRENGTH MOVES AND CHOICES

Unlike Warfare, where various stratagems and deceptions may fool an opponent, Strength is clear and straightforward. For example, attempting to fake superiority is silly. If you've got the better Strength, you can break your adversary. And vice versa. In other words, superiority in Strength is a sure thing.

One other point. Just as characters with Warfare get fast reflexes along with their military training, so characters with high ranking in Strength get martial art-style expertise along with their muscles.

Grab or Grapple. The Strength version of an all-out assault. You count on being a lot stronger, and being able to establish a controlling hold.

If you don't have Strength superiority, and you attempt a Grapple, you're going to be pretty helpless. Whatever the other wrestler chooses to do will likely succeed, and you won't be able to do much to stop it.

If you and your challenger are a fairly even match, the Grapple is probably not going to succeed. It's even risky, if the enemy uses the time to go for your throat.

If you've got clear superiority in Strength this is bound to succeed (except for wierd cases where you're battling something with too many arms, or too slippery to hold). Once the control is obtained, the controller has several options:

Pick 'em up and carry 'em. This is what Gérard did, carrying Corwin over to the cliff and dangling him up-side-down.

Swing your partner. Occasionally controlled victims are useful in other parts of combat. As shields, battering rams, and even as thrown projectiles.

Go for the Knock-out. The idea is to either whack 'em, or strangle 'em, until they go unconscious. Once a victim is in the power of superior Strength, they are pretty much helpless against blows and holds.

Make 'em hurt. The idea is to inflict pain, pure and simple.

Break 'em. Damage is pretty much optional. Choices include things like sprains and broken bones, all the way up to actions more grotesque...

Go for the Throat.

Something to do only if you think you're in a close battle, and you're willing to take it to the death.

Corwin did this on a couple of occasions, relying on his brute Strength to win out. This usually turns into a mutual thing, with each trying for the fatal snap before the other.

Although it's usually a matter of sheer Strength, where the one with the greater Strength kills the weaker, there is one factor. If the fighters are closely matched, then Endurance can come into play, and the one with the weaker Endurance will run out of steam.

If it turns out that you're a lot better in Strength, then your opponent will be helpless to resist this attack. At that point you'll have a choice of actions, strangling the foe into unconsciousness, or committing murder, or giving quarter.

Wiggle Out of It.

Escape, and defending yourself, are your primary concerns.

If you're better or equal in Strength, and facing a Grab or Grapple, then you should succeed in getting free.

If the opponent is Going for the Throat, you'll only get free if you are much stronger. You'll be stuck in any kind of even match of this kind.

If you're measurably weaker than your enemy, someone Going for your Throat, all this does is buy you a little time.

If the foe is going for a Grab or a Grapple, and you have a much lower Strength rank, then Wiggling Out of It will fail utterly.

Feigned Strength Inferiority.

This is easy, since all you've got to do is pretend not to be quite so strong.

If you are considerably better than the enemy then you've got many choices. You can pretend to be just a little bit better, and win just about anything you attempt. Or you can pretend to be inferior. You can even, if you like, pretend to get killed. Of course, lying around helplessly can be a little risky, especially if somebody decides to "just make sure."

If you're only slightly stronger, or actually weaker, then you better hope that your foe is merciful. Otherwise, pretending to be weak can lead to broken bones, or a broken neck.

Using Warfare in a Strength Battle.

Getting back to fighting dirty. In a contest of Strength, there's always the option of cheating. Get a weapon, and, so long as your opponent has left you a hand free, it might be possible to get in a strike or a shot.

This can work before things are completely resolved, or in the middle of Going for the Throat. However, once someone has put you in a Grapple, you're pretty much helpless. And if you're in a Grab, then the opponent is going to be aware of your pulling the weapon, and will have a chance to react to it.

It's also possible to use Warfare continuously during a Strength battle. The success of landing punches, kicks, and other blows, just as striking with a weapon, is determined by the Warfare attribute. The problem is, if you are significantly weaker than your opponent, then none of your kicks and punches are going to make any difference. No matter how much better your Warfare, physical attacks have to be made muscle against muscle.

Switching Between Warfare and Strength.

These are the two major choices of Combat. Benedict's way, that of skill, speed and accuracy, or Gérard's way, that of brute strength. Just to illustrate the principle, here's an abstract of a few elder Amberites and how they'd fare against each other, just using their best combination of Strength and Warfare.

The characters are as follows:

Benedict, definitely first in Warfare, and fifth in Strength.

Bleys, we'll rate him second in Warfare, and third in Strength.

Corwin, put him right under Bleys, third in Warfare, and fourth in Strength.

Julian, rank fourth in Warfare, and second in Strength.

Gérard, we'll call him first in Strength and fifth in Warfare.

So, in Warfare we've got Benedict, Bleys, Corwin, Julian and Gérard. The ranking in Strength is almost the opposite, with Gérard, Julian, Bleys, Corwin, and Benedict.

Once characters get their hands on each other in a Strength battle, whoever has the best Strength rank wins.

Let's look at how things come out differently when character's start with a sword duel. The difference is that Warfare controls the swordplay, but it's still possible to switch the battle to Strength by grabbing an opponent.

Corwin versus Gérard: Corwin's Warfare can wipe out Gérard, that's clear so long as the battle stays with swords. But Gérard isn't all that much worse than Corwin. If Gérard chooses to take a hit, buying a handhold on Corwin in exchange for the wound, then he can switch things to pure Strength. In this case Gérard probably beats Corwin.

Corwin versus Benedict: Corwin is significantly inferior to Benedict in Warfare, and he doesn't have enough of a Strength advantage to make a difference. Corwin is doomed.

Corwin versus Bleys: Corwin is outmatched in both Warfare and Strength. Unless he can bring in some other factor, Corwin gets cut up badly by Bleys.

Corwin versus Julian: Corwin, clearly superior in Warfare, has the drop on Julian. Julian can't switch the fight to Strength unless Corwin makes a major error.

Bleys versus Gérard: Bleys has Gérard in Warfare, and should be high enough in Strength to prevent being taken hand-to-hand.

Bleys versus Benedict: Benedict should win this, but if he's too cocky, and if Bleys can force the Strength issue, it could swing the other way.

Bleys versus Julian: Bleys, being superior in both Warfare and Strength, should beat the snot out of Julian.

Gérard versus Benedict: Benedict with vastly superior Warfare, will keep out of Gérard's grasp. Benedict wins.

Gérard versus Julian: Julian, ahead in Warfare, and really close in Strength, beats Gérard.

Benedict versus Julian: Benedict's Warfare cuts up Julian.

PSYCHE BATTLES

I turned to make my wishes known to my officers, and the power fell upon me, and I was stricken speechless.

I felt the contact and I finally managed to mutter "Who?" through clenched teeth. There was no reply, but a twisting thing bored slowly within my mind and I wrestled with it there.

After a time when he saw that I could not be broken without a long struggle, I heard Eric's voice upon the wind:

"How goes the world with thee, brother?" he inquired.

"Poorly," I said or thought, and he chuckled, though his voice seemed strained by the efforts of our striving.

"Too bad," he told me. "Had you come back and supported me, I would have done well by you. Now, of course, it is too late. Now, I will only rejoice when I have broken both you and Bleys."

I did not reply at once, but fought him with all the power I possessed. He withdrew slightly before it, but he succeeded in holding me where I stood."

If either of us dared divert his attention for an instant, we could come into physical contact or one of us get the upper hand on the mental plane. I could see him now, clearly, in his chambers in the palace. Whichever of us made such a move, though, he would fall beneath the other's control.

So we glared at each other and struggled internally. Well, he had solved one of my problems, by attacking me first. He held my Trump in his left hand and his brows were furrowed. I sought for an edge, but couldn't find one. People were talking to me but I couldn't hear their words as I stood there backed against the rail...

Nine Princes in Amber

The third combat arena, after Warfare and Strength, is that of Psyche.

Making Mind to Mind Contact

The first rule of Psychic combat is making contact. The nastiest, most powerful Psyche in the universe, facing off against a pathetic wimp of a human, can't do spit until there's a contact between the minds. Here are some, but not all, of the ways of plugging one brain into another.

Physical Contact or Close Eye-to-Eye Contact.
Physical contact must be flesh to flesh. A handshake, a touch on the back of the head. Clothing, armor, or any artificial covering blocks the contact. Separation, or the insertion of a barrier, will instantly sever a mind link that relies on physical touch. When physical contact is broken, the Psychic contact is as well.

Eye-to-eye contact must be very close, with the characters no more than an arm's length apart. The contact will be broken as soon as something comes between them, or when either one looks away.

Trump Contact.
The classic approach. Any Trump contact forms a junction for Psychic manipulation.

The main trick is to get the contact in the first place. If someone resists a Trump contact it generally takes a really over-powering Psychic force to break through. A Human ranked character could resist one of Chaos Psyche. One of Chaos rank could resist all but the greatest Psyche. And anyone of Amber level or better could resist even the greatest Psyche of any one individual (though their resistance could be broken by a coordinated attack by two or more characters).

Power Contact.
Touching someone with an extension of the Logrus or the Pattern is the equivalent of making physical contact. However, it should be noted that this kind of contact is far from subtle. In terms of what is needed to break through a character's resistance to contact, it's the same as a Physical Contact.

Magical Contact.
A magical mind link, an essential ingredient in many magic spells, is sufficient to create a Psychic connection. Once established, the link can be used from either end. Those artifacts and creatures with the ability to make ranged mental contact do so with the same kind of magic link.

Psychic Shadows.
Quirky Shadows exist where Psychic contact is as natural as speech. Each of these Shadows operates in a unique manner. Some such that Psyche is only possible at close range, equal to that of the spoken voice, others where Psychic contact is possible with anyone, anywhere. The type of contact, whether it is like a Trump, touch, or Magic, also depends on the nature of the Shadow.

Mental Moves

Once contact is established, there are several possible ways of dealing with the exposed mind. Each attack has its advantages, and its hazards. The more difficult the Psychic attack, the more levels of relative advantage the attacker will need.

Mindlock or Psychic Freeze.

At it's most basic, a Psychic attack is designed to simply lock up the brain of the opponent. This does the victim no particular damage, but it forces them to devote all their attention to defending themselves.

While in a Mindlock, the victim is unable to move except with slow and careful deliberation. In other words, they are helpless against physical attacks. A victim of a Mindlock is still capable of thought, and of using those powers that require the use of the mind alone.

Only someone with a clear superiority in Psyche can Mindlock another.

Psychic Assault.

Forget subtle, the attacker drives their Psyche like a weapon, seeking to rip into the victim's brain and do as much damage as possible. If successful, physical damage can take place. Eventually, if continued long enough, this can lead to the death of the victim.

Even a marginal advantage in Psyche will succeed in a Psychic Assault, if carried on long enough. However, if the mental combatants are relatively equal this kind of battle could easily go on long enough for Endurance to become a factor.

Psychic Suggestion.

A strong Psychic attack can subvert the will of the victim, putting them in a state of mind where they are willing, even eager, to cooperate with their attacker. Suggestions made after this attack is successful carry the weight of truth. However, the actual proposals should be framed so that they at least seem reasonable to the victim. Outrageous commands, that go against the victim's natural view of things, will likely cause the Psychic battle to start all over again.

Mindrape or Leech Information.

This "attack" is actually a peek into the victim's secret thoughts. In the grip of this assault, the victim will be unable to conceal their true thoughts and feelings. An attacker gets information by asking questions, or communicating images, and then "hearing" the victim's true reactions.

In order to succeed, when the opponent is resisting, the attacker must have a clear advantage in Psyche. However, a simple demand, like "What Is Your Name?" might be slipped in unexpectedly, and be successful even when the Mindrapist is inferior in Psyche.

For example, when Corwin had a death grip on the Chaos thing, Strygalldwir, their eyes locked together. Strygalldwir, probably realizing he was doomed, choose to gain Corwin's name, even though it hastened his own death. Strygalldwir might have been only a Chaos level Psyche, far inferior to Corwin, but he succeeded because of the unexpected timing of his Mindrape, and the fact that Corwin was concentrating on their physical battle.

Psychic Domination.

This is the most severe kind of Psychic assault, where an attempt is made to completely take over the victim's brain. If successful it means the attacker can move the victim as a puppet, even moving into the victim's brain (although leaving their own body helpless). Once "moved in," the attacker can manipulate the victim's body as if it were their own, relying on the victim's natural reflexes and powers. However, because the actual Psyche of the victim must remain locked away, the attacker has no access to the victim's thoughts or memories. If the Psychic contact is somehow severed, the dominator will instantly go back to their own body, and the victim will be freed.

Only someone with a significant superiority in Psyche can succeed in a Psychic Domination.

Time and Combat.

Combat is resolved fastest when the characters are mismatched.

For example, a character ranked very high in Warfare need take only a single stroke to dispose of someone ranked Human. Against one of Chaos rank the first stroke would be only a serious wound. If the enemy were Amber rank, then even the first wound might be slow in coming.

At the opposite end of the spectrum, when two characters are equal in rank, then Combat can take as long as it takes for one of the characters to be exhausted.

Speed & Timing Problems.

Some events happen in an "instant." These include a sword stroke, the pull of a finger on a trigger, a Power Word, and the releasing of a fully completed spell (one that requires no Lynchpins).

In general, Strength is the fastest of the Attributes when characters are in a hand-to-hand clinch. Otherwise, reaction time is determined by Warfare, so a better Warfare rank determines a character's reflexes.

Finally, nothing can work faster than thought, so mental combat is fastest of all. However, there must be mind to mind contact already set up. The process of making mental contact takes time and concentration, which can make Psychic battles much slower than Strength or Warfare.

Phasing Combat.

The hardest thing to keep track of in Combat is when players are doing different things. Phasing from one group to another, and back again, so everyone gets a fair chance at the action, is more of an art than a science. Practice is the only solution.

Example of Player versus Player Combat. Here we see what happens when the player characters start fighting each other. Choices have to be offered to everyone. Another problem is that other players tend to interfere, so the Game Master has to keep careful track of the time element.

G/B	PSY	STR	END	WAR
0/0 Cindy - DORELL	A	4th [9]	A	2nd [45]
7/0 Beth - YVONNE	3rd [23]	2nd [20]	A	3rd [17]
1/0 Mick - FARLEY	5.5 [11]	3rd [12]	3rd [9]	4th [12]
0/1 Willy - GARVIN	1st [52]	C	A	A

BETH: Dorell, I am sick unto death of you pushing me around!

CINDY: Yvonne, your whining is near to causing me anger. Were I you, I'd be careful how I twist my tongue lest someone hack it from my mouth.

BETH: That's it! Prepare to learn your lesson!

CINDY: Come and try it, Mistress, and you'll regret the day you were born.

MICK: Hey, Game Master! We've got a fight here.

GM (who has been running a different encounter with Alex and Willy): Just a second. Okay, what's up?

CINDY: Dorell is about to give Yvonne her comeuppance.

BETH: In a pig's eye! (to GM) I pull my sword.

GM: Okay, the fight seems to be on...

CINDY: Foolish wench! You will regret your insolence!

BETH: And I throw my sword on the ground, moving to tackle Dorell.

MICK: I'll step in between the two of them.

GM: Gutsy! Who are you facing?

MICK: Dorell.

BETH: Has Farley got his back to me?

GM: Sure. He's blocking your way and talking to Dorell.

BETH: Then I grab him by the back of his shirt with one hand, the backside of his pants with the other and pick him up.

GM: No problem, he's completely helpless.

MICK: Hey!

GM: Nothing you can do about it chum, you're in close quarters, and you turned your back to her.

BETH: Now I throw him, hard as I can, off to the right.

GM: Farley takes a little trip through the air... Cindy, what is Dorell doing?

CINDY: I draw my sword, and point it to the base of her throat.

GM: Easily done, Yvonne is close enough for sword range, are you going to lunge? Or just advance?

BETH: I hold up my empty hands.

CINDY: Drat! Dorell feels honor bound to toss away her sword.

GM: Are you doing that? Putting up your sword?

CINDY: Yes.

BETH: Great, I move in to grab her.

GM: Cindy, what is Dorell doing?

CINDY: Staying back, dropping my sword belt, and anything else that's dangling from my body, or likely to interfere with my movement.

GM: Okay. Beth?

BETH: I already said, I want to grab her.

GM: Are you moving in blindly? Rushing Dorell?

BETH: Hmmm... No, I'll be cautious, she could have a blade...

CINDY: If I were going to stab you, I wouldn't have thrown away my sword! (to GM) Am I ready?

GM: Yes, you feel you're ready.

CINDY: Then I'm going into a martial art stance, keeping my distance, but getting ready to start hurting her.

MICK: Hey, what about me?

GM: Let's see, after Yvonne tossed you aside... I guess you're okay. What are you doing?

MICK: There's something fishy about this. I want to Trump Garvin.

GM: Okay, you've got Garvin's card out. I'll get back to you when you make contact. Cindy, Yvonne is moving to seize Dorell, so what are you doing?

CINDY: I'm going to punch her right in the face!

BETH: When she does that I'll grab her arm.

GM: Dorell punches Yvonne, but it doesn't seem to slow her down. Beth, you're too slow to grab Dorell's arm. What are you both doing?

CINDY: Dorell will keep backing away, but every chance she gets she's going to punch, kick, or trip Yvonne.

BETH: I'm just going to keep moving in. I want to get my hands on her, no matter what it takes.

GM: Dorell is backing away, delivering punishing blows, and keeping away from Yvonne's grip. (turning to Willy) You're getting a Trump contact from Farley, are you taking it?

BETH: Wait a minute! How does he know who's calling? Every time I take a Trump call it turns out to be someone I'm trying to avoid!

WILLY: We initiates of the true art know these things... Yes, Garvin opens his mind to Farley's trump contact.

GM: Okay, Farley and Garvin, you've got full contact. (to Willy) Farley's mind is weak and vulnerable, do you wish to do anything with this opportunity?

MICK: Wait a minute...

WILLY: No, we'll just talk.. Farley, dear cousin, why have you summoned me?

MICK: Yvonne has challenged Dorell to a duel, and I think there's something fishy about it. Can you check things out?

WILLY: I agree, the duel sounds suspect. Bring me hither.

MICK: "Hither?" Cheeesh! (to GM) I'll bring him through.

GM (Interrupting Willy's reply): Okay, Farley and Garvin are now together. Since you've got a conversation going, I'll let you two chat, while I get back to the fight. Okay?

MICK: Sure, we can handle it.

GM (turning back to Beth and Cindy): Dorell is still pounding Yvonne, who is still advancing. Are either of you doing anything different?

CINDY: No.

BETH: Do I still feel okay?

GM: Yup, you're fine. Dorell, on the other hand, is starting to run out of room to retreat...

MICK: Wait a second! We want to try something first.

GM: Hmmm. Yeah, you've had time to talk it over, what are you doing?

MICK: It's Willy's play.

GM: So, Willy, what is Garvin doing?

WILLY: I want to see the magical aura, if any, that surrounds Yvonne.

GM: How are you doing it?

WILLY: (passes a note to the GM) The usual.

GM: Of course. You find a strong presence hovering over Yvonne. It seems to be a creature of magic...

WILLY: Is it of Chaos, Pattern or Shadow?

GM: You have no idea. The method you are using doesn't tell you that.

MICK: Nice try, Willy.

WILLY: You never know... (to GM) What does the entity seem to be doing? And can I tell if it is intelligent, or just a creature. And, is it natural, or some created thing?

GM: Let's see... The entity, as you call it, is feeding some kind of magical energy into Yvonne, somehow helping her. You don't know if it is intelligent, and you don't know whether it's natural or not. Anything else?

WILLY: I'm going to tell Farley about it.

GM: Fine, I'll get back to you guys. Beth, Cindy, Yvonne grabs Dorell.

BETH: Gotcha!

CINDY: You witch! (to GM) Shouldn't she be damaged by now? I'm no wimp, and you say I've been punching and kicking her for a few minutes now.

GM: Yes, you think so. You've scored quite a few hits, but she doesn't seem hurt at all. Maybe she's tougher than you thought.

CINDY: I don't buy that. I think she's cheating somehow.

Other Factors in Combat.

Amber Combat, aside from everything that has already been explained, is made more unpredictable, and more complex, because all kinds of other factors get involved. Here are some examples.

Inspecting the Ground.
When Bleys fought his way up Kolvir, killing and killing and killing, part of the reason he was able to do so was because of the ground. He was fighting on a narrow stairway, with a cliff wall on one side, and a sheer drop-off on the other. There was no way an opponent could sneak in behind him, or flank him, so he only faced one or two at a time.

Later that day, in the same battle, Corwin loses all of his army. Alone, in the open, where he can be attacked from any direction, he is overwhelmed.

Given the choice of ground, Amberites will make sure it's to their best advantage. For example, Gérard, when he wrestled Corwin, picked a bowl-like spot where Corwin could not escape.

High ground, confined spaces, walls, or obstacles can all be brought into the Combat situation.

Good Stuff, Bad Stuff and Zero Stuff.
No matter how good a character may be, a point of Bad Stuff can turn things around. Likewise, a character with a point of Good Stuff can usually get out of the most disadvantageous Combat.

Take a look at page 120. You'll see an illustration of three characters on a stairway. A barrage of arrows comes from stage left.

The top character, Carolan, having a few extra points of Good Stuff, faces the arrows without hesitation, even pausing to grab one out of mid-air.He doesn't collect so much as a scratch.

In the middle, hiding behind his shield is Morgan. At Zero Stuff his character is stuck in the position of having to make choices. Rather than depend on luck, the Game Master asks, "how are you defending yourself?" and "are you cowering behind your shield or looking over it and using it actively?"

Finally, at the bottom, we see Gwynt, struggling with his Bad Stuff burden. If an arrow can hit him, it will.

Good Stuff and Bad Stuff are never used up. They come into use in every Combat, giving the Game Master the opportunity to swing around any event that might be dependent on luck or chance.

Allies, Enemies and Other Interested Parties.
Things in *Amber* rarely happen without witnesses and manipulators.

Each player character is likely to have Allies and/or Enemies. Each can, if things look right, play a part. Using agents, advanced forms of the Powers, or other abilities, the on-lookers often meddle in affairs of Combat.

Again, let's look at *Amber* as Zelazny has defined it. Here we have characters who have hated each other for thousands of years. Who challenged each other, fought, and had the power to kill, yet never actually caused the death of any sibling. How could this be?

Most probably it has to do with the fact that everybody is always watching everybody else. And, when things come down to that final, fatal stroke, either the potential killer decides there might be witnesses, or one of the witnesses actually interferes.

Weapons.
If it's handy, if you can get it moving toward the enemy, then you can call it a weapon. Amberites love throwing things at their opponents. Small hard things are usually projectiles. Big heavy things are missiles. Fluids, dirt and powders are good for heaving in the eyes. Cloaks, rugs, or large flexible things can be used to entangle, or throw over the other guy's head. In other words, Amberites use whatever they've got.

However, some weapons are better than others.

Swords, Bladed & Edged Weapons. There is a special catagory for this kind of weapon. "Extra Hard" do no more damage than usual, but are resistant to breakage and can sometime cut through defenses. "Double Damage" weaons tend to increase damage, so a "Slash" might become "Life-Threatenin." Deadly Damage" weapons can take a "Slash" into a "Maim" or increase "Life-Threatening" into "Mortal." Arrows and crossbow bolts are considered edged weapons.

Impact & Force Weapons. Bare hands and feet are all that a character needs, but a weapon can increase the damage from an impact. Clubs, maces and flails are impact weapons. Another advantage of impact weapons is that they are very effective against armor since they don't really require penetration.

Gunfire & Beam Weapons. High technology weapons have much more damage potential than hand-based weapons. Fortunately for most characters, technology is something that only works in limited regions of Shadow.

Blast & Burn Weapons. There are powerful magical weapons that do a great deal of damage. High powered technological weapons, like explosives and flame throwers also fall into this catagory. Likewise, some magical spells do damage equivalent to blast and burn.

Inflicting Injury.

WARNING: SOME OF THE DESCRIPTIONS OF WOUNDS AND DAMAGE ARE PRETTY DISGUSTING!

If you find something that bothers you, stop reading and skip to the section on Armor. The descriptions of damage get progressively worse from level to level.

In Amber wounds are part of the story.

Wounds are clues. Clues that point to the superior combatant in a contest of skill. Clues telling a character that things are not going well. If a character gets a scrape, that means the next blow might be a cut or a slash. A stab will likely be followed by something more serious.

Here are eight levels of injury.

1. Non-Impact Hints and Near Misses.
A bullet whizzing by an ear is a really good way of telling someone to keep their head down. Likewise, in just about any

form of Combat there are less-than-damaging ways of clueing in the player. Near misses can be hits to the weapon, or the weapon's guard piece, or cuts that damage the character's clothing and possessions. Attacks adjusted by a character's Good Stuff, or the enemy's Bad Stuff, are often these kind of near misses.

2. Scratches, Bruises and Scrapes.

Also includes bruises, muscle strain, nose bleeds, broken fingernails, and other unimportant injuries. These wounds never get in the way of the character's actions, and are only really noticeable when the character has the leisure time to tend to them.

Swords & Such. Swords and other edged and pointed weapons can hit at the wrong angle, or with insufficient force to do real damage.

Impact & Force. In hand-to-hand Strength battles, there are a lot of these kinds of injuries, particularly when one character smashes another against walls, floors and other objects.

Gunfire & Beam. Burns and clips from the close passage of ranged weapons will hurt, but do no real damage.

3. Cuts, Pricks and Sprains.

An open wound, bleeding, and painful. While it doesn't have any impact on the character's other attributes, it is an indicator that all is not going well.

While blood loss from a cut could eventually drain the character into a life-threatening state, it's pretty unlikely. Usually the blood coagulates (forms a scab), and stops the flow, before the loss is serious. Even uncontrolled loss from a wound like this would take an hour or more to seriously weaken someone of Human Rank Endurance.

A wound could alternatively be some kind of sprained joint, like a sprained ankle or wrist, that will impair their Warfare and Strength in the affected area.

Swords & Such. In sword battles the most common wounds are to the sword hand and arm, then the shoulder, and finally, rarely, there are hits to the body or the head. The trouble with the hand and arm wounds is they usually keep bleeding as long as the character keeps fighting. It's distracting, and the psychological edge can sometimes push a character into retreating.

Impact & Force. A victim, in the hands of someone of superior Strength, will have their limbs twisted to the point where they become useless with the pain. Dashing a character against a solid surface will result in massive, painful bruises.

Broken bones also cause blood loss. When a bone breaks, there is damage inside and all around the site of the breakage. At this level the breaks are of small bones (finger, toes, small bones in wrist or ankle) so the blood loss is minimal.

Gunfire & Beam. A graze from bullets or beamed weapons will cause a painful open wound.

Blast & Burn. Close passage of high energy weapons and near explosions will cause painful burns on the character. This kind of near miss can also set clothing, hair, or possessions on fire.

4. Slashes, Stabs and Other Serious Blood Loss Wounds

Now things are getting serious. Without first aid of some

kind, the character will start feeling the loss of blood. Even slamming a hand, or a bit of cloth on a gushing wound will help the situation. No one, not even an Amberite, can remain conscious if they lose too much blood.

All characters go through the same stages in blood loss. First there is a feeling of weakness, then a narrowing of vision. After that a character feels unbalanced. Finally the character blacks out as the body goes into shock, and circulation fails from blood loss.

Swords & Such. Blades can rip open bodies to such a degree that major blood channels are cut. This means a character has only a limited time to patch things up before blood loss is serious.

Impact & Force. Heavy duty impacts can cause major fractures. These result in serious loss of blood inside the body. It's one of the reasons that people tend to go white when they have a serious break.

Gunfire & Beam. Shots at this level tend to pass through the body, leaving great gaps for blood to pour out. High caliber gunpowder weapons often punch small entry wounds, but leave larger, life-threatening exit wounds.

Blast & Burn. Explosions and fire inflict sufficient damage for nasty second degree burns (blistering) or third degree burns (burnt down through the skin). Characters of a Human Rank Endurance won't heal from these wounds without hospitalization.

5. Life-Threatening Wounds.

This is a major sword cut, a knife stroke, or a bullet wound that passes through a critical part of the body. Blood loss will be very serious.

A serious wound will generally slice open a significant artery or enough of the body's veins to cause a lot of blood to escape the body.

Normal people can survive this kind of blood loss for up to seven minutes. This doesn't change according to the person's size or shape. Bigger people have a bit more blood, but also bigger arteries that pump blood faster. Smaller people might have less blood (though not all that much less), but they also have smaller arteries. So seven minutes is pretty much standard.

Swords & Such. Thrusts deep into the body, or a hit on a major artery in a limb, cause sufficient damage to threaten a character's life. Blood loss is hard to stop without skilled medical help. Such wounds can kill in just a few minutes. Those with Human Rank Endurance will pass out almost immediately, and Chaos Rank won't remain conscious for more than a minute or two.

Impact & Force. Impact great enough to cause a concussion, or major rending of the spinal column. Characters, even those with great Endurance, usually pass out from the pain. Internal bleeding is immediate, but even Human Rank Endurance characters can survive for a few hours. Movement, other than possibly crawling, becomes impossible.

Gunfire & Beam. Impact is in the body, or possibly the head, and is great enough to rip apart internal organs. Survival for those of Human or Chaos Rank Endurance is impossible without skilled medical care. Even those with Amber or better Endurance are threatened.

Blast & Burn. All exposed flesh is seriously burned, and even the parts protected by clothing will be affected. If the eyes were facing the source, then the character will go blind.

Characters with Human or Chaos Rank Endurance will die unless they have immediate medical help.

6. Maim or Sever.

The severing of a hand, the putting out of an eye, something that permanently (well, not permanently for most Amberites, but you get the idea) disfigures the character.

Swords & Such. Swords usually do their damage to hands and arms, cutting them off, or puncturing eyes.

Impact & Force. Brawny opponents, of high Strength, do have the capacity to tear the limbs from their opponents.

Gunfire & Beam. Any serious blast, from a bullet or a beam, can tear away a limb. This is usually a foot or a leg, but can be a character's arm or hand.

Blast & Burn. Major burns can turn extremities like hands or feet into useless stubs.

7. Mortal Wound.

When Brand was released from his imprisonment, the knife wound to his kidney would have been fatal had he not received immediate medical attention. Neglected, he would have surely died. Characters with Mortal Wounds instantly have their attributes reduced.

Swords & Such. The weapon damages either the heart, lungs, throat or another internal organ. Characters, no matter what their Endurance, are stricken and unable to stand up. Surgical treatment is needed or the character will surely die.

Impact & Force. The rib cage is crushed. Only major surgery and modern medical maintenance devices, or magical means, can save the character.

Gunfire & Beam. Same as a sword, except the damage is more extensive, and the wound is larger.

Blast & Burn. Tossed like a rag doll, the character suffers massive breakage throughout the body. Only characters with better than Amber Rank Endurance can even survive the initial blast. Burns and cuts cover the body, and broken shards of bone penetrate internal organs.

8. Death Blow.

A direct, completely fatal, blow to the heart or brain. Dealing death is usual only when one character completely overmatches another.

Of course, there's always murder. A deliberate killing stroke to a helpless victim. If you've got anyone unconscious, paralyzed, mind-locked, or otherwise out of commission, then killing 'em is easy. So long as it is understood that it would be cold-blooded murder.

Swords & Such. Direct, precise thrust through the heart or the brain. Death is instantaneous.

Impact & Force. A major impact breaks the neck of the victim, severing the spinal cord, and causing immediate death.

Gunfire & Beam. As with a sword, the bullet or beam destroys either the brain or the heart.

Blast & Burn. Character is shattered or blown apart.

Reinforcing the Hurt.

In *Amber* we don't have any specific gauge of how much a character is hurt or wounded, any more than there is any such thing in real life. However, just as in real life, characters will tend to notice certain signs from their bodies, telling them that injuries must be fixed. Here are a few samples:

- "All at once you feel this prickly, pins and needles feeling running down your limb to your wound. What are you doing?"
- "Ooh, you suddenly see little sparkly things right in front of your eyes. What are you doing?"
- "Everything seemed to go dark for a moment. You realize that you sort of lost it there, and you're not standing up straight any more. What are you doing?"
- "You stagger, losing feeling in the area, and then the pain comes back sharp and new. You feel like cursing or just gritting your teeth, or maybe even screaming a little bit. What are you doing?"
- "Abruptly you open your eyes, and you're not sure how long you've been unconscious. You still feel totally drained. What are you doing?"
- "Your skin feels cold and clammy, and you're sweating an awful lot. What are you doing?"
- "With a clunk, you hear something hit the floor. You look at your hand and you realize it's now empty. Not that you can recall what you were holding. What are you doing?"
- "You realize you don't have the vaguest idea of what you are doing, where you are, or what's going on. What are you doing?"

Armor Deflections.

Armor can do two things. First, it can deflect a blow, saving the character from all but minimal damage (even if the character's armor deflects a bullet, there's still the likelihood of getting a bruise). Second, armor can slow down or reduce the impact from a blow, lessening the damage.

For example, Julian's white lacquer armor is impervious to bullets. Since it covers most of his body, hits from bullets tend not to bother him. Cutting at Julian's armor with a sword is also pretty useless. The attacker may succeed in striking Julian, but the armor will prevent the weapon from doing any damage.

Conventional armor (equivalent to "Resistant to Conventional Weapons") will deflect all but the best blows. Therefore a character would have to deliver something serious, like a life-threatening wound, before the weapon would penetrate the armor. Even then, the effort of punching through the armored material would soften the force of the impact, so the wound would be less.

Still, armor only protects what it covers. For example, most characters avoid wearing heavy gauntlets, because it interferes with their sword work. A good pair of armored gloves, such as those used by Eric or Corwin, will protect the character from the hand wounds that swords often inflict.

Death.

Like anything else in Combat, Death is a piece of the story. It shouldn't be used unless it has some point. No *Amber* Death goes unpunished, or unexamined.

If it works out that a player character kills someone, then the perfect death is one where the player character feels immeasurable guilt, where they suspect outside influences and where a fair fight, in retrospect, looks more and more like murder. In short, each death should open the door for future stories in the campaign.

Amberites are tough suckers. They are very, very hard to kill. Almost impossible to kill.

Which brings up an interesting question. How do you kill an Amberite? Here are three basic possibilities. Murder, bloodshed, and misadventure.

1. Murder. A helpless character, bound or unconscious, can be killed without a whole lot of effort. Murder also takes place under the cover of Combat.

2. Bloodshed. In other words, dying in combat.

In combat Amberites can be killed. It's just awfully tough.

Take a sword fight as an example, based, of course, on each character's Warfare Attribute.

If a player character is ranked as a wimpy Human in Warfare, then death comes very, very easily. So easily, that a skilled opponent may deliver the fatal blow without realizing it. Likewise, if a character is Chaos ranked and if the opponent is of very high rank, then, again, there may be a quick death.

It's just that very few Amberites have this kind of weak Warfare. And those that do are rarely stupid enough to get into sword fights.

When combatants are closely ranked, the fight is never a matter of a quick death blow, but is more in the nature of a gradual fencing for position, information, and advantage. In this situation, once the inferior in a combat recognizes that position, it doesn't make a whole lot of sense to stand around being slain.

3. Misadventure. Usually boils down to either suicide, or "accidental" suicide.

It's always possible for a character to walk directly into death. Jumping off a cliff.

One popular method of suicide is taking dumb risks in combat. For example, staying in a combat situation where things are going badly. Or leaping to an all-out attack before adequately measuring the opposition's relative skill.

One all-time favorite is for a Logrus Master to teleport to the center of the Pattern. Or, likewise, for an Amberite to teleport directly into Primal Chaos. Both are quick ways to go. And, they save on funeral costs (no coffin needed).

Combat and the Art of Visualization.

All of the technical rules that are listed above are really an afterthought in the design of Amber. From 1985 through 1990, all the various Amber game masters had no such rules at their fingertips.

Yet they used no dice. And the combat was satisfying for the players. And often memorable.

So how did they manage to moderate combat in Amber?

You could put a lot of names to the process. Visualization. Artistry. Applied imagination.

Basically the Amber game master develops a "feel" for the combat. There are so many variables, that it's rough to try to logically analyze exactly which is most important in every case.

Instead, game masters use their experience, and their knowledge of the Amber universe, to judge combat intuitively.

What am I saying?

That this entire section, all these rules, tables, and explanations, they are all baby steps. They'll teach a game master how to "walk" through Amber combat.

If you want to "run" Amber, at some point, you'll have to leave these rules behind.

You'll know when you're ready.

Combat versus Mortality.

In any role-playing combat system there is a trade-off. Usually the trade-off is between realism and playability. In *Amber* we've avoided that particular trade-off. Combat can be as real, or as playable, as the Game Master likes.

However, there is another trade-off, that between mortality and justice. In a purely just role-play, each Combat would be judged on its own merits, so the outcome would always be based on absolute, measurable factors.

However, there is a drawback to a purely just game. It tends to kill characters.

Consider that in the entire *Chronicles of Amber* only one character of the blood of Amber, Eric, actually died in Combat. Even his death was suspect, since he may have been killed not by the Combat, but by the complication of his wearing of the Jewel of Judgement.

That kind of mortality, where Amberites rarely die, should be true in role-playing *Amber* as well. The players' investment in their *Amber* characters is just too great for them to die easily. It just doesn't make any sense for someone to put in two or three (or eight, or eighty) hours into a character, and then have it wasted on a whim.

Amber's combat system reflects the kind of problems faced by Zelazny's characters.

Corwin knew he could best certain people. Any Shadow dweller was so much meat for his grinder. Most of his brothers were not regarded as a threat.

On the other hand, Corwin knew that, in a fair fight, he was doomed if he fought Benedict. Against Eric and Bleys, he wasn't sure. And, against total strangers, those exhibiting strange powers, he could never be sure of the results...

So victory isn't rare. One character can best another. It's just that the loser usually has the power to escape even when totally defeated in battle.

As a Game Master you will sometimes be faced with a choice. Whether to run things justly, or whether to warp things to allow players to survive. Stick on the side of mercy, and, just to keep things "fair" give the same mercy to the non-player characters.

Closing Combat.

To sum up Combat, let's take another look at the *Chronicles*, this time when Zelazny relates Corwin's last real fight, when he's ragged and tired, and meets up with Duke Borel of the Courts of Chaos.

'Lord Corwin of Amber!'

He was waiting for me as I rounded a bend in the depression, a big, corpse-colored guy with red hair and a horse to match. He wore coppery armor with greenish tracings, and he sat facing me, still as a statue.

'I saw you on the hilltop,' he said. 'You are not mailed, are you?'

I slapped my chest.

He nodded sharply. Then he reached up, first to his left shoulder, then to his right, then to his sides, opening fastenings upon his breastplate. When he had them undone, he removed it, lowered it toward the ground on his left side and let it fall. He did the same with his greaves.

'I have long wanted to meet you,' he said. 'I am Borel. I do not want it said that I took unfair advantage of you when I killed you.'

Borel... The name was familiar. Then I remembered. He had Dara's respect and affection. He had been her fencing teacher, a master of the blade. Stupid, though, I saw. He had forfeited my respect by removing his armor. Battle is not a game, and I had no desire to make myself available to any presumptuous ass who thought otherwise. Especially a skilled ass, when I was feeling beat. If nothing else, he could probably wear me down.

'Now we shall resolve a matter which has long troubled me,' he said.

I replied with a quaint vulgarism, wheeled my black and raced back the way I had come. He gave chase immediately.

'Coward!' he cried. 'You flee combat! Is this the great warrior of whom I have heard so much?'

I reached up and unfastened my cloak. At either hand, the culvert's lip was level with my shoulders, then my waist.

I rolled out of the saddle to my left, stumbled once and found my footing. The black went on. I moved to my right, facing the draw.

Catching my cloak in both hands, I swung it in a reverse-veronica maneuver a second or two before Borel's head and shoulders came abreast of me. It swept over him, drawn blade and all, muffling his head and slowing his arms.

I kicked him then, hard. I was aiming for his head, but I caught him on the left shoulder. He was spilled from his saddle, and his horse, too, went by.

Drawing Grayswandir, I leaped after him. I caught him just as he had brushed my cloak aside and was struggling to rise. I skewered him where he sat and saw the startled expression on his face as the wound began to flame.

'Oh, basely done!' he cried. 'I had hoped for better of thee!'

'This isn't exactly the Olympic Games,' I said, brushing some sparks from my cloak.

The Courts of Chaos

THE MECHANICS OF *AMBER*: A GAME MASTER'S TOOLBOX

Here is a toolbox full of tricks, methods, tactics and strategies designed to help the Game Master conduct an *Amber* campaign.

SUCCESS & FAILURE

The number one question put to any Game Master is, "does it work?"

Whenever player characters try to do anything, from opening a door to creating a universe, it's up to the Game Master to decide whether or not the character is successful. And player characters are only the beginning, since the Game Master has to answer the same question, silently, about anything attempted by any of the dozens of non-player characters operating behind the scenes.

Does it work?

In *Amber* the answer is almost always yes. Characters, player characters or otherwise, almost always succeed at everything they try.

Of course most things are just the little things we do every day. Characters are usually successful at walking, talking, reading, eating, playing and sleeping, just like people in real life. When it comes to using the Powers, like Pattern or Magic, most of the things that characters try will succeed.

There are exactly three exceptions, three cases where characters can fail. They are when a character has Bad Stuff, lacks the ability, and/or is opposed by some other character.

Bad Stuff. Any character with Bad Stuff will occasionally experience the difficulties associated with ill fortune. This is an on-again, off-again kind of thing. How much failure is inflicted on characters depends on how much Bad Stuff is involved.

Inability. Any action can fail if the character lacks the qualifications, over-reaches their capacity, or if the character is trying something for the first time.

At the point of character creation, the players are allowed to fill in the details of their character's life and skill experiences. The player always deserves the benefit of the doubt, so if any part of their background could provide a needed skill, assume that it does. However, if nothing in the character's history indicates the opportunity to learn the skill, then assume that the player character never picked it up.

For example, a player may not have specified computer training. However, if the character's background includes a couple of years of study at a modern university in a fairly sophisticated technological Shadow, then it's fine to assume that the character must have picked up some computer training. Perhaps not enough to re-program a supercomputer, or to build a security-cracking program from scratch, but enough to know the basics of exploring a data base and getting on line on a network.

As far as inability to use Powers, characters often fail as they experiment with new aspects. Look, for example, at an expert auto mechanic. The mechanic can fix most problems, but will run into difficulty when faced with a car imported from a different Shadow reality. It's not that the mechanic lacks the skill, it's just that dealing with a totally new car requires a bit of experimentation, with a good chance of occasionally screwing up.

Garvin, Willy's character in our example, has Advanced Pattern Imprint. Suppose that somehow Willy hears of something called "Pocket Shadows." Here's how the Game Master might handle it.

Cindy: So now we've got the Crystal Ball. Now what? What do we do with it? If we just leave it out in Shadow, it'll just get snatched. If we take it with us, we might be followed.

Willy: I've got an idea. (to GM) Is it possible for me to make a Shadow Pocket? As a place where we can stash the Crystal Ball?

GM: I don't know. What are you going to do?

Willy: Hmmm. It seems that bringing Pattern to mind is always the first step whenever I want to play with my power. I bring Pattern to mind.

GM: Okay, that's going to take you awhile. Cindy, are you doing anything?

Cindy: No, I'll just stand guard.

GM: Are you going to keep holding the Crystal Ball?

Cindy: I'll set it on the table, but I'll keep my eye on it and see if any of the images reappear.

GM: Okay, Cindy is staring at the Crystal Ball. Willy, you bring the Pattern up to mind. Now what are you doing?

Willy: Well, as I said, I want to make a Shadow Pocket.

GM: You've never done it before. So how are you going to try doing it now?

Willy: Let me look at the Shadow I'm in now.

GM: Examining this Shadow, you find that the immediate area, everything within a few miles, seems rather ordinary. You and Dorell are the only things of substance. The Crystal Ball is not of this place, and it seems to have some kind of strange energy. If you like, you can start looking outward, checking the rest of this Shadow.

Willy: No, I don't need to see the rest of it. Though, now that you mention it, can I see if the energy of the ball seems to be leading anywhere?

GM: No, it seems self-contained, with no outside connections, here or through Shadow.

Willy: Back to making a Shadow Pocket. I'm going to manipulate the Shadow stuff, using it to make a pocket.

GM: I don't follow you. What do you mean, making a pocket?

Willy: Well, I sort of want to grab the fabric of Shadow, like it was a piece of cloth, and pull it open, so there's like a gap between layers.

GM: Yes, it seems you can do that. Cindy, you see a spark of green light appear in the center of the Crystal Ball.

Cindy: Really? I'll look closer.

Willy: Wait, does the light have anything to do with what I'm doing?

GM: You have no idea. You're concentrating on the Pattern. You feel like the Shadow is giving way a bit, as if the fabric of reality is coming apart. Are you going to rip open a pocket?

Willy: Yeah, that's what I want, a Shadow Pocket.

GM: Cindy, the light is getting brighter inside the ball, are you still concentrating on it?

Cindy: Well, yeah, but I'll also take a look around, just to make sure Garvin is okay.

GM: Everything seems to have gotten a lot darker and a lot quieter, except for Garvin. He's covered in light, his eyes open, but focused on something you can't see. It's eerie, but you think you can almost see right through Garvin. What are you doing?

Cindy: Dorell screams!

GM: Willy, you hear Dorell whispering something. Things seem to be coming along, and you'll have that pocket ripped into the Shadow with just a bit more effort. What are you doing?

Willy: I don't hear her scream?

GM: Just a whisper.

Willy: Since I'm concentrating on the Pattern, I'll hold the Shadow rending for a second and take a peek around.

GM: Where there was one Shadow, it seems there are now two. One with you in it, and the other with Dorell and the Crystal Ball. She does seem to be screaming over there. You realize that you haven't actually heard any sound from her, that the whispering is something you were picking up with your Psyche. There's a strange gap between the two pieces of Shadow, corresponding to the hole you are creating. What are you doing?

Willy: I don't see anything wierd about the ball? No light?

GM: Nope.

Cindy: I'm going to try to grab Garvin.

Willy: What do I think is happening to the Shadow? Why is it breaking in two parts?

GM: Just a second. Cindy, your hand passes right through Garvin. When it comes out your hand is covered in frost, chilled to the bone. Everything gets even darker.

Cindy: I'm going to grab the Crystal Ball, and start walking out, Shifting Shadow away from here.

GM: Willy, you're going to use your Pattern to figure out how the Shadow is changing?

Willy: Yes.

GM: If you do that, you'll have to stop concentrating on the pocket, which means it might close up again.

Willy: Forget the pocket. I want to know what is happening.

GM: Dorell Shifts away, just in time, as the Shadow seems to fall out of existence. Willy, as soon as you stop concentrating on the pocket, it's like everything dissolves and the entire Shadow self-destructs around you. You are falling into a void, what are you doing?

Willy's character Garvin didn't exactly fail. He tried creating a Shadow Pocket in totally the wrong way and ended up destroying an entire Shadow. Had he tried something else, or experimented more thoroughly, he might just as easily have succeeded. And, having created a Shadow Pocket successfully once, Garvin could feel confident that the process could always be repeated without failure.

It's also important to note that the way things worked out here is not the way they'd work in another campaign. Willy's Game Master has a version of *Amber* with fragile Shadows, easily ripped apart by a careless Advanced Pattern Initiate. In another campaign, with another Game Master, this might turn out to be exactly the correct way to make a Shadow Pocket.

Opposition. The most common reason for a player character's failure is that they are being opposed.

When a character says, "I plunge my sword through the creature's neck," it automatically succeeds if the creature is helpless. However, most of the time, the creature is not cooperating. A character's attempt at murder is usually contested by some other character's attempt at escape. They can't both succeed, so the result is a matter of the Game Master's judgement.

Contests in *Amber* are often waged without the characters even being aware of it. Player characters will sometimes be opposed by hidden elder Amberites, or off-stage Lords of Chaos. In these cases, the player characters could easily see their failure as due to luck (Bad Stuff), or inability.

Example of Player Character's Failure. In the following example, three characters attempt to pick the lock that bars their way.

GM: The door is locked.

Beth: No it isn't. Using my Pattern, I determine that it's probably open, but that I turned it the wrong way. I'll turn it the other way, and open the door.

GM: No you don't, it's still locked.

Beth: I thought I could manipulate Shadow stuff?

GM: Well, it seems this lock is real enough to resist your use of Pattern. What are you doing?

Beth: So I bash it in...

Ted: No! We don't want to leave any trace of our passage.

GM: Well?

Beth: Don't look at me. Yvonne doesn't know the first thing about picking locks. Except for the old-fashioned way, of course.

Ted: What do you mean, the old-fashioned way?

Beth: Just grab hold of it, smile, and use a little muscle.

Alex: Give me a try at it. My character Harick has had plenty of experience with this kind of thing.

GM: Okay, what are you doing?

Alex: I'll snap off a piece of wire, bend it into the right shape, and unlock the door.

GM (consults the character summary, and sees Harick's eight points of Bad Stuff): For some reason your piece of wire feels really spongy. It just keeps wiggling around, without affecting the lock.

Kevin: Let me in there. I can get this thing open...

GM: Alex? Are you stepping aside?

Alex: I might as well. Nothing ever seems to go right for me...

Kevin: Gee, I wonder why? I'll examine the door's lock.

GM: Hmmm. I don't remember anything about Roderick studying locks. Where did you get this experience?

Kevin: Hey, wasn't it part of my standard training out in the Courts of Chaos?

GM: As you look at the lock, you're reminded of the time you took off to learn skate boarding on Shadow Earth. It seems you missed quite a few of your classes, and a whole year's worth of exams. You seem to think lock-picking might have been part of that.

Kevin: I've already got Logrus up to mind, I'll insert a Logrus filament into the lock and feel around.

GM: The filament recoils in shock as it touches some remnant of Pattern in the lock. Not quite enough of a jolt to make your Logrus fall apart, but you didn't hold the contact for more than an instant. What are you doing?

Kevin: Yeah, it's a lock all right! Ariel, why don't you give it a shot?

Ted: Okay, at least I know something about locks. I'll check it out.

GM: That's right, you're quite an expert as I recall. Hmmm. The brand here is unfamiliar, but it looks like your number six lock pick should do the job.

Ted: Great! I open up the lock.

GM (this time, consulting Ariel's Warfare Rank, the Game Master realizes that lock's builder was much higher.): It doesn't work! Somebody has customized this sucker, putting in some kind of double failsafe tumblers.

Ted: Uh oh! We must be dealing with a real professional here.

GM: What are you doing?

Ted: Well, how long do I figure it would take me to get this thing open, if I just keep working on it?

GM (again, seeing that Ariel is completely out-classed): This thing seems devilishly clever. It's far beyond your experiences. You figure you need at least twenty minutes just to get a good idea as to how it really works...

Beth, playing Yvonne, has no doubts that she can get through the lock, using her strength. Of the others, Alex's character, Harick, fails because he is cursed with Bad Stuff, and Kevin's character Roderick admits to having never learned the skill. That leaves Ted's character, Ariel, who should succeed. What stands in Ariel's way is the maker of the lock, an elder Amberite, who would use tactics (putting complex locks together is as much tactics as anything else) and dexterity. Both the tactics and the dexterity are governed by Warfare. Ariel's Warfare is vastly out-ranked, which means that the lock may still be picked, but that it will be a long and arduous proces.

DECISIONS VERSUS RANDOMNESS

So-called *random* events seem to be part of our day to day lives. It's just that, upon examination, there are few things that are really left up to chance. No matter what the choice, event, action or environment, a Game Master can eventually figure out a logical, non-random solution.

Unfortunately that word, *eventually*, means that Game Masters often don't have the time to resolve things logically. They've got to fall back on a few role-playing tricks.

Using Good Stuff, Zero Stuff & Bad Stuff.

Since the player characters made their own "luck," with Good Stuff and Bad Stuff, the Game Master can make decisions based on how much they've got. In general, the more Bad Stuff, the worse things will be for the character, at Zero Stuff things are neutral, and Good Stuff characters have the universe smiling on them at all times.

Choosing Random Events.

Random events usually involve a range of possibilities. Randomness could be just a choice between two outcomes, or the almost infinite variety of personalities that reside in any given human being.

Determining the range, the number and form of the

possibilities, is not a random thing. No, a Game Master figures out the possibilities and limits what the random events will be in any situation.

For example, on an urban street in a city of Shadow Earth the range of encounters varies from beggars to bandits, children to the aged (though limited to those both old enough and young enough to get around on the street), and all the various normal city folk. The range would not usually include wizards, dragons or aliens.

What is seemingly random is a matter of choosing from the range of possibilities. On the city street the possibilities are known. Any one encounter, policeman or newspaper vendor, is something that could be random.

Yet why should it be random? A novelist or screenwriter doesn't pick a random encounter. No, they pick whatever "random" event is most interesting, or the one that works best to move the story along. Why should they want any boring events? Or purposeless events? Or events that repeat themselves?

It's the same in role-playing. All the Game Master has to do is visualize or list the possibilities, and then select the one choice that seems to work best for the current role-playing situation.

Leaving Choice Up to the Players. If, and only if, there seems to be more than one attractive possible event, then let the "decision" fall on the actions of the players.

Note that it isn't player "choice," it's player "action." After all, the players don't know that they are making the decision for you.

Looking at the possible events, figure out which actions would branch outward, leading to a single event from among a range of choices.

For example, suppose a couple of player characters have wandered into a strange city. There are two important events happening that night, one of which is a secret blood sacrifice, where the identity of their tormentor might be revealed, and the other is the departure of the ship that carries the Crystal Ball they seek.

Now assume that two elder Amberites are trying to influence events. Each has agents in the city, and each agent has a plan for pulling the characters to one of the events. Now it becomes a matter of choice for the players. Will they seek out a comfortable inn where they can refresh themselves, and where one of the agents will be informed of their presence? Or will they directly seek the Crystal Ball, in which case they may find the other agent, set to watching the Street of the Jewelers.

The players select the event that they'll witness by their actions.

An Example of Letting the Player Decide. Let's go back to one of our previous examples, where Ted's characters Ariel is dealing with a lock that is challenging his Warfare.

Seeing that Ariel's Warfare rank is straight Amber, and that the builder of the lock has over forty points in Warfare, The Game Master decides that even with twenty minutes to explore the lock, there's still a fifty-fifty chance that Ariel will fail. Here's how it might work:

Ted: Hasn't it been twenty minutes? Have I figured out how to pick this lock?

GM: Yes. After careful probing with your lock picks and probes you figured out that this lock is so tricky that it would require somebody to insert a key halfway, turn to the right, then push it all the way in, and then turn to the left. What are you doing?

Ted: Great! I open up the lock.

GM: You turn it to the right, you turn it to the left. You hear a loud click from the lock mechanism. Now what are you doing?

Ted: I'll open the door.

GM: Apparently not. It's still locked. Now there's a ticking noise coming from inside the door.

Beth: What?

Kevin: Bad vibes, dude!

Ted: Hush up, everybody. I assume I've still got my pick in the lock. What can I figure out about this situation?

GM: It seems that everything is normal. The lock is designed to set off some kind of trap or alarm, you're not sure which. Whoever set this thing up, set it up so the user of the key would have to make a final decision, whether to make another right turn, or to continue turning left. What are you doing?

Ted: I have no idea. Can I back out of this?

GM: Yeah, you could probably turn everything back the way it was. So long as you're fast enough to do it before whatever is clicking decides to do something else.

Ted: Mmm...

GM: It keeps clicking. Are you turning that last tumbler right, or left? Or are you just going to wiggle it back and forth?

Ted still has a range of options open. Everything from picking between right and left, to trying to undo his work before the clicking stops, to asking Beth to step back in and use some muscle.

Which was the right choice, right or left? Well, there's a couple of things the Game Master might have done. First, just before this point, the Game Master could arbitrarily decide on *left* as the correct answer.

Another answer is that the Game Master can leave it to the character's Good Stuff and Bad Stuff. That way, no matter whether Ted picks right or left, basically a matter of luck, success will depend on the amount of Stuff invested in the character.

Manipulators & Puppeteers.

Things that we think of as random here on Shadow Earth are easily manipulated by those who have the powers of Amber and Chaos.

Weather, for example, is something an Amberite controls easily while walking through Shadow. Crowds will be as thick or as thin as an Amberite wishes. And finding a parking space isn't a problem for an Amberite.

GAME MASTERS & POWERS: ROLE-PLAYING WITH THE GLOVES OFF

There are no limits in *Amber*. Player characters can, if they're ambitious enough, or careless enough, destroy the whole campaign. It's just habit with me. I've always liked the idea of player characters being given sufficient power to blow themselves to kingdom come.

So the characters here have no particular limits.

Yet, at the same time, player characters can't do everything. At least not right away.

This gives the Game Masters another tightrope to walk. You've got to be able to tell a player that anything is possible, but that it doesn't work the way it's being tried right now.

It would be impossible to cover all the possibilities. Phage Press is planning a series of "Amber Master Guides," each devoted to explaining the possibilities inherent in a particular Power. For example, *An Amber Master's Book of Trump* will cover such exotica as "Designer Trump," "Trump as Weapons," "Trump Traps," and "Automatic Trump," explained, in character, by the players who create and use them.

PATTERN

Player Definitions. Sometimes characters are "destined" to have various encounters while travelling through Shadow. These can be meetings with their elders, or things relating to the story in progress. The Game Master can leave the details of the location of these encounters to the player, by asking questions about their choices in moving through Shadow.

GM: Mick, as I recall Farley is mounted up on his horse, and he's ready to head off into Shadow.

Mick: Yup.

GM: Where are you going?

Mick: I'm not sure exactly, I just want to find out something about the guys in pink armor.

GM: So you're looking for a Shadow where there might be some answers?

Mick: Yes, that sounds good.

GM: Alright, You start in Forest Arden, at the point where Shadow can be manipulated. What are you changing first?

Mick: Does it matter?

GM: No, all roads will eventually lead to what you seek, I just want to hear some details.

Mick: Okay, around the corner the path will turn into a yellow brick road with a bright green city off in the distance.

GM: Whoa! Too much of a change for the edge of Forest Arden. Try giving me a little detail.

Mick: How about a green mushroom?

GM: Better. You come into a clearing dotted with green mushrooms. Next?

Mick: The tree trunks will get a little more mossy, and the sky will get cloudy.

GM: About half an hour later you are surrounded by mossy trees, green mushrooms and travel under dark clouds.

Ahead you see a man in pink armor on a huge pink horse. What are you doing?

This could have easily ended up on the yellow brick road, or in the green city. If the Game Master isn't satisfied with one location, all they have to do is keep questioning the player until a better place comes along.

Shadows of Desire.

Any character can find exactly the Shadow they desire. When this is done by a player character, the Game Master should get a written description of the Shadow from the player. Game Masters should also be aware of the unconscious aspect of a player character's desires. The character's problems, hopes, dreams, fantasies, and fears, all the things that can show up in quizzes or in role-playing, can also show up in a character's Shadow. For example, if a character is having a problem with a particular elder Amberite, then one, or more, Shadow versions of that character might be found in the Shadow.

Shared Shadows. One interesting variation on a Shadow of Desire is a Shadow sought by two or more characters at the same time. This can create some interesting side-effects since the contents of the Shadow will reflect both characters' desires, conscious and subconscious.

The Danger of Walking the Pattern.

It's sometimes tough to get across to the players that walking the Pattern is a dangerous, life-threatening, exhausting activity. As a Game Master you know that with one misstep the character can die. Therefore you might want to encourage players to have alternate means of escaping the potential death-trap of a Pattern Walk.

One way to do this is, early on in your campaign, is have a character be in danger mid-way through walking the Pattern. A figure could appear before the character, blocking the way. Or some outside force could agitate the Pattern, causing the sparks to fly higher, and the resistance to increase to the point where the character can no longer move forward. Of course, this first time, there ought to be an elder Amberite available to save the character at the last second.

Since there is no turning back from a Walk on the Pattern, and no stepping off, there are only a limited number of escape routes. Trump is one possibility, and another is having an Advanced Pattern Initiate around to assist. The point is, player characters can, and should, try to safeguard their passage.

Taking Chances. Characters who lack the appropriate Attributes, or who are weakened, exhausted, or otherwise in bad shape, should be warned. After all, their characters know how tough a Pattern Walk can be. If they're not ready, then the Game Master should say, "you know that if you try walking the Pattern right now, in your current state, you will probably die!"

TRUMP ARTISTRY

There are a lot of unanswered questions about Trump. Corwin, our first narrator, only used Trump, and had only a vague idea of how they worked and their ultimate potential. Merlin, the second narrator of the Amber series, is a Trump artist. However, Merlin never actually shows us the creation of a Trump. We know he made a batch, including his own, Dara's and Martin's, and we get to see him make a Trump Sketch, but he reveals little or nothing about the secrets of Trump Artistry. So we end up with a lot of questions to be answered by each Game Master.

What is the Power of the Trump?

Early on, it's Corwin's opinion that Trump are based on some aspect of the Pattern. His confusion is understandable, since every Trump he ever saw depicted either someone with the Blood of Amber, or some subject connected to Amber. However, having found out that Trumps and Trump Artists are quite common in the Courts of Chaos, we have to come to the conclusion that Pattern is not the basis of Trump. If not Pattern, then what?

Logrus as Trump Power. It seems unlikely, but still possible, that Trump are based on the Logrus. After all, the only Amberites who we can definitely identify with making Trump are Dworkin and Brand, both of whom have shadowy connections with the Courts of Chaos. Perhaps it's even possible that Trump are some creation based on Chaos Creatures, with each card being prison to some Psychic critter.

Magic as Trump Power. Again, this doesn't seem real likely. Magic is too ephemeral, too easily dispelled. However, there may be a large dose of Magic involved in the creation of Trump, again because all the known Trump Artists could very easily be Sorcerers or Conjurers.

The Reverse Image as the Source of Trump Power. With the Unicorn as the back of all of Dworkin's Amber Trump, it seems possible that the image is more than mere decoration. If this is the case, then it's possible that the Trump of the Courts of Chaos depict the Serpent of Chaos, or some other powerful icon capable of "fueling" the power of the Trump.

Trump as Innate Power. The obvious answer to the Trump's power is that it is a separate power. This solves one problem, but creates a bunch more questions. If Trump is a major power, does it have a representation, like Pattern or Logrus? If so, where can the source of Trump power be found?

Trump as the Power of Creation. Could it be that we're looking at things backwards? Perhaps it was Trump Artistry that preceded Order and Chaos, Pattern and Logrus. It could even be that Primal Chaos was created by the drawing of the Logrus, and that the Primal Pattern itself is nothing but a Trump image.

How are Trump Duplicated?

It doesn't seem logical that Dworkin painted each card, in each deck of Trump. Certainly he must have had some short-cut, some way of printing a number of Trump decks. The question is, how?

Since Trump Sketches can use ordinary materials like pencils and paper, or stone and a stylus, then it would seem that the image is most important. Still, a Trump Sketch doesn't have the true power of an actual Trump card.

Conjuring Trump. If Trump are somehow Magic objects, or some kind of Powered items, then the only way to duplicate them will be with some form of Conjuration. In this case the Conjurer will probably need to have Advanced Trump Artistry as well.

Printing Trump. First answer this question; Are Trump special because of their special image, because of the cards themselves, or both? If Trump work only because of the artistry involved, then anyone could make a deck of Trump, starting with an original deck and good color printing or color photocopy technology. If only the cards need be special, then add to the list of supplies a stock of special Trump-sensitive material.

Trumping a Trump. Here's a wierd possibility. Maybe it's possible to use a Trump for a contact, not with the subject of the card, but with the card itself. Then, having "trumped" the card, perhaps one can, like a magician pulling cards out of thin air, pull out endless duplicates of the card. If this were possible, then it must certainly be a very special ability, worth ten or fifteen points as an addition to Trump Artistry. An alternative method may allow a character to duplicate an entire deck, concentrating, shuffling the cards and gradually "growing" the deck until it has been fully duplicated.

What is the Link Between the Card and its Subject?
It would seem that being the subject of a Trump card confers some special qualities, though what these may be is uncertain. It's possible that making a Trump of someone or something gives them an extra bit of reality.

Trump Window to the Mind. One of the scary things about Trump is that they create an opening to the subject's mind. For example, Caine, as he quietly spied with the Trump, could read the surface thoughts of the subjects, *even when they were not in Trump contact!*

What are the Limits to Creating a Trump?
Brand created a Trump of someone he didn't even know, just from descriptions given to him by other people (although it may have taken him years of effort). Perhaps that was possible because of Brand's extraordinary Psyche, an Attribute that might have allowed him to build up a Psychic impression of the person, just from the second-hand impressions of other people. We know it's possible for any Trump Artist to make a Trump of anyone they know well. Does that mean a Trump Artist can successfully capture the image of someone they met only briefly?

Making a Trump from an Image. It's unlikely that any Trump artist could make a Trump based on a painting, or a photograph of a subject. That's because there would be no Psychic impression of the person, and that's probably essential to creating a Trump.

What does it take to Force Trump Contact?
Characters can generally refuse a Trump contact. However, it is possible to break through a character's defenses, forcing the contact. It's up to each Game Master to decide just how much Psychic superiority is needed for such a feat.

Group Trump Contact. A popular experiment is duplicating the feat in the books, when Corwin leads a batch of family members in a mass Trump contact, recovering their brother Brand. However, when Corwin led his pack of family members on a group Trump contact, remember that Brand wanted to be contacted. The power of the group was used to overcome an environmental problem, the Trump barrier around Brand's prison. It's also worth noting that at least one, and possibly two, of the characters in the group were actively fighting *against* the contact (something that none of the others noticed). Using a group to break through a character's "go away" barrier could be another matter altogether.

The Great Trump Trick.
Trump, when used from player character to player character, works entirely too well.

There should be a major disadvantage to using Trump. The drawback should be that the character should have no idea who is calling until some kind of contact has been made.

Which means that the only way to avoid the other characters you want to avoid, or avoid having your privacy (or mind, or body) invaded by some stranger, is to refuse to take Trump calls. This should be a blind refusal, because most player characters have no way of knowing who is calling.

Except, of course, that one player can hear another player, right across the table, tell the Game Master about the Trump call. So, naturally, the listening player knows who is calling.

So, how does the Game Master foil the player's unauthorized listening?

Well, there's no point in haranguing people into not talking, or into "playing in character."

Instead, use trickery.

Here's how it works.

First, make sure when you prepare the game, that you carefully look over the list of non-player characters and determine what kind of possible Trump calls they might want to make to the player characters. Will Bleys Trump his son, telling him of some new development in a plot? Would King Random, concerned about Castle Amber's security, recall some of the player characters away from their adventure? Has some villainous Lord of Chaos, or lost Amberite, come across the Trump of a player character, and is looking forward to matching Psyche? Or, could there be an assassination attempt in the cards?

For each player character, there should be a list of some pre-prepared options.

Then, add to the list of possible Trump calls as the game continues. Be sure to record the exchanges of Trump, when characters give their Trump to others, and accept new Trump cards for their own deck. It's also important to keep track of each contact with non-player character Trump Artists, each of whom could create new Trump cards and/or Trump Sketches based on the player character.

Prepared with a list of the possible calls that the player characters could receive, you will now be ready for the first player-to-player Trump call.

Players are not likely to make the Trump calls when their characters are in the same room, or in sight of each other. If so, there's no need for intrigue, and the Game Master can just allow the connection.

Start with the player character using the Trump. As a safety measure, just in case the receiving end is resistant, be sure to "push" the role-playing of the caller a bit. It goes something like this:

Ted: I'll pull out Harick's card and Trump him.

GM: Hmmm. You're getting some resistance. Are you going to keep trying?

Ted: Why aren't I getting through?

GM: Gee. Could be the distance through Shadow, or maybe the nature of the place where he is. Or he could be busy. The question is, are you giving up, or do you want to push it?

Ted: Yes.

GM: Okay, I'll have to get back to you.

Once you've got a commitment on one end, as above, you've got to put the caller on "Game Master Hold," take a moment to consult your list of potential Trump calls. Now move to some other slice of the action. You could go direct to the player getting the Trump, but it's usually best to wait awhile, and to take care of other business first. After a bit, you finally get to the player character who is the object of the Trump call.

GM: So, Alex, what is Harick doing?

Alex: I'll take the Trump call.

GM: What Trump call? Are you planning on using your Trump? I thought you were busy hauling on the ropes of the mainmast, controlling your voyage through Shadow, and trying to keep your sailboat from drifting in the storms.

Alex: I'm not getting a call?

GM: No, I'll tell you when you get one. What are you doing?

Alex: Well, how's the *Goldenrod* doing? Are we taking on more water? Are we keeping on course?

GM: So far, between your Strength, and the skill of your small crew, you've managed to keep things going. Any change in plans?

Alex: No, I still want to make land by morning.

GM: All right, it's about two hours later, the storm has quieted from gale force, down to gusts of wind, and you see a ship, brightly lit with torches, about a mile behind you.

Alex: Damn. Is it moving toward us?

GM: Yes, and it seems a lot faster than your ship.

Alex: I'm continuing to Shift Shadow, is it staying with us?

GM: Seems to be keeping up with no problem.

Alex: I want to get my spyglass, and climb up the mast, so I can take a better look.

GM: Okay, you're up near the crow's nest, one foot braced on the wet mast, and reaching up for a rope that keeps snapping in the wind, and you feel the itching of a Trump contact. Are you taking it?

Alex: Now? Wait a minute! Can I climb up first?

GM: Not if you want to block the contact. Concentrating to keep out a Trump contact takes all your attention. It seems pretty strong and persistent. Are you taking the call? Or staying rigid and resisting?

Alex: Arggh!

GM: While you're deciding, your foot slips off the mast and you are now hanging on with one hand and with your other foot looped around some rope. What are you doing?

Alex: All right! I'll take the contact!

Ted: Great! Harick, I need...

GM: Ted, wait a minute. The call is not from Ariel.

Alex: It's not? Who is it?

GM: You're not sure. Whoever it is, they seem to be looking at you from a distance, probably from that other ship!

After the players have had an experience or two with this kind of Trump switch, they'll lose their eagerness for every "friendly" Trump contact.

It's even nastier when one player makes a call, and the Game Master immediately asks the player at the other end if they'll accept it. That's an opportunity to bring out some prepared Trump contact that reads like this; "Something inhuman makes contact. It is reaching out for you, attempting to break into your brain. What are you doing?"

Coded Calls. Another, related, Game Master tip. Some players may get the cute idea of "coding" their calls. They'll try pulsing them, or calling with three short contacts and one long. Remember that a lot of folks out in the Amber universe are listening in, and the energy from the initial Trump call is a lot louder than the communication once the call is established. Sooner or later, somebody really nasty is going to catch on to the code.

Random Trump Selection. Players tend to spring this on their Game Masters frequently, especially when things are pretty desperate. In response to "What Trump are you using?" players will say, "I'll just pull one out at random..." or, "the top one off my deck..." In some cases, when the player character has a hefty dose of Good Stuff, this makes perfect sense. Being lucky, they'll draw a card and hope it's the perfect one for the occasion. However, it's best to discourage this kind of thing, asking something like "Are you sure you want to leave it to chance?" or saying "You pull the last card you've used, the one you keep trying because you get no response."

Trump Decks, Standard and Otherwise.

Corwin describes the deck he retrieved from the Library at Castle Amber as complete. Of course, that's complete as far as Corwin knows. He found other cards in Dworkin's cave that were totally unfamiliar.

Number of Cards in a Deck. There's no exact answer here, because Corwin (or Zelazny) never mentions it. Most Tarot card decks, which are presumably the Shadow Earth reflections of the Trump, have seventy-eight (78) cards, arranged in the following way:

1. Twenty-Two Major Arcana. These are the cards that represent major ideas or concepts, ranging from The Fool, to The Magician, to Death, to the Wheel of Fortune. They are (usually) each labelled with roman numerals, except for the Fool, which is either unnumbered or marked with a zero.

2. Fifty-Six Minor Arcana. Arranged into four suits of fourteen cards each. As Corwin says, they are arranged into Wands, Pentacles, Cups and Swords. Which, if you take a look at a standard deck of playing cards, translates into the suits of Clubs, Diamonds, Hearts and Spades.

3. Sixteen Court Cards. Each suit has four royal figures. The King, Queen and Jack (called a Page in Tarot), plus a Knight.

Corwin describes the cards that depict the various elder Amberites as *Greater Trumps*. Whether he's talking about the Major Arcana or the Court Cards is unclear, but it makes more sense for them to be Court Cards.

SCRYING WITH TRUMP

And some nights I dwelled upon the playing cards. The missing Trumps had been restored to the pack I held. One of them was a portrait of Amber itself, and I knew it could bear me back into the city. The others were those of our dead or missing relatives. And one was Dad's, and I skipped it over quickly. He was gone.

I stared at each face for a long while to consider what might be gained from each. I cast the cards several times, and the same thing came up on each occasion.

Nine Princes in Amber

Corwin's casting of the cards is an attempt to perform a scrying with the Trump. It's a form of fortune telling, and one of the Amber family's favorite games.

How Trump Scrying Works From the Character's Point of View.
Since the cards are icons, pictures of the major forces and characters of the *Amber* universe, mystically connected with their subjects, then the cards can be used to show how those people and things are in conflict. This appears as an abstract picture, with each card representing one or more of the forces at work in the universe. It would seem that the more talented the reader of the Trump is in Psyche, the clearer and more accurate the reading will be.

How Trump Scrying Works from the Game Master's Point of View.
Trump fortunes are a way for the Game Master to leak suggestions, hints and background to the player characters. Trump readings should be ambiguous and vague, so the Game Master doesn't have to reveal anything really important. It's almost as if the universe itself were speaking to the character, but in a "voice" that is expressed only by the pictures on the cards. They say a picture is worth a thousand words, and, in a Trump reading, each image could sure be interpreted in a couple of dozen different ways.

The twisted element of manipulation by elder Amberites can also be involved in Trump Scrying. Cards that appear in a reading could be directed by an unseen character, as a way of urging player characters in one direction or another.

Trump Readings, Detailed and Vague.
Most of the time the Game Master will just present the results of Trump Scrying at the spur of the moment. If something comes to mind, some indicator, omen or premonition, fine. If not, then the cards can always give a result of "situation hazy, try again later." Game Masters also have the option of preparing the occasional complete spread, giving the players a snapshot of some major conflict in the campaign, where each card is presented in detail, with a number of alternate meanings and explanations.

Casting Fortunes with Trump.

Using a deck of Trump as a fortune-telling device, a process that Corwin calls "casting the cards," is done in the following way. First there is a shuffling of the cards, all while the *querant*, Tarot-talk for somebody looking for an answer from a Tarot reading, thinks about a specific question or just about the situation in general. If there is more than one character involved, then each character cuts the cards after the shuffle. The cards are then dealt out, one at a time, face up, in a special arrangement. This can range from the simple "six card spread," described below, to complex patterns involving every card in the deck. Once the cards are laid out, those present attempt to read the cards.

Interpreting the Spread. Each place in the spread has a different meaning, and each card has several possible meanings. Depending on where a card appears, it could represent the specific character painted on the card, or something in the background, or a trait exhibited by the character. Not only could a card have a different meaning depending on where it appears, its meaning could also be affected by the placement of the other cards in the spread. Figuring out all these variables can take hours, and rarely lead to any solid conclusions.

Six Card Spread. This is the spread used to get a Psychic snapshot of the current situation from the querant's point of view. The first card is flipped over in the center, followed by a card to the right of the first, then one to the left of the first, then one above, then one below, and finally, one crossing the first (overlaid, but sideways).

Meanings of the Cards in the Spread. Depending on its position, each card tells a part of the overall story. Here's what each card means in a Six Card Spread, along with a sample of how it might be interpreted if Caine's card appeared in that spot:

1. *The Querant's Present Situation.* "This is the card that covers you," is how a Trump reader would describe it. If there are several things going on related to the character, then this card is an indicator of which matter is being described.

If Caine's card were in this position, it could refer to Caine being the one the characters are trying to help. Or, because Caine is associated with the sea and with ships, it could mean that the reading has something to do with a naval problem.

2. *The Likely Outcome of the Current Affair.* "This is the course that lays before you." The card describes the likely outcome of the current situation, if nothing much changes.

If Caine's card were in this position, it could mean that the characters will run into Caine, or that he will win out in the current contest. On the other hand, it could mean that the ending of the current affair will end as Caine ended the Patternfall War, by assassination.

3. *The Querant's Pursuer.* "That which hounds you." Usually a character, but it can represent either a person, or some aspect of the person. For example, Benedict in this position could mean that Benedict is pursuing the character, that military events are chasing the character, or that the character is being dogged by his or her need to improve their Warfare.

If Caine's card were in this position, it could be that Caine is chasing the querant. Or, if the querant were being pursued by one or more of Caine's offspring, whose cards are not in the deck, this could be a clue as to their origins. On the other hand, since Caine used the Trump as devices for spying, maybe it means that a Trump spy is keeping tabs on the querant.

4. *That Which Inspires the Querant.* "That which gives you comfort." A lofty goal, a moral code, the inspiration of a loved one, or somebody the character respects.

If Caine's card were in this position, it could mean that the querant's ultimate goal is to rescue Caine from some danger or imprisonment. Or perhaps the querant's true desire is to be as sneaky as Caine, or as willing as Caine to end the treachery against Amber.

5. *That Which Seeks to Manipulate.* "That which hides its face." This could illustrate, or at least hint, at whoever is pulling the character's strings. It could be somebody specific, or it could refer to a general place like Amber or the Courts of Chaos.

If Caine's card were in this position, it could mean that Caine is manipulating the querant, or the whole scenario (quite likely, since we're talking about one of Amber's finest manipulators). On the other hand, since Caine once faked his own death, maybe this is saying that the true villain, or true manipulator, is someone who is assumed to be dead and is really alive.

6. *The Pivot.* True Solution or Next Step. "And this is the card that crosses you, the key to your dilemma." The most important card in the reading, and the one that contains the main hint. As a Game Master tool, it allows you to point the character in the direction of their next logical step. Usually it's kind of ambiguous, but sometimes it directly names a character, or a place, of great significance.

If Caine's card were in this position, it could be saying that a contact with Caine would be the next logical step, or that Caine is the key to the whole affair. Or that Caine's methods of hiding, waiting, and watching, would be the best strategy for the querant.

Examples of Game Mastering a Trump Scrying. Here are a few responses to players attempting to read their fortune with a deck of Trump. We'll start with how a couple of player characters might start the process:

Ted: I know, let's try casting fortunes with our Trump.
Peggy: Sounds okay to me. I'm completely baffled anyway. Maybe we'll get some kind of idea about where we should go.
Ted: I'll shuffle my cards.
GM: Are you using that strange pink card you picked up from the citadel?
Ted: Hmmm...
Peggy: Bad idea, bad idea. I don't like that thing. Who ever heard of a Trump of somebody wearing clown make-up. I say we leave that one out.
Ted: Yeah, but if we leave it in, maybe we'll get an idea of how it fits in to the overall picture.

After the players decide on whether or not to use the strange card, they deal out a Six Card Spread. Here are six of the possible responses that a Game Master might give.

1. *"The Spirits are Restless Tonight."* The Game Master doesn't have a clue as to what to say, or doesn't think the players should be trying to take such an easy way out:

GM: You lay out the six cards, study them for awhile, and don't see anything significant. Maybe you should try again later.

2. *"You Have a Premonition of Doom, but a Cheerful One."* The Game Master wants to tell the players they should get moving, because there is a sense of urgency, but isn't willing to give any details:

GM: The whole reading seems to indicate that there is trouble afoot, that you shouldn't trust anyone, and you should watch your back. There's only a vague sense of anything concrete.
Peggy: In Amber that's sort of common knowledge, isn't it?
GM: Yes, but this reading seems to indicate that something is about to happen very soon. You see that the picture of Castle Amber itself is in the "likely outcome" place. The problem is that it's reversed...
Peggy: What does that mean, reversed?
GM: It means that the card is up-side-down. That means really big trouble, and it could mean that if you don't act quickly Amber itself could be damaged or destroyed.

3. *"Yes, a Face Appears..."* The Game Master is willing to give the player characters a broad hint of what is going on and who might be behind it all:

GM: Everything seems to point to one person, the key to your dilemma, the one you need to contact. In the number six position, representing the true solution to your problem, you see the card of Corwin.

4. "I See a Mystery..." The Game Master chooses to focus on a particular problem, giving the players a couple of ways of looking at things:

GM: Hmmm... This looks like a pretty clear reading. Some of it actually makes sense. For example, in the manipulation spot, which refers to the manipulator, you see Julian's card.

Ted: Julian! He's the one who sent us out on this wild Unicorn chase. That makes sense. What about the rest of the cards?

GM: They seem kind of vague, except for one, they are all minor cards.

Ted: What about the one?

GM: What seems weird is that the Dworkin card appears in the third position, the one that's supposed to describe who is following you.

Ted: It's more likely to be a person in that spot, isn't it?

GM: As far as you can tell.

Peggy: What do we know about Dworkin?

GM: He's probably your great-grandfather, he's the creator of the Trump, and most of your relatives think he's mad.

Peggy: No, I'm thinking of something else. Yes! Dworkin said that he really was the Pattern, and that the Pattern really is Dworkin.

Ted: Huh. So Dworkin's card might actually mean the Pattern itself.

Peggy: Yes, and then this all makes some sense...

5. "Yes, I Don't Know the Question, but the Answer is Definitely Yes!" The Game Master has plenty of Trump scrying ideas. It's just the complexity involved in the campaign, with all the strange things going on, and its huge cast of characters, make it hard to decide which fortune should be used. In this case the Game Master decides to try to get the players to ask for something specific:

GM: What question would you like to have answered by this reading?

Ted: I thought we had to ask questions like that before we spread out the cards?

GM: Usually, that's a good idea. It's just that this looks like it might be the answer to a yes or no question. So what would your question be?

Ted: I guess we want to know whether the Unicorn is leading us, or running from us.

GM: Can you put that into a question that can be answered yes or no?

Ted: How about, should we continue to follow the Unicorn?

Peggy: Better yet, how about, does the Unicorn want us to follow her?

GM: Ted?

Ted: Yeah, that sounds good. Does the Unicorn want us to follow her?

GM: You can't be sure that the reading is actually referring to the Unicorn, but the answer to your question is definitely yes. What are you doing now?

Peggy: I think we should do another reading, just to see if we can find out more about the Unicorn.

Ted: That might be a problem. (to GM) Is there a card that represents the Unicorn?

GM: There's no card for the Unicorn.

Ted: I've got an idea. I'll take my card, turn it Unicorn face up,

and then shuffle it into the deck. Then I'll try another reading and see if the Unicorn shows up.

Peggy: Good idea!

GM: Are you doing another reading?

6. "I Feel a Presence Enter the Room..." A Trump reading is also an opportunity for an elder Amberite to be revealed. In this example, Osric has been manipulating events, keeping close tabs on the player characters, and may even be setting up the cards in this particular reading:

GM: Wow! Your reading looks like a who's who of Amber. There are three images of elder Amberites, and Peggy's character, Iresa, makes an appearance.

Peggy: I do? Where is Iresa in the reading?

GM: Center stage. Iresa is in the first card slot, meaning the whole problem revolves around you.

Ted: Hey, maybe this will say something about Iresa's missing Mom!

GM: From the other cards, it does look like the reading is about Iresa's personal situation. You also see Julian as the manipulator.

Peggy: That makes sense, he's been pushing us around a lot lately.

GM: In the second position you see the card of Brand.

Ted: *Brand* is the likely outcome?

GM: Or something associated with Brand. Unless you do something to change things.

Ted: I don't like the looks of that. All we know about Brand is that he wanted to destroy the whole universe and create a new one that would worship him.

Peggy: Sounds like a real pervert. That's Julian and Brand, you said there were three elder Amberites in the reading. Who is the other one?

GM: Hmmm... I'm not sure if you even know who this guy is...

Peggy: If it's a card in the standard deck, I'll know who it is. Julian went through the whole deck with me, and he put names to every one of the family.

GM: Fair enough, I forgot about that. Yes, the card is Osric's.

Ted: Osric? He's dead!

GM: Right. Well, Osric's card is in the sixth place, meaning that he's supposed to be your True Solution. From everything else in the reading, Osric's card seems to point to a living person. Could it be referring to one of Osric's kids?

Peggy: I don't think he's dead. I think his being alive makes perfect sense.

Ted: Yes. Too much sense.

Peggy: I'll put my hand on his card and try to make a contact.

GM: Smoothly, like oil spreading over water, Iresa feels a cool sensation. It's like what your fingers feel when you touch a Trump card, but it's covering your whole body. Ted, you get a creepy feeling. What are you doing?

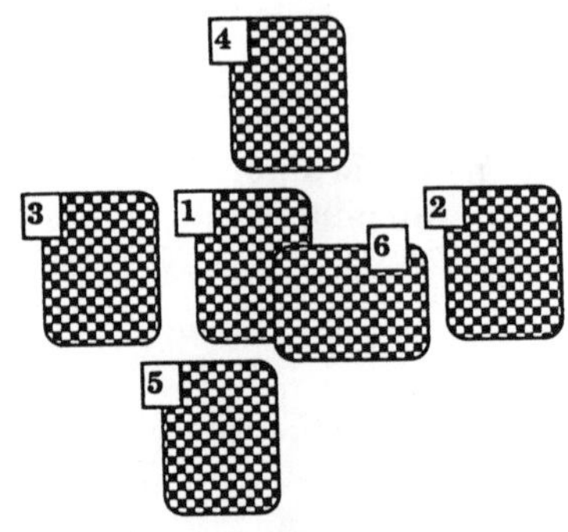

SHAPE SHIFTING

To keep things balanced the Game Master should always warn the player as they embark on a dangerous, self-destructive course of action.

The temptation to push the boundaries of safety seem irresistible for Shape Shifters. Too often they forget that they are changing the all-too-fragile container of their souls.

It's the temptation of power that keeps the player characters from crossing the line of safety. Plus the fact that the line is very broad. A character, succeeding in crossing the line once, and returning to a body and mind restored, will be doubly tempted next time, and may go even farther.

Shape Shifting Gone Wrong.

Eventually, if players push their Shape Shifting, they'll go over the line into some kind of personal horror story. The mistakes are almost never life-threatening, but they are agony for players. Still, getting burned by Shape Shifting is another great way to build up the player character's personality through adversity.

Losing Personality in Shape Shift. One of the most common problems facing a Shape Shifter is losing their own personality, their own sense of self. Usually this is a result of the character attempting to imitate another character, or trying to take on the form of some psychically powerful entity. When this happens a player may lose control of the character, making it necessary for the Game Master to take over. In the best of cases, the player character's persona returns to control whenever the body goes unconscious. However, it is possible that the "foreign" personality can take over for extended periods.

Losing the Power to Shape Shift. This is a particularly dangerous and debilitating problem that can take one of two forms.

Loss of Shape Shift Skill. The character's mind can lose the ability to control the Shape Shifting mechanism of the body. The character will then have to relearn Shape Shifting through trial and error, a process that will probably take a couple of weeks.

Loss of Shape Shift Power. A more serious condition, the body's cells lose their special Shape Shifting property. The body becomes immutable and unchanging. Eventually, as normal healing and regeneration take place, the cells will recover. How long this will be depends on the character's Endurance. Those with Human Rank will need a year or more of restoration. Chaos Rank characters will be back in Shape Shift form in about a month. Having Amber Rank means being able to go back to Shape Shifting within a week. Higher Ranks are faster yet.

Loss of Both Skill and Power. The body's cells will have to recover completely before the character can start exercising the skill of Shape Shifting again. Add the recovery time and the relearning time together and that's how long it will be before the character is back to Shape Shifting normally.

Involuntary Shape Shifting. In some cases the player may end up conditioning their character's body into reacting in certain ways automatically. Most of the time this is good. For example, if a character falls from a great height, consciously thinking about making a change isn't nearly as fast as letting the body do the change automatically.

Infection with Primal Chaos. Player characters who push their Shape Shifting too far, inflicting incredible stress on their body, run the risk of their shape shifting going berserk. Every living thing must have a bit of Primal Chaos as part of their makeup. Without chaos there is no birth, growth or aging, only stasis. Shape Shifters are those who can control the element of chaos that dwells within every living thing.

Primal Chaos Cancer. Characters who have pushed themselves too far, through exhaustion, fatigue and starvation, can find that they lose control of some of their body's cells. This is almost always discovered too late for any preventative measures. Characters will find themselves constantly drained as both body and Shape Shift abilities attempt to contain the cancer. Both "tumors" and wild cells roving through the body, will Shape Shift with incredible speed, adapting to consume other cells, and multiplying every few hours. If left unchecked, Primal Chaos Cancer will eventually attack the character's vital organs, appear on the skin, and will generally start to eat the character alive.

Total Primal Chaos. If allowed to completely run its course, a Primal Chaos Infection will not necessarily kill the character. Instead every remaining cell of the character's body will become a separate tiny Shape Shifter. Effectively the character becomes a mass of constantly changing cells, and looking like an amorphous blob. No thought will remain, because the neural connections will be severed. However, even in this state it's possible for the character to be restored to normal form.

Curing Primal Chaos. Specific cures are up to each individual Game Master. One possibility is the use of Advanced Pattern, though this is incredibly destructive as the Primal Chaos literally explodes through the body. There are also "Cancer" specialists out in the Courts of Chaos trained in dealing with this kind of problem, though their methods are rumored to be barbaric and torturous.

Advanced Shape Shift. The risks associated with Shape Shifting never go away. The possible ways a character can fall into trouble with Shape Shifting just multiply as the power increases.

The most important advancement in Shape Shifting is the ability to alter the character's own mind. This is done in one of two ways. The most dangerous method is for the character to *imitate* another character, literally becoming a complete copy, and losing the character's own identity.

The other method is to Shape Shift into the persona of another character. That means the Shape Shifter is just copying the other character's Psyche, personality and mental state, while the Shape Shifter's real character personality remains in control. This is dangerous, but more because of the strange perceptions rather than because of the chance of getting "lost."

SORCERY

...I resolved, though, to lay in a decent supply of spells the first chance I got, both offensive and defensive, on the order of the one I had primed against my guardian entity. The trouble is that it can take several days of solitude to work a really decent array of them out properly, enact them and rehearse their releases to the point where you can spring them at a moment's notice—and then they have a tendency to start decaying after a week or so. Sometimes they last longer and sometimes less long, depending both on the amount of energy you're willing to invest in them and on the magical climate of the particular shadow in which you're functioning. It's a lot of bother unless you're sure you're going to need them within a certain period of time. On the other hand, a good sorcerer should have one attack, one defense and one escape spell hanging around at all times. But I'm generally somewhat lazy, not to mention pretty easygoing, and I didn't see any need for that sort of setup until recently. And recently, I hadn't had much time to be about it.'

Merlin's explanation in
Blood of Amber

Magic Spell Casting.

Yes, spells are powerful. The problem is, spells are also a royal pain.

As a Game Master, the trick is to make the player aware of the time, effort and trouble involved in keeping the spells "hung..." Consider asking:

- "Do you want to go to the party, or are you going to refresh your spells?"
- "Are you going to take the Hell-Ride with the rest of the group or are you going to take care of your spells?"
- "You're right in the middle of weaving all the refreshed spells together again, when a trump call comes on you with the force of major headache #47. Are you going to try to hold the spells together, opening up your mind to the call, or drop the spells and put your mind to work on Trump?"
- "It's coming up on your final examinations. You've been maintaining your spells all term, even though you could have really used the time to study your texts. Now it seems pretty clear, either you drop the spells and cram for the next two weeks, or flunk out of school. Which will it be?"

As Game Master you can also zap the player through months, or even years, of play-time, especially if the character is out in Shadow or in the Courts of Chaos. Then, just as a test, see just how long the character is willing to let the fun and interesting things in life pass in order to maintain a bunch of spells. The only characters who will really keep up a full rack of spells will be those who are real drudges.

Serious Drawbacks of Spell Casting.

Defining Shadow. Each spell must be tailored for the magical environment where it will be cast. If the character has Pattern or Logrus the quality of Shadow will be obvious. Characters without these powers, and having no other way of evaluating the quality of Shadow, can be stuck with a rack of spells and no way to properly cast them. This is one of the reasons why Shadow Mages, awesomely powerful and capable of casting a virtually limitless array of breath-taking spells in their home Shadow, are helpless elsewhere. In another Shadow, where the ambient rules of magic are different, none of the Mage's spells will work.

Magic versus Power. In the face of any real power, Magic pales. Spells are simply ineffective against Pattern, Logrus or Trump, unless the spell itself contains some form of Power.

DESIGNING NEW POWERS

Take a look through the descriptions of the elder Amberites and you'll see some extensions of existing Powers, and even a few things that seem like new Powers altogether.

Zelazny's characters do things that are mysterious and sometimes downright baffling. Sometimes that's because the characters are expert users, and in their centuries of experimentation they've figured out how to manipulate the Powers in extraordinary ways. Other characters may be able to do wondrous things because they have multiple Powers and have found ways of using them synergistically, so that their effects are multiplied instead of merely added together. Finally, some characters display implausible Powers simply because there are levels of attainment beyond those covered in this book.

New Powers are a grand opportunity for every Game Master. Not only is it possible to come up with your own Amber, but it's also within your power to come up with "exalted" versions of any Power beyond anything described here.

Inventing New Powers.

One of the great strengths of Amber is that it is limitless. If Pattern was created, as Dworkin put it, out of an island in the darkness beyond Chaos, then what other realities await?

Take Zelazny's recent books, the Merlin saga. Here we come across characters who have absorbed Powers offered by "Broken Patterns" and the "Keep of the Four Worlds," and even Merlin's Trump Construct, *Ghostwheel*. Phage Press plans to develop these Powers fully in the upcoming book *Shadow Knight*, but it's also possible for each Game Master to come up with their own versions of these unknown Powers.

Before working up the specifics on any new Power, or any advancement level of an existing Power, think about the answers to the following questions.

- **How does the Power feel?** As when summoning the Logrus, does the new Power fill the character with madness? Or just a feeling of dread, or a reddening of the eyes? Does the Power display a distinctive color, light, or scent? When invoking the Power is there a sense of great mystery? Or maybe contentment?

- **What makes the Power work?** Is there some Primal Source, such as Chaos or the Pattern? Does it tap into the life energy of the wielder? Or is it drawn from a particular environment?

- **How is the Power gained?** Is practice required? A personal sacrifice? Some act or journey? Or does a character have to be born with some special quality?

Calculating New Power Basic Costs.

It's not precise, but one the best ways of computing the cost is by comparing the new power with the standard ones.

Realm of Power. This defines the area Created by the Power. Is the Power restricted to a particular area, or unlimited?

None. The Power only exists where a character invokes it. Shape Shifting is a "Locus" power because it affects only the character who possesses it. Cost: 0 Points.

Locus. A single location contains the Power and is affected by it. This can be anything ranging from a tiny island in the void to an entire world, but with no Shadows cast or created by the Power. The Power is available at full strength only at the Locus. Cost: 1 Point.

Focus. Like Pattern or Logrus, the Power extended out infinitely in all directions, stopped only by interactions with other Powers. However the "focus" of each Power is quite clear, just as the Primal Pattern is the generator of Pattern, and as Primal Chaos is the central point for Logrus. The strength of the Power is at its peak at the center and gradually falls off as one travels outward. Cost: 2 Points.

Infinite. The Power seems boundless and effective everywhere. Trump is an example of an Infinite Realm of Power since it seems to work everywhere equally. Cost: 4 Points.

Power vs. Power. The other measure of a Power is how well it stacks up against other Powers. Each Power should be evaluated against each of the five other Powers listed below. If your campaign includes other Powers, then you'll want to add a listing of how the new Power compares with those Powers as well.

Helpless Against Existing Power. The new Power is incapable of challenging the old Power, both defensively and offensively. For example, a Power "Helpless Against Logrus" would be unable to affect any character using the Logrus, and would be wide open to a Logrus tendril attack.
Against Pattern. Cost: Zero.
Against Logrus. Cost: Zero.
Against Trump. Cost: Zero.
Against Shape-Shifting. Cost: Zero.
Against Magic. Cost: Zero.

Weak Against Existing Power. Against another Power the new Power can almost always be overcome. If the Power has a focal point, then that is the only place where it even has a chance of competing.
Against Pattern. Cost: 4 Points.
Against Logrus. Cost: 3 Points.
Against Trump. Cost: 2 Points.
Against Shape-Shifting. Cost: 1 Point.
Against Magic. Cost: Zero.

Equally Matched Against Existing Power. The two Powers, old and new, are equivalent in strength. A contest in one or the other's focal point would probably result in a victory for the "home team."
Against Pattern. Cost: 8 Points.
Against Logrus. Cost: 6 Points.
Against Trump. Cost: 4 Points.
Against Shape-Shifting. Cost: 2 Points.
Against Magic. Cost: 1 Point.

Strong Against Existing Power. Although not overwhelming, the new Power can eventually defeat the old Power. This is most difficult at the old Power's focal point, and easiest at the new Power's focal point.
Against Pattern. Cost: 16 Points.
Against Logrus. Cost: 12 Points.
Against Trump. Cost: 8 Points.
Against Shape-Shifting. Cost: 4 Points.
Against Magic. Cost: 2 Points.

Dominant Against Existing Power. The new Power can overcome the existing one anywhere they meet, either offensively or defensively.
Against Pattern. Cost: 32 Points.
Against Logrus. Cost: 24 Points.
Against Trump. Cost: 16 Points.
Against Shape-Shifting. Cost: 8 Points.
Against Magic. Cost: 4 Points.

Multipliers to New Power Cost.

After buying the Realm of influence and the five comparisons with other Powers, the total is multiplied by whatever abilities that the new Power confers.

Communication. As with Trump, the power confers the ability to communicate. This assumes a Psychic communication and also provides a conduit for mind-to-mind combat.

Communication with Others Sharing the Power. Works only between common users of the new Power. Cost: *2 Points.
Communication with Empowered Targets. As with Trump, where the subject is allowed only a Psychic connection, and limited to a particular class of targets, across a particular range. Cost: *3 Points.
Omniscient Communication. Anyone within the realm of the Power can be contacted by an initiate of the new Power. Cost: *4 Points.

Transportation. A Power's capacity to transport characters from one point to another. Shadow walking is considered to be part of Shadow Manipulation, not transportation.

Through Communication Channel. Once a Psychic connection is made with the Power, travel can be made in either direction. Cost: *2 Points.
Transportation Through Power Points. Character can teleport from any Power locus to any other. For example, an initiate can travel to the Power's central locus, or to any other initiate or item that contains the Power. Cost: *3 Points.
Focal Point Teleporter. As with Pattern, an initiate can travel to the central Focus of the new Power and use it to teleport anywhere within the range of the Power. Cost: *4 Points.

Defensive Use. How well the Power works as a defense depends on the "Power vs. Power" rating. Defensive use of a Power almost never works against purely physical attacks from swords, bullets, etc...

Power Realm Defense. The Power only defends characters who are actually in the Power's central focal point. Cost: *2 Points.
Invoked Defense. The Initiate must somehow concentrate on the Power, or manipulate an artifact based on the Power. Cost: *3 Points.
Automatic Defense. Initiates are always "charged" with the new Power, and are constantly defended by the Power. Cost: *4 Points.

Offensive Use. Powers can also be used as weapons. As with Power Defensive use, the effectiveness of a Power's Offense depends on its "Power vs. Power" rating. Offensive use is always a pure energy phenomena.

Offense at Power's Focus. Attacks only those who directly come in contact with the Power's source or focus. For example, Pattern is usually only destructive to those who touch it. Cost: *2 Points.

Invoked Offense. The Initiate concentrates on the Power, or manipulates an artifact based on the Power, and projects the energy of the Power. Cost: *3 Points.

Attack Within the Realm. Initiates can wield the Power as a weapon and direct its energy at any target within the area influenced by the Power. Cost: *4 Points.

Power Manifestation. Usually Powers are subtle and do not manifest themselves except in the performance of their abilities.

Power is Inward Only. The initiate character can manipulate the Power within their own body. Cost: *1 Point.

Power Extends to Energy. Some kind of projections allow for the Power to be used as raw energy. Cost: *2 Points.

Power Extends to Physical. Just as the tendrils of the Logrus can be used to touch, grab and bend physical objects, so the new Power can be used to manipulate objects. Cost: *3 Points.

Shadow Manipulation. Can the Power be used to create Shadow? Destroy it? Or just to manipulate it into different forms? Except for Trump, all the other Powers are able to play with Shadow stuff. Creating, and destroying, Shadow is certainly within the realm of the Powers of Pattern and Logrus.

Item Manipulation. Using the new Power a character can change ordinary Shadow things. Cost: *2 Points.

Self Manipulation. Initiates of the power can manipulate their own body. This is the primary multiplier for the Shape Shifting Power. Cost: *3 Points.

Shadow World Manipulation. The new Power can be used to manipulate whole Shadows, just as Pattern Initiates can walk through infinite Shadow. Cost: *4 Points.

Invocation. How does a character invoke the Power? With standard Pattern Imprint, or Power Words, or Shape Shifting, the Power is built into the character, so there is no invocation required. However, Logrus and Advanced Pattern require that the symbol be invoked in order to utilize the Power. The Power of the Jewel of Judgement can only be used while in possession of the object itself.

Requiring Invocation makes a Power cheaper. Cheaper yet if some token or artifact is required. It should be more expensive for a Power that can be used without preparation.

Within Power Only. The new Power can only be used when the initiate is in direct contact with the focal point. For example, Pattern's teleport function only works when a character actually walks a Pattern. Cost: *1 Point.

Object Oriented. Like Trump, the Power is contained in some kind of artifact (or creature), and can be used only with the item. Cost: *2 Points.

Invoked. Characters with the new Power must "bring it to mind," a process of a minute or two, before it can be used.

Logrus and Advanced Pattern are examples of Powers that must be invoked. Cost: *3 Points.

Automatic. The initiate can use the Power without any special preparation or artifact. Works just like Pattern, where Initiates can Shift Shadow simply by doing it. Cost: *4 Points.

Improving Existing Powers.

The basic and advanced versions of the Powers presented earlier in this book are only a beginning. Every Power can be improved or expanded. Sometimes simply by increasing the amount of control available to the character, and sometimes by coming up with a totally new approach to using the Power. For more examples check out the "Exalted" Powers of the elder Amberites. Here's one sample.

"Reversed" Pattern Imprint. If we assume that the Primal Pattern is "Clockwise," that is, that one walks it by entering and walking from right to left around the center. Could there also be an identical Pattern where one walks "Counter-Clockwise," entering and walking from left to right around the center?

The reverse Pattern would generate an entirely new, "flip-side" of the universe. A character possessing this expansion of Pattern would be capable of going to the mirror universe and manipulating the Shadow in that place. Of course, the contents of the Counter-Clockwise Amber are entirely up to the Game Master.

Cost: 15 Points (added to Pattern, for a total of 65 Points, or to Advanced Pattern for a total of 90 Points)

Extending Artifact and Creature Potential.

Just as elder Amberites and other non-player character have improved Powers available, so they can also put together creatures and artifacts with qualities and powers not available to player characters. Here are just a few examples.

Stupendous Vitality. Confers a strength beyond that of any natural creature. Humans with Stupendous Vitality have their Strength raised to above Amber Rank, to somewhere in the Attribute Auction. Costs 8 Points.

"Exalted" Vitality. The level of a Gérard, capable of defeating virtually anyone in a contest of strength. Costs 16 Points.

Combat Supremacy. Items specializing in a particular combat will surpass the reaction time of an Amberite, reaching into the lower ranks of the Attribute Auction. Costs 8 Points.

Impervious to Damage. No natural force or material can damage the sword, nor can it be affected by Magic or any lesser Power. Costs 8 Points.

Destructive Damage. Rips through opponents, causing massive damage. Penetrates any type of armor, including armor "Impervious to Damage." Costs 8 Points.

CAMPAIGN BUILDING

AMBER UNDER CONSTRUCTION.

Here's where each Game Master gets to design their own unique version of the *Amber* universe.

Why?

Why not just use Zelazny's version of *Amber*? Why go to all the trouble of building a whole complex universe, with different versions of all those complicated characters?

There are three reasons.

Surprise! *Amber* is role-playing based on surprise. Players enjoy it best when they must uncover each of *Amber's* many secrets step-by-step, so each Game Master's *Amber* should provide the opportunity to discover the unknown. Putting the answers down in black and white, even if it were possible, would take a lot of fun out of the role-playing.

Amber stays fresh and new, delighting even jaded long-time players, when each Game Master builds their own version.

Zelazny Doesn't Say. One of the main problems with using Zelazny's version is that we don't know enough about it. He's not finished building his *Amber*. Other books are boiling out of him, books based on his own secrets, his developing insights, and his changing view of this fascinating universe. A Game Master can't run a campaign without knowing the secrets of the universe. To properly run a campaign in Zelazny's version of *Amber*, you'd need to know his secrets. And he's not telling.

The Players Shape the Universe. Each Game Master starts with a different cast of player characters. We can all start with the same roots, Corwin's saga in *The Chronicles of Amber*. Those first five books are primarily about the elder Amberites, and only two of the next generation appear, Merlin and Martin, players of relatively minor supporting roles.

Zelazny's following books are centered around his version of the next generation. Your *Amber* is going to have a different cast, the characters generated by your players.

Player characters in *Amber* are not minor figures. They have the power of universal creation and destruction. Each player character with the blood of Amber has the capacity to destroy the Primal Pattern, wiping out the roots of creation, and plunging the universe into chaos. The liquid that can perform this feat runs through the veins of any character with Pattern Imprint.

Even their minor actions can have major consequences.

Build your own universe, Game Master. Take Oberon's children, take Pattern, Logrus and Trump, take the grandeur of Castle Amber and Forest Arden, the wierd perversity of the Courts of Chaos, and the infinite range of Shadow, and make the universe anew. Start with the events of the Patternfall War and then let your mind range over the possibilities.

Will the Courts of Chaos be a major threat, massing armies to assault their ancient enemy? Or will the Courts of Chaos be thrown into disarray, involved in bitter feuding and beset by coup and revolution? Or, is it possible that the Courts of Chaos were never much concerned with Amber after all, and that only a small faction had gotten involved in the border skirmish known as the Patternfall War?

The elder Amberites have sworn fealty to their new King Amber. Yet could their constant fighting have really come to an end? Is one cabal scheming to take the throne? Or are separate groups of plotters each arming, plotting, and conniving to replace the all-too-young new king? Have recent events unbalanced any of the siblings, taking them to new heights of madness, or turning them to a path of evil? What long-lost brothers and sisters, seeing a new king in Amber, will return from their Oberon-imposed exile for a new try at the throne?

GAME MASTER'S GUIDE TO CAMPAIGN BUILDING

Here's where the Game Master learns all about putting together *Amber* Campaigns. Before getting to the nuts and bolts, there are three concepts that we have to review.

First, there's the whole idea of *Story Composition*. Any role-playing is the performance of a story. Any competent Game Master should be able to tell the difference between a good story and a bad one.

Then there is the *Idea Gathering* stage of things. If you're interested in role-playing, you've probably got plenty of ideas. This section, with hundreds of *Amber* story ideas, should really get sparks flying in your imagination.

Take those ideas, and add the *player characters*. Here's where you want to see all their strengths and weaknesses, and all the ways that they might interact with Amber and the elder Amberites in interesting ways.

Adventure and Campaign Construction comes dead last, after you've looked at all the possibilities. If everything comes together right, any Game Master can hold an intriguing, involving, unique campaign that'll knock the socks off the player characters.

STORY COMPOSITION

'Good evening, Lord Corwin,' said the lean, cadaverous figure who rested against a storage rack, smoking his pipe, grinning around it.

'Good evening, Roger. How are things in the nether world?'

'A rat, a bat, a spider. Nothing much else astir. Peaceful.'

'You enjoy this duty?'

He nodded.

'I am writing a philosophical romance shot through with elements of horror and morbidity. I work on those parts down here.'

'Fitting, fitting,' I said. 'I'll be needing a lantern.'

He took one from the rack, brought it to flame from his candle.

'Will it have a happy ending?' I inquired.

He shrugged.

'I'll be happy.'

'I mean, does good triumph and hero bed heroine? Or do you kill everybody off?'

'That's hardly fair,' he said.

'Never mind. Maybe I'll read it one day.'

'Maybe,' he said.

The Hand of Oberon

Before we get to the ins and outs of building adventures and campaigns, let's first take a look at what we're trying to accomplish.

As Game Masters we're all trying to do the same thing, provide our players with some fun. We want to give them a few hours, or a few hundred hours, of enjoyable entertainment.

Some of that fun will come automatically. Get any bunch of role-players together for a few hours and, chances are, they'll have fun. Give them some really interesting characters and that'll provide even more laughs and excitement.

Get together with your role-players, throw their characters in with a batch of hapless monsters, a couple of villains and a puzzle or two, and, bingo, the result is role-playing.

It's not a role-playing campaign. It's just role-playing.

To turn role-playing into something more intense, the Game Master has to do some work.

The hard part is creating the *story*.

According to the dictionary, a story is a narrative. In other words, the telling of some series of events. For the purposes of building a campaign let's talk about the pieces that make up a story.

Beginning, Middle and End.

Break apart the idea of a story and one thing you come up with is the concept of a *beginning*, a *middle* and an *end*.

Pretty obvious, huh?

Of course a story has to start somewhere, things have to happen, and then, sooner or later, it's got to stop. Isn't that obvious?

No.

Each part, the beginning, middle and end, has an important relationship to the other parts. The beginning sets things in motion, but contained in the beginning has to be the seed of the ending of the story. The ending must be what stops the story, and the thing that explains the beginning. And, while a lot of other things can happen in the middle of a story, the only important parts of the middle are the things that cause the problem in the beginning to get to the solution at the end.

Stories that miss having a clear beginning, middle and end are not satisfying.

What if you heard a friend tell you, "What a morning I had! It all started when I pulled out of the driveway this morning, and I had to swerve to avoid an elephant!"

Wow! you'd think, an elephant! This is sure to be a great story.

So your friend continues by saying, "then, when I got down to the shopping mall..." and proceeds to describe shopping at a bunch of stores, looking for the right kind of light bulb. Fifteen minutes later your friend stops talking, and you still haven't heard anything interesting.

Wait, you'd say, what about the elephant?

See, you knew that the elephant was the beginning of the story. When your friend never got back to telling you about the elephant, you knew you'd been cheated.

The moral is, if you're going to tell a story that starts out with an elephant, it sure better end up explaining the elephant.

As role-playing Game Masters, especially when running a long campaign, with a lot of players, it's easy to get the stories mixed up. One character's story may start with a family fight, another with the destruction of a favorite Shadow, and everybody might be involved in another story concerning a plot against Amber by a member of the Courts of Chaos.

Your job is not to leave out any elephants.

When you introduce an element to the campaign, especially an interesting element, then make sure you follow through.

Sure, this is a role-playing campaign. The players may never go looking for that elephant again.

The point is, *don't introduce the elephant unless you know why it's there, and what you're going to do with it.*

That's the whole point to having a beginning, a middle,

and an end. The beginning sets up what makes the story interesting, the middle offers an obstacle to figuring out how things work, and the end is the explanation.

An elephant story begins with meeting an elephant. The middle is the character confronting the problem (how do you track down the owner of a lost elephant?). The end is when you have a satisfactory answer to why the elephant was there in the first place, and you've also solved the problem of what to do with the elephant.

The Hook.

The elephant, in the example above, is the hook. It's the real *beginning* of a story. Any good story starts with some kind of hook. The character takes the hook, follows the attached line, and, if the story is put together properly, ends up at the other end of the line, at the end of the story.

The trick is getting the character, or, in the case of role-playing, getting the player to take the hook. One way is to bait the hook, enticing the character to take a bite. Sometimes it comes out of nowhere, and the story starts with a character surprised to find themselves hooked.

In *Amber*, the hook is usually some kind of intriguing problem. A missing relative, a puzzling Shadow, or a mysterious artifact. Or, to make things a little more obvious, an enormous beast that suddenly appears and eats the Pattern, making every single Shadow go "poof." Any of these things hook the player into following the line of the story. Don't disappoint, make sure the other end of the line, no matter how many tangles or snarls or false lines there may be in between, eventually leads to the reason for the hook.

Conflict.

The middle of any story is the part where the characters have to bump heads with their problems.

We all hear boring stories. We hear them all the time. They're the telling of the little events that happen in our lives constantly. What happened today in school, how I found the store that carried what I was looking for, and where I got sent on an errand by my boss. These are all stories.

Problem is, most of them lack conflict. We're telling the story to relate information, share a hardship, or just to be sociable.

Stories get a lot more interesting when there's conflict. There's a big difference between "I walked to school instead of taking the bus," and "I walked to school so that bully wouldn't beat me up."

Both cases are the beginning of a story. The second one is a beginning with conflict built in. Is the teller going to dodge the bully all day long? Or will there be a fight? Or something unexpected?

Stories with conflict are unpredictable. They have drama, because the audience can't guess the outcome. Stories with conflict are perfect for role-playing settings because the player characters can make their own endings, but not without uncertainty.

In *Amber* conflict is sometimes difficult to pull off. The characters are just too powerful for many potential stories. Characters with Pattern just aren't threatened by most terrible monsters, forces of nature, frightening villains, or omnipotent gods, because they are usually confined to a single Shadow.

Real conflict in Amber is other Amberites, or solving the dark enigma of how things really work.

Character Development.

Somewhere between the beginning and the end of a story, it's possible for something to happen. Something called *character development*.

Not every story has character development. A hero, like the Lone Ranger, sometimes defeats the villain without changing at all. At the end of the story the Lone Ranger is pretty much the same, and so is the villain. Unfortunately, after a while, these stories get kind of boring.

Character development doesn't mean just getting bigger or more powerful. Character development is all about the changes that take place in a character's personality. It's when a character grows and gains hope, or finds disappointment and despair, loses faith or discovers inspiration. Something has to happen to the character's direction, goals, outlook and attitude.

It's all about growing up and learning new things. What we all have in common is the changes we go through as we discover the world around us. We start out as little children with childlike innocence and a willingness to return every affection tenfold. Reality creeps into our lives, we find that people can lie, and disappoint, and we become more cynical. As teenagers, after years of adolescent frustration and degradation, we eventually find love, or true friendship, and it changes our scornful skepticism into new hope. Betrayed by a friend, we close up. Observe the gentle side of a bitter enemy, and we find hope again.

In *Amber* the idea is to subject the player characters to stress, to give them an opportunity to change. If they can find out something about themselves, or about their relationships with other characters, that's pretty good. If the player character changes somehow, in mood or attitude, that's better. However, if a character becomes real, so real that the player feels what the character feels, that's the best.

Closure.

The difference between a good story and a great one can be the ending. Does the role-player, or reader, or viewer come away satisfied? Their satisfaction is a mark of good closure.

Loose Ends. Actually, since we're talking about role-playing campaigns, there should be a lot of loose ends. After all, that's how to get the players coming back for part two. Still, if a story offers a core problem, then that problem has to be addressed for the story to really end. The players will be happiest if the problem is solved, but that's not the only possible ending.

An ending is just as valid if the players are defeated, humiliated, and/or driven from their homes. In this case the end of the story involves something like the players in full retreat, with one of their number turning back to the enemy, shaking a fist, and saying, "This is not the end! We will never forget!"

Another strong possible ending is a stalemate. The villain is still loose, the problem is still unresolved. In order to have closure, the players finish the story by discovering what their problem was and where the line from the hook began.

In other words, the ending has to explain the beginning.

Planning for Closure. When putting the ideas for a

campaign together the Game Master should be always be thinking of how the thing is going to end. Not the exact end, but the range of possible endings.

Corwin asked, "does good triumph and hero bed heroine? Or do you kill everybody off?" A Game Master of stories, unlike a writer of stories, can't force a particular ending. However, a good Game Master can be prepared for a range of possible endings.

In *Amber* closure is usually character oriented. A story ends when the players discover the villain, or figure out the problem. Closure, in *Amber*, is best when it also provides insight into how the universe works.

Moral Resolution.

A story can have a good structure, with interlinked beginning, middle and end, an involving conflict, characters that develop and change, and closure, and still be missing something.

Moral Resolution, the sense of a story revealing something truthful about the universe, is the most difficult thing to build into a story. Dickens' "A Christmas Carol," and Aesop's Fables will be with us forever, while lesser stories are forgotten, because they manage to get across some universal insight, some peek into the underpinnings of the world.

In *Amber* moral resolutions usually have to do with people. Characters come to recognize their failings, or the folly of misjudging, or hurting the innocent. Or, to put it another way, goodness triumphs, evil yields, and the pure of heart prevail.

CAMPAIGN IDEAS FROM ZELAZNY'S *CHRONICLES OF AMBER*

Roger Zelazny's concepts in the *Chronicles of Amber* are truly inspired. There is no limit to where a Game Master can take a campaign from the countless ideas that are scattered about his universe. What follows are three sets of guidelines to the ideas of *Amber*.

First, a discussion of the books themselves, and how to get the most out of them.

Second, a compendium, in alphabetical order, of idea-rich concepts taken from the books.

Third, another list, but this time of ideas from Zelazny's unanswered questions.

Reading Between the Lines: Digging Out Ideas from Zelazny's *Chronicles of Amber.*

The very best way of gleaning ideas for an *Amber* campaign is by reading the books. There's no better way to understand how an *Amber* campaign can be built then by reading through the whole series. Here are some of the things to watch for while reading, or remembering, Zelazny's

Chronicles of Amber.

First Person. Everything in *The Chronicles of Amber* is told in first person, in the voice of Corwin. That means that we see the Amber universe through his eyes only, not through the "truth" told to us by an impersonal narrator. Corwin's perceptions are flawed, and a serious limit on the accuracy of the *Chronicles*. As you read, learn to look behind what Corwin says, for the things he misses, or misinterprets.

For example, before Corwin recovers his memory, we are told things about Amber that are untrue. Early on, Corwin thinks that Random is his full brother and Eric is a half-brother. It turns out to be the other way around.

There may also come the time when a player may argue with the Game Master, citing some passage from the *Chronicles*. "That's not the way it is," they'll complain. Counter that with the fact that everything in the books is only Corwin's opinion. Corwin makes many statements of fact that he latter recants, or contradicts. Fact is, Corwin is sometimes wrong.

Corwin Lies. Be suspicious of what Corwin says. Remember, he is speaking to his son, but also to a son of the Courts of Chaos. That means he may be telling tales to an agent of Amber's great enemy. No doubt Corwin accurately recounts those things for which there are witnesses, but he occasionally covers up his own actions.

It's possible to read the entire series from another point of view altogether. History, as they say, is written by the victors, and Corwin is the victor in the Patternfall War.

Corwin himself said it best:

So I simply said one of the great rite truths: 'There is generally more than one side to a story.'

The Courts of Chaos

The Great Questions. Each Game Master's version of Amber is unique. One of the ways to "customize" Amber is to find your own answers for the unanswered questions. The questions are scattered throughout the books, voiced by Corwin, or sometimes left unsaid.

It's a good idea to jot down the questions as you find them. Yes, sometimes a question will seem to be answered by the end of the series. Think about it again a week later, or a month later, and you may find yourself questioning the answer. In *Amber* nothing is as simple as it seems.

Every Action has Consequences. One of the great lessons to be learned from the *Chronicles* is how every action of Corwin's can have bitter consequences. Whenever Corwin makes a mistake, or acts out of weakness, he is usually made to pay a price. Converting that sense of actions and consequences to role-playing can make a campaign great.

Amber is Deep. To put it another way, don't trust your first reading. I've been reading the *Chronicles* since the first book was published back in 1970. I reread them at least three times even before I got involved in role-playing. Since I started working on *Amber* in 1985, I've read them again, cover to cover, over thirty times. *And I still keep finding things I didn't know, and I'm still surprised!*

IDEAS BASED ON ZELAZNY'S *CHRONICLES OF AMBER*

First start by taking a little trip through the Glossary. Every entry provides an element for campaign building. Then, take any of Zelazny's Amber books in hand, open to any page, and start reading. Ideas for entire campaigns are scattered everywhere. The following ideas are just a few of the possibilities.

The Abyss.

Somewhere near the Courts of Chaos, perhaps serving as a border, perhaps as a hyperactive moat, there exists the biggest hole of them all. It's another of the mysteries of the Amber universe.

Does it border on Shadow? If so, how many Shadows? All of them? Some? Most? Few?

If the Abyss borders on more than just one place, can it be used as some kind of shortcut? Or as a highway?

What is the Abyss? The birthplace of everything? The source of Chaos? Or just a cosmic garbage dump?

Is the Abyss a real place, like Amber or Chaos? A place of greater reality? Less? Of greater chaos? Or perhaps of deathless stability?

There are only two things we know for sure. First, that after Brand and Dierdre plunged into the Abyss, everyone took them as lost forever. Period.

The second thing we know about the Abyss is that the Unicorn climbed out of it shortly after Brand and Dierdre took their dive. Around the Unicorn's neck was the Jewel of Judgement, which had just fallen in with Brand.

Which leads to the inevitable questions. If the Unicorn can get something out of the Abyss once, can she do it again? If the Unicorn did it, can others enter the Abyss and leave again?

Amber Lands.

Zelazny has only described a small piece of the world of Amber. All we know of the true world called Amber is from one city, a castle, a range of mountains, a large forest, and the surrounding sea.

That's like trying to visualize the whole of Shadow Earth when all you've seen is Scotland's Edinburgh, or Seattle, Washington, USA.

Since Castle Amber is located on the real world of the Pattern, probably at least as large as our planet Earth, what happens in the rest of the world could have major importance.

Did Dworkin and Oberon once range across the world of Amber, exploring and conquering their new domain? Or did Oberon's children once use the world as their playground before they walked Pattern and took off to Shadow? What might they have left behind? Children? Treasures? Artifacts mystic and arcane? It leaves a whole world to explore.

Castle Amber.

Amber's Castle is vast and old. You'll find no map of it here, for it's secrets are hidden in every Game Master's campaign. I imagine it to be a sprawling mass of anachronistic wings and towers, each part contradicting the architecture of every other part, the work of many hands. Others see it as a single shining tower, matchless in beauty. It could be like one of those Bavarian fairy castles, as ornate as lace and structured as a wedding cake. Or picture it as a fortress supreme, bristling with crenelated walls, ramparts, gateways, and battlements.

Inside Castle Amber. As with any building that is both a building of government and a family home, there are different areas, each with a different feel.

Those who visit Castle Amber in search of its government will see the administrative suites of the officials, the various Halls of State, the waiting chambers, and possibly the Throne Room itself.

Visiting dignitaries will see a different Castle Amber, with its Dining Halls, ranging from the formal to the informal, the sitting rooms, the lounges, the game rooms, and the penthouses reserved for guests. For relaxation these guests will also see the gardens, the ramparts, the courtyards, and the stables.

Amber's royal retainers, from kitchen workers, to attendants in livery, see yet another Castle Amber. They see the back rooms, the working areas, the delivery doors and the places where all the goods and treasures of the Castle are stored and maintained. For them, Castle Amber is a great town complete with kitchens, laundries, workrooms, and washrooms, with special chambers for artisans of every craft. Just as Oberon's family has living chambers, so too do the retainers, passing their apartments and dormitories down through the generations.

Castle Amber's defenders, the guardsmen, live among the royal retainers but see the Castle from a different perspective. For them there are the defenses of Amber, its walls and towers, gates, bars, portcullis, drawbridges, and prison cells. Practice yards, the blacksmithy, and every armory makes up their world in Castle Amber. They patrol the Castle's every entry and exit, from the dungeons to the high walls, and for them the Castle is a hard and rugged place.

Finally, there is the view of the royal family of Amber. Oberon has reserved for his family private areas where the public may not go. The King's area alone could fill a large house, as it comes complete with bedrooms, drawing rooms, waiting halls and private dining facilities. Each family member has a lodging of two or more rooms, laid out and decorated according to their own taste. Common family rooms include the library, a variety of sitting rooms, several eating areas suitable for groups of various sizes, for the various meals of the day, and chambers appointed as lounges, smoking rooms and intimate parlors. Somewhere, if one where to poke around, there are also the old nurseries and playrooms, no doubt with old toys and dolls carefully packed away and preserved for the next generation of royalty.

Even this listing of the contents is not complete, for Castle Amber is filled with secrets. Hidden hallways, rooms intended for clandestine meetings, false walls dotted with peepholes, crawl spaces, and concealed stairways are everywhere. Some rooms have been sealed for millennium, others are simply forgotten. In fact, it is rumored that the Castle is itself a living thing, capable of shifting its stone, of moving hallways from place to place, and of having windows that look out upon vistas entirely foreign to Amber.

The Dungeons and Caverns. No known Amberite has ever fully explored all the cavities under Castle Amber. In the dungeon there are many different areas, including a range of accommodations for those who have fallen out of favor with the Court. No doubt there are torture chambers and places of execution as well.

Numerous tunnels cut deep into Mount Kolvir. Take one of these tunnels to the seventh side passage, and you'll end up in the Pattern Room (a room that is locked, with the key hanging in the hallway *outside* the Pattern itself). Could a tunnel lead, like some ancient bolthole, all the way to a cavern by the sea? Or is it possible that one can walk a tunnel through to Dworkin's cave in the realm of the Primal Plane? Could it be that one tunnel extends all the way to an ancient homestead, perhaps the Barimen estate in the Courts of Chaos?

If you visit the guardsroom in the dungeons of Castle Amber, you'll not want for interesting tales. They will tell you of guards lost for days or even years. Of guards wandering for an hour and coming back aged ten years, and of guards lost for years and returning with no sign of the passage of time. They'll tell you of battles with strange creatures, and conversations with ghosts, and fleeing from dark forces that dwell in the nether regions of the dungeon. They can also talk about locked gates and doors that have never been opened in their collective memory, yet seem to hold back growling creatures, and even stranger things. Still, guards are guards everywhere; their duty is boring and their tales always grow in the telling.

City Amber.

'We were both wonderfully drunk that night, and it seemed but a brief while that you talked - weeping some of the time - telling me of the mighty mountain Kolvir and the green and golden spires of the city, of the promenades, the decks, the terrace, the flowers, the fountains...'

Ganelon, recounting Corwin's description of Amber
The Guns of Avalon

I heard her humming a tune: 'The Ballad of the Water Crossers,' the song of Amber's great merchant navy. Amber is not noted for manufacture, and agriculture has never been our forte. But our ships sail the shadows, plying between anywhere and anywhere, dealing in anything. Just about every male Amberite, noble or otherwise, spends some time in the fleet. Those of the blood laid down the trade routes long ago that other vessels might follow, the seas of a double dozen worlds in every captain's head. I had assisted in this in times gone by, and though my involvement had never been so deep as Gerard's or Caine's, I had been mightily moved by the forces of the deep and the spirit of the men who crossed it.

The Hand of Oberon

Amber, like many a sea port city, is a bustling, lively place.

The economic layout seems to follow the land itself. The higher the location on the slope to Mount Kolvir, the more expensive the land, and the more prosperous the inhabitants. Conversely, the lower the area, and the closer to the sea, the sleazier it gets.

It's also very much a navy town, centering around the ships, both military and commercial. Even socially this tends to be the case, with high status coming from those having high ranking positions in either fleet.

Corwin's New Pattern.

...Cassis, and the smell of the chestnut blossoms. All along the Champs-Elysées the chestnuts were foaming white...

I remembered the play of the fountains in the Place de la Concorde...And the Rue de la Seine and along the quays, the smell of the old books, the smell of the river... The smell of the chestnut blossoms...

Why should I suddenly remember 1905 and Paris on the shadow Earth, save that I was very happy that year and I might, reflexively, have sought an antidote for the present? Yes...

...And, as the Pattern in Rebma had helped to restore my faded memories, so this one I was now striving to create stirred and elicited the smell of the chestnut trees, of the wagonloads of vegetables moving through the dawn toward the Halles. ...I was not in love with anyone in particular at the time, though there were many girls - Yvettes and Mimis and Simones, their faces merge - and it was spring in Paris, with Gipsy bands and cocktails at Louis'. ...I remembered, and my heart leaped with a kind of Proustian joy which Time tolled about me like a bell. ...And perhaps this was the reason for the recollection, for this joy seemed transmitted to my movements, informed my perceptions, empowered my will...

...Poppies, poppies and cornflowers and tall poplars along country roads, the taste of Normandy cider. ...And in town again, the smell of the chestnut blossoms...The Seine full of stars...The smell of the old brick houses in the Place des Vosges after a morning's rain...The Bar under the Olympia Music Hall...A fight there...Bloodied knuckles, bandaged by a girl who took me home...What was her name? Chestnut blossoms...A white rose...

The Courts of Chaos

Thus does Corwin set out to create a new Pattern, thinking of Paris in 1905. He succeeds, but we never see what the place is really like (for all I know, we live there now). If you like, use the quote above as hints for the universe that Corwin's Pattern might generate.

Corwin's Pattern is a wonderful setting for a brand new *Amber* campaign. There are already many versions of the place, each a unique creation.

One way to set it up is to have Corwin return there, and be father to one, or more, or the entire group of player characters. This is handy for players who have yet to read the *Chronicles*. Those characters who come from Corwin's Pattern will not be at any disadvantage, since they'll only know of their own home and not of Amber.

With Corwin's Pattern the Game Master is also freed of many other restrictions. Create the place, and all its many Shadows, exactly as you see fit. It may even be that you'll want to announce to the players that you are running a campaign based on a universe of your own creation, using this book as a reference for Powers and the like, but using nothing of Zelazny's universe. Then, if you wish, you can keep it as a separate thing, evolving your own story line, or, if it seems appropriate, you can have the players "discover" their connection with the Amber universe.

The Courts of Chaos.

...It was as if I stood at the end of the world, the end of the universe, the end of everything. But far, far out from where I stood, something hovered on a mount of sheerest black - a blackness itself, but edged and tempered with barely perceptible flashes of light. I could not guess at its size, for distance, depth, perspective, were absent here. A single edifice? A group? A city? Or simply a place? The outline varied each time that it fell upon my retina. Now faint and misty sheets drifted slowly between us, twisting, as if long strands of gauze were buoyed by heated air. The mandala ceased its turning when it had exactly reversed itself. The colors were behind me now, and imperceptible unless I turned my head, an action I had no desire to take. It was pleasant standing there, staring at the formlessness from which all things eventually emerged... Before the Pattern, even, the thing was. I knew this, because I was certain that I had been here before. Child of the man I had become, it seemed that I had been brought here in some distant day - whether by Dad or Dworkin, I could not now recall - and had stood or been held in this place or one very near to it, looking out upon the same scene with, I am certain, a similar lack of comprehension, a similar sense of the forbidden, a feeling of dubious anticipation. Peculiarly, at that moment, there rose in me a longing for the Jewel I had had to abandon in my compost heap on the shadow Earth, the thing Dworkin had made so much of. Could it be that some part of me sought a defense or at least a symbol of resistance against whatever was out there? Probably.

The Hand of Oberon

Thousands of places, another infinity of Shadow, uncounted numbers of powerful Lords. The Courts of Chaos could easily dwarf Amber.

The critical thing is that there is a much larger royal family in the Courts of Chaos. Amber is a new court, with only Oberon's children and grandchildren. The Courts of Chaos have been breeding for millennia. Worse yet, with the time distortion, every week in Amber could be another year in the Courts of Chaos, and their population that much greater.

The Politics of Chaos. This is another unknown. For all we know, the Patternfall War was a backwater battle that hardly affected the true center of the Courts of Chaos. Indeed, it may be that Dara, and her son Merlin might be from some backwater, only getting involved in the conflict with Amber as a means of gaining power. On the other hand, it could be the opposite. Perhaps the royalty of Chaos were severely damaged by their loss to Amber, and now are threatened with second class status. Either way, agents and spies from the Courts of Chaos will probably be inserting themselves in and around Amber at every opportunity.

The Diplomatic Mission. This works two ways. It seems likely, with the end of the Patternfall War, that diplomatic missions have been exchanged between the two great powers. Player characters can face possible conflicts with Lords of Chaos wandering the halls of Castle Amber, or could even be sent on missions straight into the depths of the Courts of Chaos.

Creatures of Chaos.

Their markings were precisely those of Siamese cats, only these were the size of tigers. Their eyes were of a solid, sun-bright yellow, pupilless. They seated themselves on their haunches as I turned, and they stared at me and did not blink.

They were about thirty paces away. I stood sideways between them and the stretcher, my blade raised.

Then the one to the left opened its mouth. I did not know whether to expect a purr or a roar.

Instead it spoke. It said, 'Man, most mortal.'

The voice was not human-sounding. It was too high-pitched.

'Yet still it lives,' said the second, sounding much like the first.

'Slay it here,' said the first.

'What of the one who guards it with the blade I like not at all?'

'Mortal man?'

'Come find out,' I said, softly.

'It is thin, and perhaps old.'

'Yet it bore the other from the cairn to this place, rapidly and without rest. Let us flank it.'

I sprang forward as they moved, and the one to my right leaped toward me.

My blade split its skull and continued on into the

shoulder. As I turned, yanking it free, the other swept past me, heading toward the stretcher. I swung wildly.

My blade fell upon its back and passed completely through its body. It emitted a shriek that grated like chalk on a blackboard as it fell in two pieces and began to burn. The other was burning also.

But the one I had halved was not yet dead. Its head turned toward me and those blazing eyes met my own and held them.

'I die the final death,' it said, 'and so I know you, Opener. Why do you slay us?'

And then the flames consumed its head.

The Guns of Avalon

Giant cats. Masked women with no faces. Horned Demons. Grass that grabs at feet, and leeches strength and life away. Living things from the Courts of Chaos can be a bit different from the life forms in and around Amber.

In Amber, there is a clear dividing line between the few who wield real Power and the rest of Shadow's living things. To put it another way, either you've got Amber Blood, or you don't.

Things aren't so distinct in the region of Chaos. The Lords of Chaos, counterparts of the Princes of Amber, are just the top of a gradual continuum of Power. Nearly all creatures of Chaos can Shape Shift, many can perform Magic, and, given time, even the most bestial can control some aspects of the Logrus, accessing its power from time to time.

On the Amber side of the tree of Ygg, only Amberites are truly real. On the Chaos side, the difference in reality between a lowly creature and a High Lord of Chaos is just a matter of degree.

Lesser Ones. These creatures are the servants of the Lords of Chaos. They can take any form, though they are fairly limited in their power. Their Attributes tend to be either Human or Chaos Rank. If one has an Amber Rank in one Attribute the other three are likely to be Human Rank. While they can move through and manipulate the Shadows near to the Courts of Chaos, they are generally unable to affect the Shadows on the Amber side of Ygg.

Few of the Lesser Ones are naturally intelligent, at least to the point of being able to speak with humans. However, intelligence is something that a Lord of Chaos can impose on a Chaos Creature, forcing it to Shape Shift into a creature that can listen and obey, understand and report. Lesser ones can also be made capable of weaving and casting spells, but only those taught to them by Lords of Chaos.

Disembodied Creatures of Chaos. Some bizarre forms of Chaos Creatures are entities made of pure Psyche and Logrus. They are capable of using at least some of the powers of a Logrus Master. Having no bodies of their own, they reach out to the minds of physical creatures, engage them in a Psychic battles, and seize their bodies as hosts.

Weaker versions of these bodiless creatures can be dominated by others, and then put into the bodies of animals. For example, a horse, combined with a disembodied Chaos Creature, has several advantages as a mount. For example, the Chaos Creature can keep the horse moving and fighting after it would normally die, and even after all its vital organs have stopped working. Weaker versions generally have a Psyche of Chaos Rank.

Greater disembodied creatures will sometimes bargain for a position in a characters body, negotiating their considerable services in exchange for a "body ride." Of course, the honesty and reliability of any such creature is questionable. The greater ones have a minimum Psyche of Amber Rank.

Disembodied creatures can also have Magic, ranging from Power Words to Conjuration. However, the creatures cannot use Magic while in pure Psychic form. They must have a physical body in order to release Power Words, cast spells, or conjure.

The Crown of Amber.

'I stared at the crown of Amber upon the crimson cushion Caine held.

'It was wrought of silver and had seven high points, each topped by a gem stone. It was studded with emeralds, and there were two huge rubies at either temple.'

Nine Princes in Amber

The intrigues will likely continue, even after Random is made King of Amber. Aside from any treacherous attempts at removing Random, there will be the time-honored tradition of debating the succession. Just who comes next in line, once Random is removed? Putting the player characters in these positions makes for some interesting role-playing opportunities. Here are some of the possibilities.

Children of Random. Although Martin has a strong claim to the throne, especially with his Rebma royal blood, he is also suspect for a variety of reasons. Other children of Random may have been born first (in the usual way, or because time is playing its usual tricks). Based on Corwin's discussion of line of succession, whoever is born in wedlock (which Martin was not), has a stronger claim. Likewise, someone born after Random became King might have priority.

Children of Eric. In many kingdoms, when an elder brother dies, and the younger takes the throne, it is the children of the elder who are considered the rightful heirs. This could work out quite well, since there are still bitter feelings toward Eric from some quarters. A child of Eric might campaign for formal recognition, so that Eric will not be forgotten if for no other reason.

Children of Corwin. Although Corwin was never crowned, there are many who feel that he was the rightful first in line. Which means that his children should be the logical successors. Merlin, of course, has a very strong claim, but his Chaos connections may disqualify him.

Other Children of Elder Amberites. Since

just about any of Corwin's siblings could make a case for their own right to the throne, it seems reasonable that they would also make the same claims for their own offspring. After all, just because they swore fealty to Random, doesn't mean they feel that any child of Random is entitled to a dynasty.

Dworkin.

'...Get out of here!'

I was about to protest that I hardly feared any physical violence he could muster, when his features began to flow like melting wax and he somehow seemed much larger and longer-limbed than he had been. Seizing the light, I fled the room, a sudden chill upon me.

Corwin's flight from Dworkin,
The Hand of Oberon

Dworkin is more than a character. He is a primal force in the *Amber* universe. It's hard to imagine a campaign element with more potential. He created Pattern and Trump, consorted with the Unicorn, and either fathered or is relative to every player character.

Dworkin's Mind. Doesn't it make sense that every dream, every whim, every passing thought of Dworkin's might have world-shaking consequences?

Suggesting disturbing or upsetting things to Dworkin can have real consequences. For example, a player character speculating with Dworkin on the nature of reality, should be very frightened if Dworkin takes seriously the proposition that they do not exist.

Dworkin's view of time might be rather twisted as well. He could speak of things that have not yet happened, or not remember the recent past. His instructions to player characters can likewise be contorted, since his perceptions are not normal.

The Reflection of the Pattern. Dworkin isn't just the creator of Amber. He is Amber. He is the universe. He is the Pattern. That means that whatever ails the Pattern will also be reflected in some way in Dworkin. When Brand stained the Pattern with Martin's blood, Dworkin was driven to bouts of madness. Likewise, any threat that a Game Master inflicts upon Amber can show up, in twisted miniature, in Dworkin.

Possibilities are to have Dworkin fade in and out of reality. Or to have him regress to his pre-Pattern self, becoming once more a Lord of Chaos and unable to recall his own offspring. Dworkin could take any beastly form, and be more frightening and powerful than any monster. If Dworkin were somehow captured or controlled by an enemy of Amber, there would be little they could not do.

Dworkin's Chamber:

I cast my eyes about the place. There were bookshelves on all four walls. No windows. Two doors at the far end of the room, right and left, across from one another, one closed, the other partly ajar. There was a long, low table covered with books and papers beside the opened door. Bizarre curios occupied open spaces on the shelves and odd niches and recesses in the walls - bones, stones, pottery, inscribed tablets, lenses, wands, instruments of unknown function. The huge rug resembled an Ardebil. I took a step toward that end of the room and the lantern sputtered again. I turned and reached for it. At that moment it failed.

I growled an obscenity and lowered my hand. Then I turned, slowly, to check for any possible light sources. Something resembling a branch of coral shone faintly on a shelf across the room and a pale line of illumination occurred at the base of the closed door. I abandoned the lantern and crossed the room.

I opened the door as quietly as I could. The room it let upon was deserted, a small, windowless living place faintly lit by the still smoldering embers in its single, recessed hearth. The room's walls were of stone and they arched above me. The fireplace was a possibly natural niche in the wall to my left. A large, armored door was set in the far wall, a big key partly turned in its lock.

The Hand of Oberon

Grove of the Unicorn.

Another little known place is the Grove of the Unicorn. Situated between Kolvir and Forest Arden, it is the place renowned for sightings of the Unicorn. It is also where the path begins that took Corwin and Random to the Primal Pattern. If the Unicorn is truly a being of some different reality, of another dimension, then the Grove is where others of its kind may appear. Or where player characters may find themselves slipping into other realities.

The Gunpowder of Avalon.

Corwin shifted the balance of power, in Zelazny's *The Guns of Avalon*, by bringing workable guns to Amber. It had always been known that technology would not function in Amber, and that there was no compound that works there as gunpowder.

So Corwin's innovation has probably changed the equation.

Or has it?

First off, we don't know what was in Corwin's compound. Corwin described it as jeweler's rouge from Shadow Avalon. However Corwin also disappeared, with the right digging implements, just before the "powder" was complete. His story was that he needed diamonds to pay for things. Since most Amberites rarely want for treasure, it seems possible that the diamond-hunting story was just a cover, and that Corwin's little trip might have involved digging up some secret ingredient, without which the powder would be useless.

Another possibility is that Corwin went off to find an Amber substitute for primer. Modern, cartridge style bullets need two explosive compounds elements. Gunpowder, and also primer, a substance used to detonate the gunpowder.

Pattern & Function. One interesting question is why gunpowder didn't work in Amber to begin with. Is it just coincidence that neither gunpowder, nor electricity, nor engines, nor even modern medicines perform in Amber? What if it's not coincidence, but design? Suppose that the Pattern itself can be slightly altered, or tuned? Couldn't Dworkin, or Oberon, have deliberately set things as they are? It would make perfect sense, since it prevents Amber from having to engage in a weapons race with the Courts of Chaos or some other force. Given that it is possible, then anyone with the Jewel of Judgement might be able to tune the Pattern, adjusting details of Amber's environment. Oberon, when he repaired the Primal Pattern, had the perfect opportunity to fix things in this way.

Ramifications. On the other hand, perhaps Corwin's discovery could have untold repercussions. A chemical analysis of the powder could lead not only to the development of a whole new range of weapons (Amber with cannon? hand grenades? missiles?), but also of engines based on similar combustion principles (Amber cars and trains?).

Either Corwin's powder works, but is rare and precious. Or Corwin's powder works, and is in the process of changing the face of Amber. Or somehow Amber has changed, so that the powder no longer works there.

The Jewel of Judgement.

'Why does he want so badly to attune it? So he can raise a few storms? Hell, he could take a walk in Shadow and make all the weather he wants.'

'A person attuned to the Jewel could use it to erase the Pattern.'

'Oh? What happens then?'

'The world as we know it comes to an end.'

Random asking Corwin about Brand in
The Hand of Oberon

The nature of the Jewel of Judgement is unknown. It could be a construct of Dworkin's. Or something found in the Abyss. Or, as Zelazny has it in the Merlin Saga, an eye stolen from the Serpent of Chaos.

None of these answers is entirely satisfactory.

The image within the Jewel, so we are told, is like a pattern rendered in three dimensions.

Most remarkable is that the Jewel of Judgement can be used as a tool for inscribing a new Pattern. One supposes that it can also be used for altering or erasing an old Pattern.

The following Power is an addition available to anyone who correctly attunes themselves to the Jewel of Judgement. This can be done by walking the Pattern while holding and focusing on the Jewel, or with the assistance of someone who is already attuned. If attunement is performed by a character in the course of a campaign, the ten (10) points are instantly subtracted from the character, in the form of a loss of Good Stuff or increase in Bad Stuff.

ATTUNEMENT TO THE JEWEL OF JUDGEMENT

10 Points.

Only a character attuned to the Jewel of Judgement may call upon its powers. The attunement also links the character with the Jewel. In some cases this is good, as the Jewel will attempt to protect an attuned wielder. However, the Jewel also can be destructive to those who are attuned, drawing away too much of their mortal energies. Every use of the Jewel of Judgement results in strain and fatigue in the wielder. Drawing power direct from the Jewel can offset this life force drain somewhat, but it can never take the place of rest and separation from the Jewel.

Jewel of Judgement Abilities.

Attunement Link. Characters who have attuned to the Jewel of Judgement establish a Psychic link. This allows them to sense the Jewel's presence in a place, or, if the character has a powerful enough Psyche, the direction or location of the Jewel. It also means that someone who is attuned can control the Jewel from some distance away, manipulating its functions, or blocking someone else from using it.

Weather Control. Using the Jewel of Judgement a wielder can summon or banish storms, manipulate lightning, and control the winds. This can be done in Amber, or, looking through the Jewel's Pattern, at any point in Shadow. The Jewel can also be used to control Shadow Storms, not necessarily dispelling them, but at least redirecting them.

Molding Shadow. Dworkin warned Corwin of the hazard of leaving the Jewel of Judgement in Shadow. As Dworkin said, "It is best kept near the center of things," and "It tends to have a distorting effect on shadows if it lies too long among them."

Presumably this also means that the wielder of the Jewel can deliberately mold Shadow in a variety of ways. It may also mean that entire Shadows can be created and destroyed with the Jewel.

The Portable Pattern. The act of walking the Pattern, and then teleporting from its center, can be reproduced using the Jewel. Beyond that, it may be possible to use the Jewel to teleport objects away from the wielder, to another place in Shadow.

Temporal Control. Another unknown and unexplored aspect of the Jewel is its control over time. A side-effect of this control is the speeding up experience that a wielder of the Jewel will experience if they've maintained the contact for too long. From the point of view of the wielder, everyone else seems to be moving too slowly.

The Ladies of Oberon.

Like many Kings, Oberon went through a succession of women. And, tiring of their demands, or their jealousies, he put them aside.

What about all those wives of Oberon? Did they really die, or has Oberon merely put them aside, hidden somewhere in Shadow? Or were any wise enough to remove themselves to a place of safety, before their fickle lord could dispose of them? If alive, could any of them yet be still active in affairs of Amber, with their children as their agents?

What powers might these ladies have? Do they still control their children, consult with them, manipulate them?

Here, probably in order of importance, and appearance in Oberon's life, are the women that Corwin has mentioned.

Cymnea, mother of Benedict, Osric and Finndo. A strong possibility exists that Cymnea may be a Lady of the Courts of Chaos. It's possible that she dwells in the Courts even yet.

Faiella who bore Eric and Corwin, is said to have died giving birth to Dierdre. Perhaps, perhaps not. Like all rumors of death in Amber, this is suspect.

Clarissa, the fiery redhead who begot Fiona, Bleys and Brand, was probably a sorceress to be reckoned with. She and Oberon had a stormy relationship, and it's entirely possible that she has simply made herself scarce to avoid dealing with the old king. Perhaps, with his absence, Clarissa will reappear in the Court of Amber.

Rilga is the mother of Caine, Julian and Gerard. One of the most recent of Oberon's wives, Rilga is seldom mentioned. She may still be living in City Amber, quite openly.

Moins, a noblewoman of Rebma, is mother to Llewella. It's interesting that Moins had enough political clout to persuade Oberon to recognize Llewella as an "official" heir to the throne of Amber.

The list does not end there. In latter books Zelazny also refers to *Paulette* (mother of Random), *Dybele* (Florimel's Mom), *Lora, Kinta, Deela* and *Harla*, all as consorts or wives of Oberon. Fact is, Oberon was a lusty guy, and he had thousands of years to fool around. Any woman that enthralled him must have been, and likely continues to be, a character worthy of the intrigues of Amber.

Lords of Chaos.

Every so often Corwin will describe one of the Lords of Chaos in great detail. Unfortunately, Corwin then goes on to remove them from the lists of potential encounters. This he does by killing them, usually with Greyswandir, but occasionally with his bare hands.

Among Corwin's more memorable victims were Strygalldwir, an impressive gent with an equally impressive set of antlers. He also had leathery wings, a rune-carved sword, and a talent for magic.

Lord Borel, a High Lord of Chaos, was an even more impressive opponent, who made the mistake of challenging Corwin to a fair fight. Needless to say, Corwin did not fight fair. Borel died.

Still, Lords of Chaos make for impressive opponents. They generally have Logrus Mastery, Shape Shifting, and some Magical Powers. As a general guideline, a Game Master might construct a Lesser Lord from one hundred points or less, a Lord of Chaos in the hundred to two hundred range, and any High Lord of Chaos from a pool of more than two hundred points.

Lost Children.

Corwin claimed that Oberon's offspring numbered fifteen sons and eight daughters, of whom an indeterminate number were dead. Take away the acknowledged nine Princes and four Princesses, and that leaves six men and four women unaccounted. For examples of how a Game Master can make use of the missing sons and daughters, take a look at *Finndo* and *Osric* in the section on Amberite profiles.

These misplaced characters are probably the most powerful campaign elements available to a Game Master. Hidden in Shadow, or behind Patterns of their own making, or in darker realms, the lost siblings can be authors of plots and schemes that are centuries in the making.

A useful way of portraying the missing ones is by fairly subtle hints. For example, in the combat section you'll see an example where a player character, John Speck's character Godfrey, fenced with Reanna, an elder daughter of Oberon. He knew little about her other than her name, but he did remember the wine that she offered him after their bout. Later on, at a family banquet, where many of the elder Amberites were in attendance he served a few bottles of the wine (it's amazing what you can find in the wine cellar of Castle Amber, if you take the time to look). The reactions of the other Amberites ranged from ignorance to shock to immediate withdrawal and preparation for battle. Those reactions, from Dworkin and Bleys and Finndo spoke volumes to the players about the importance of Reanna, and her relations with the other Amberites.

The Patternfall War.

Corwin reports the war is over.

Is it?

Can the great conflict between Chaos and Order, between the Court of Amber and the Courts of Chaos, ever truly be resolved?

Another great opportunity for role-playing is the continuing threat of reactivating the war. Player characters can be put in the role of peace makers, attempting to patch up differences between the courts. No doubt there are agents and factions on both sides who are hot to create incidents and to revive hostilities. Investigating and uncovering such plans is another opportunity for role-playing.

Shadow/Hellriding/Shadow Shifting.

...Do we make the Shadow worlds? Or are they, independent of us, awaiting our footfalls? Or is there an

unfairly excluded middle? Is it a matter of more or less, rather than either-or?

The Courts of Chaos

Here's one of the great philosophical questions posed in *Amber*, in Corwin's own words, "We had spent much of our time wandering in Shadow, or in our own universes. It is an academic, though valid philosophical question, as to whether one with power over Shadow could create his own universe. Whatever the ultimate answer, from a practical point we could."

So, can you?

Can you create your own universe? Or in the infinity of possibilities will you always come across the universe that you imagine?

The whole process of moving through Shadow is a mystery to Amberites. Are there just uncountable Shadows awaiting discovery, or do those with the power of the Pattern actually create something in the Shadows they seek?

Supernatural Creatures.

I was being pursued by a manticora.

The last time I had seen its like was on the day before the battle in which Eric died. As I had led my troops up the rearward slopes of Kolvir, it had appeared to tear a man named Rall in half. We had dispatched it with automatic weapons. The thing proved twelve feet in length, and like this one it had worn a human face on the head and shoulders of a lion; it, too, had had a pair of eaglelike wings folded against its sides and the long pointed tail of a scorpion curing in the air above it.

The Hand of Oberon

Scattered throughout the *Chronicles* are mentions of all sorts of supernatural creatures. Corwin battles Werewolves, tells of battles with Centaurs, escapes a Sea Serpent, pats a Hippogriff, lets a wheezing Dragon pass by, and drinks ale with evil little people (Leprechauns? Fairies?). All these creatures seem to have, in whole or part, the mystical powers that legend grants them.

Where these creatures come from, and the source of their power, is up to each Game Master. Perhaps they are formed from Shadow itself. Or perhaps they are remnants of a reality that existed before Pattern. Or maybe, like the Unicorn, they simply wander in from realities totally foreign to the Amber universe.

Tir-na Nog'th.

Floating above Amber, but only in moonlight, Tir-na Nog'th is purely a place of dreams. It is sought because it contains its own Pattern.

Yet Amberites also seek it out for signs of that which is to come, as a place of portent and omen. For example, it is said that Brand visited Tir-na Nog'th, and then decided to get rid of

Corwin before the events of *The Chronicles of Amber* ever started, because he saw that Corwin would be the cause of his defeat.

For a Game Master Tir-na Nog'th offers a chance to reflect the pain of all the wrong decisions a player character may have made. They'll meet those they have wronged, or those who died through their mistakes or inaction, and see how their world might have been made better, or worse.

The prime example of a painful experience in Tir-na Nog'th was inflicted upon Corwin. He spoke to only one of the silvery ghosts, the woman he had known as Lorraine. What could she say that would wound him more than that he could have saved her, and Eric, and a kept a sense of unity in Amber, if only he had not turned his back on her?

'We had that argument,' she said. 'You followed me, drove away Melkin, and we talked. I saw that I was wrong and I went with you to Avalon. There, your brother Benedict persuaded you to talk with Eric. You were not reconciled, but you agreed to a truce because of something that he told you. He swore not to harm you and you swore to defend Amber, with Benedict to witness both oaths. We remained in Avalon while you obtained chemicals, and we went to another place later, a place where you purchased strange weapons. We won the battle, but Eric lies wounded now...'

Sign of the Unicorn

Shadow Storms.

Shadow Storms, like thunder storms and hurricanes, rip across a landscape cutting a swath of destruction. The difference between Shadow Storms and mundane violent weather is that Shadow Storms rip across the boundaries between Shadow, tearing apart the orderly separation of Shadow. The Storms in their passage tear things from one place and cast them into another, or change the stuff of Shadow, or simply destroy that which they touch.

The source of the Shadow Storms is completely unknown. Each Game Master is free to calculate their cause, as well as coming up with characters and Powers that are capable of controlling them.

Which leaves many questions for the Game Master to answer:

• What is the effect of a Shadow Storm on a creature of substance, such as an Amberite or some Logrus Master?

• What lies behind these storms? Are they merely the result of the ebb and flow of the energies that course between Pattern and Logrus? Or is there a more sinister purpose?

• What are the limits of Shadow Storms? Do they feed themselves on the Shadow they destroy? Or can the only continue as long as they are fueled?

Trump.

'Do you know how many sets there are?'

'Well, everyone in the family has a pack or two and there were a dozen or so spares in the library. I don't really know whether there are any others.'

Sign of the Unicorn

Control and distribution of Trump decks is a crucial factor in setting up a campaign. By linking a group together with Trump, the role-playing remains group based, even when all the individuals are scattered across Shadow. When player characters are lost out in the wastelands of Shadow, or stuck somewhere, Trump is often the only way for them to return. Characters who carry Trump decks have another dimension of options for movement and communication. For the character who is lost or trapped, sometimes being contacted by another player character, holding their Trump, is the only way out.

In addition to the importance of the usual Trump, there are several other variations on Trump that can be turned into story elements for campaign. Here are just a few of the possibilities:

Unknown Trump. A great way of foreshadowing the involvement of an some unknown character is by introducing their Trump into the campaign. A Trump discovered in the hands of an assassin, or in some significant place, can serve as a *hook* to involve the player characters in yet another story line.

Unknown Trump can be one of those cards found in a standard deck, but that are largely ignored (such as those of Osric and Finndo), or of a new character altogether.

Trump Backs. An important question every Game Master has to answer is the meaning of the image on the back of the Trump. Dworkin's Amber cards all have an image of the Unicorn, rampant, on the back. Does this mean that the power of the Trump is somehow drawn from the Unicorn? Or is it just a way for Dworkin to place his own mark on the cards?

Trump also exist, the product of many hands, in Courts of Chaos. Do they have a single image, like the Great Serpent of Chaos? Or does each school of Trump Artists have a common logo or mark? It's even possible that each and every Trump Artist might have a unique creative icon.

A strange Trump back is another campaign or story element that is sure to keep the players guessing.

Full-Sized Trump. Although Trump are nearly always presented as cards in the *Chronicles*, we know that Dworkin can sketch a Trump image anywhere. He even managed to inscribe one on a stone wall with a sharpened spoon. It's interesting to speculate that one or more of the thousands of paintings and tapestries hanging about in Castle Amber may have Trump qualities as well.

Trump Traps. It's certainly possible that a Trump could be created specifically as a trap. Such a Trump could draw the holder into a hostile environment, a Shadow from which conventional escape is impossible, or into the jaws of some waiting monster. It's also possible that a Trump might be designed to remove a character from the physical universe altogether, suspending them in the region between Trump contacts. Such characters might be trapped in the force of Trump itself, and might only be released through manipulation of the card that formed the gateway.

The Unicorn.

'Corwin! Look!' It was a whisper, and Gerard's hand closed on my elbow as he spoke.

I followed the direction of his gaze and froze. Neither of us moved as we regarded the apparition: a soft, shimmering white encompassed it, as if it were covered with down rather than fur and maning; its tiny, cloven hooves were golden, as was the delicate, whorled horn that rose from its narrow head. It stood atop one of the lesser rocks, nibbling at the lichen that grew there. Its eyes, when it raised them and looked in our direction, were a bright, emerald green. It joined us in immobility for a pair of instants. Then it made a quick, nervous gesture with its front feet, pawing the air and striking the stone, three times. And then it blurred and vanished like a snowflake, silently, perhaps to the woods to our right.

I rose and crossed to the stone. Gerard followed me. There, in the moss, I traced its tiny hoofmarks.

'Then we really did see it,' Gerard said.

I nodded.

'We saw something. Did you ever see it before?'

'No. Did you?'

I shook my head.

'Julian claims he once saw it,' he said, 'in the distance. Says his hounds refused to give chase.'

Sign of the Unicorn

Aside from its physical manifestation, its mythic connotations, and its spiritual symbolism, the Unicorn remains an enigma.

Is it representative of some greater reality?

Something outside the Amber universe?

Ever hear of a story called *Flatland*? It could easily apply to Amber.

Imagine a world in two dimensions, as on a single, infinite, flat piece of paper. Flat creatures roam its length and breath, never having the faintest idea that such a thing as height could possibly exist. You, a human of three dimensions, look down upon the plane like a god. And, touching the land of those two dimensional beings, you seem strangely powered able to mysteriously grow, expand, shrink, and disappear, all as your finger passes through, and out of, Flatland.

What if the Unicorn were a creature from a *four*

dimensional place? That would explain it's other-worldly aspect, and how it is able to come and go in impossible ways. Also, a four dimensional Unicorn implies other creatures with an extra-dimensional nature, or even an entire reality peopled with mythic creatures.

Ygg.

Partway down, with the fog just beginning to creep and curl about me, I spotted an ancient tree and cut myself a staff. The tree seemed to shriek as I severed its limb.

'Damn you!' came something like a voice from within it.

'You're sentient?' 'I'm sorry...'

'I spent a long time growing that branch. I suppose you are going to burn it now?'

'No,' I said. 'I needed a staff. I've a long walk before me.'

'Through this valley?'

'That's right.'

'Come closer, that I may sense you presence. There is something about you that glows.'

I took a step forward.

'Oberon!' it said. 'I know thy Jewel.'

'Not Oberon,' I said. 'I am his son. I wear it on his mission, though.'

'Then take my limb, and have my blessing with it. I've sheltered your father on many a strange day. He planted me, you see.'

'Really? Planting a tree is one of the few things I never saw Dad do.'

'I am no ordinary tree. He placed me to mark a boundary.'

'Of what sort?'

'I am the end of Chaos and of Order, depending upon how you view me. I mark a division. Beyond me other rules apply.'

'What rules?'

'Who can say? Not I. I am only a growing tower of sentient lumber. My staff may comfort you, however. Planted, it may blossom in strange climes. Then again, it may not. Who can say? Bear it with you, however, son of Oberon, into the place where you journey now. I feel a storm approaching. Good-bye.'

'Good-bye,' I said. 'Thank you.'

The Courts of Chaos

By the end of the *Chronicles* there was another tree, started from one of Ygg's branches, and planted at the beginning of Corwin's new Pattern.

What these two trees represent, where they come from, and what they might do, is unknown. Somehow the trees of Ygg have the ability to look out into Shadow, possibly all the way to Amber and Chaos.

Threats to the trees are also fairly serious. Ygg does more than simply mark the boundary, it is a boundary. Without it the rules of the two domains may pass back and forth, causing massive confusion, possibly Shadow Storms, and even threatening the stability of Shadow itself.

IDEAS FOR CAMPAIGNS BEYOND ZELAZNY'S *AMBER*

One of the best things about Zelazny's *Chronicles of Amber* is the way the universe keeps unfolding. Corwin's story starts here on Earth, opens up to reveal Earth as one of a number of infinite Shadows, and then opens up again to show Amber, the central point of all reality, complete with its two reflections. In the next book the universe gets even bigger, as we find the Black Road, part of a festering force that corrupts Shadow, and which allows things from somewhere beyond the Pattern of Amber to invade. The third book expands Corwin's view of the universe yet again, revealing the hidden Primal Pattern, upon which Amber is based. The Courts of Chaos are revealed in the next book, demonstrating that Amber is not alone as a power in the universe. Finally, in the fifth book of the series, the universe is seen to be larger yet, partly by Corwin's own creation of a new Pattern, and partly by clues of other lands altogether.

Continuing this ever-expanding tradition makes for wonderful *Amber* campaigns. As if *Amber*, with its infinite Shadows, weren't big enough, here are a few ideas for taking a campaign out into the farther reaches of the imagination.

Other Patterns.

If Corwin could create a Pattern to co-exist with Amber's Pattern, why couldn't there be others? Distant, true, and perhaps not as vibrant or real as Amber, but Patterns just the same.

Each Pattern offers yet another version of infinity, with Shadows demonstrating every possible aspect of creation.

Early Experiments. It's possible that Dworkin, or Oberon, or some other Lord of Chaos, may have created other Patterns before Amber. These could be considered "failures," and may yet remain as lonely outposts. Their contents unknown and unexplored, they could provide new territory for the player characters to scout, or sources of new threats.

Pattern Retreats. Missing elder Amberites, or other rebel Lords of Chaos, might also have created Patterns. Each could be a private haven, containing long lost characters.

Alternate Patterns.

Look at the description of the Jewel of Judgement. It is said that inside its ruby depths is a vision of the Pattern rendered in three dimensions. One two-dimensional "slice" of the Pattern forms Amber, and another "slice," inscribed by

Corwin, created a new Pattern. Who is to say that there are not an infinite number of Patterns yet to be created? Or, for that matter, as is suggested about Shadow, could it be that each Pattern already exists?

Alternate Universes.

Just as there is infinite Shadow, there might also be infinite Pattern. Stacked up, each differing from the next by only a hair, all these variations on Amber could contain the basic Pattern, Castle Amber, but with slightly altered versions of the elder Amberites and in each a different cast of the younger generation. This is the basis for Cross-Over sessions, where characters from different campaigns can come together for role-playing.

Realms Beyond Amber and Chaos.

Zelazny has described a universe with two poles, Amber on one end and Chaos on the other. However, that doesn't mean that other realities might not exist. If a character were to find the right approach, it could be that totally foreign sources of Power exist. Travel far enough, or through a twisted doorway, and a whole new realm of universes can open up.

ANALYZING PLAYER CHARACTERS

In *Amber* there are two halves to campaign building. One half is the Amber universe itself. The other half is made of the player characters. It doesn't matter how brilliant an idea may be if it doesn't fit with the characters.

Fitting Stories to Characters.

It's up to the Game Master to tailor the stories to the characters set up by the players.

Let's take a peek at *Hamlet*, the classic Shakespearean play.

Just to sum things up quickly, Hamlet, our main character, is confronted by the ghost of his father. "Son," Hamlet is told, "my murderer is the guy who just married your mother. So go kill him!"

What happens next depends on Hamlet's personality.

If Hamlet had been a gung-ho, damn the torpedoes kind of guy, he would have confronted the old lecher and skewered him. End of story.

On the other hand, Hamlet might have been the kind of cautious, bright guy for whom a ghost's advice is not something you can bank on. In which case he would have gotten himself to a monastery, a psychologist, or whatever, and freed himself of this ridiculous Freudian and Oedipal complex.

Either way, we end up with not much of a story. If you're going to entertain a mob of role-players (or, from Wm.'s point of view, a crowd of rowdy Elizabethan play-goers) for a few hours, you need a little more drama.

So Shakespeare's Hamlet is somewhere in the middle. He dithers, he talks to himself, he tortures everybody around him, and he goes looking for clues. Gradually he drives his

girlfriend crazy, makes an enemy out of his best friend, kills a couple of by-standers, and generally sets the stage for a lot of gore and bloodshed at the end of the play.

Shakespeare had it easy. He could make his characters do anything he wanted. The story came first, and it always ended exactly the way it was supposed to.

Game Masters have a rougher time. Characters in role-playing are run by players who generally don't cooperate with any pre-ordained story line. No, they do whatever they feel like doing, pretty much the way people do in real life.

Which doesn't mean that we Game Masters can't still connive the player characters into some great stories. For example, let's look at Hamlet's situation and see how it could be played out if Hamlet's personality were different.

As before, Hamlet is clued in by the ghost of his father. This time Hamlet is being played by some upstart actor who sees himself as a man of action. He takes the guards in tow, chases on down to the old man's room, and guts him then and there.

Not to be outdone, the Game Master has to react. One possibility is blood-stained Mom treating Hamlet as a lunatic, and set things up so everybody else thinks Hamlet has gone Throne-crazy. What might make it more interesting is to have some kind of evidence come to light, showing that the "ghost" was some kind of fake, staged by his step-father's enemies. Now, even though Hamlet screwed up the original idea, there's still a story that can be played out.

Take another tact. What if Hamlet is more easy-going?

This time Hamlet is played by an actor who laughs off the ghost's warning, convinced that apparitions from the nether world aren't real, and, even if they were, are never honest. Hamlet goes about his merry way.

In this case the Game Master can set it up so that the ghost was right. Somehow the King hears the ghost's story and decides that more bloodshed will be required. Hamlet, foolishly ignoring the warning, has put his Mom and his own life in danger. As Hamlet blunders in as the next target in a foul murderer's schemes we're on the track of a new story.

The moral is that Game Masters have to design campaigns with a number of alternative story lines. Even more important, the stories have to be built around the player characters.

Basing the Campaign on Player Characters.

First, inventory all the points expended by your players. Take a look at what they've spent for Attributes, Powers and all the other things that go into making up their characters.

Look at these points as money. The players are your customers, and they spend their points on your campaign.

If a player spends, say, fifty-one points on first place in Warfare, and the next bid is fifty, these players, collectively, have invested over one hundred points in that one aspect of the game.

A pair of players who invest over one hundred points in Warfare are buying a campaign where Warfare is important. Neglect Warfare, by setting up a campaign where the real villains only fight with Powers and Psyche, and you've cheated this pair of customers.

One way of figuring out if a campaign is properly balanced is to total up all the characters' points and compare it to the threats they'll be facing. If the proportions work out, it should be a well-balanced campaign.

Good Stuff, Zero Stuff & Bad Stuff.

Each character's "Stuff" should be used to help build the campaign.

Using Good Stuff. The best use of a point of Good Stuff is translating it into a friendly interest from an elder Amberite. Good Stuff is also used to set up the character's background, and the more a character has, the more pleasant, loving and trouble-free will be their earlier years.

When Bad Things Happen to Good Characters. Every character should be given threats in a campaign. Of course when the threat is against the whole universe, obviously it's going to affect all the characters. Yet every character should have at least one bad potential event built into their situation, just to keep things exciting.

Zero Stuff Characters. Even players with no Bad Stuff or Good Stuff can still have campaign elements based on their characters. Just balance out the forces that affect them. For every threat, they should have some positive force seeking to help them.

Pawns, Tools and Puppets. Another important campaign element for Zero Stuff characters is how the elder Amberites will react to them. Players who plan characters as Zero Stuff are usually those who have well worked out objectives, clear goals, and Powers that they exploit and explore well. From the perspective of the elder Amberites, those master manipulators, Zero Stuff characters are the perfect tools. After all, the best tool, is something predictable. Something that does what you want, when you want.

Bad Stuff Characters. Each point can be programmed into the campaign as hatred or distrust from an elder Amberite, as another element in a background and/or childhood filled with pain and suffering, or as betrayal of the character's secrets to other characters.

Agitators. Not only are they fun by themselves, but Bad Stuff characters can also be useful for getting an entire campaign moving. The events generated by their Bad Stuff can serve as forewarning of the major events of a campaign. For example, a child of Brand might be viewed as a competitor to some entity seeking to use Brand's legacy. In this case attacks upon that character could work to get the whole player group moving.

MAKING THE BIG DECISION

SELECTING CAMPAIGN IDEAS

One last word of advice to any Game Master. Think hard upon the basic concepts behind your campaign. Your players will likely get very attached to their characters, characters with a great deal of power and the capacity to live forever. Your campaign is likely to last a very long time.

Ask yourself. *Do you want to spend the next year of your life role-playing this campaign? These ideas?*

And a year may be an understatement. The first Amber campaign started back in 1985. It's still going strong. Many of the original story lines have yet to be completed.

Endangering the Universe.

The best opening story for an *Amber* campaign is one where the entire universe is in danger.

There is a very good reason for needing this kind of opening in a campaign. Without something really nasty to work against, the player characters tend to scatter.

Not that they won't take off on their separate ways anyway. It's just that if they don't develop some kind of relationship with each other early on, they may never get back together again. In which case you might as well schedule a series of one-on-one sessions with each player, because the role-playing group may not exist.

The player characters are incredibly powerful. Too powerful to need to work together in any mundane kind of adventure. To bring them together you need a story where there is no escape, where the threat reaches the player characters wherever they might hide, and where only a combined effort can be successful.

In other words, you've got to threaten to destroy the world (or at least all the player characters) every time a campaign starts up.

Making it Personal.

'You think that a human agency arranged that entire chain of events, up through the recovery of the Jewel?'

'I don't know about that. What's human? But I do think that someone we both know has returned and is behind it all.'

'All right. Who?'

I showed him the Trump that I held.

'Dad? That is ridiculous! He must be dead. It's been so long.'

'You know he could have engineered it. He's that devious. We never understood all of his powers.'

Benedict & Corwin,
The Hand of Oberon

In *Amber* there is no threat without a personality behind it. No matter that it could be a Black Road feeding demonic troops from the Courts of Chaos to the heart of Amber, or a vile stain crippling the Pattern. Some agent, some *character*, caused these things to come about. The same is true of clues, or matters of assistance, where behind-the-scene relatives are performing the charitable acts for their own purposes.

Figure out the antagonistic, and look at the universe from their point of view. What are their goals and objectives? What Powers are at their disposal? Who do they hate or love? Get into their head, and look through their eyes.

Then, when you, the Game Master, feel comfortable "playing" the character responsible for the threat, every action and sequence will be the result of role-playing. As the role-play progresses, keep stepping back into the roles of the manipulators. What will they learn of the player characters' actions? Will they see things clearly, or misinterpret events and intentions? How will their plans change? What will the off-scene characters do next?

It turns into a kind of internal juggling act after awhile. The Game Master not only has to keep track of how the non-player characters react to the player characters, but also of how the non-player characters react to each other. Eventually a good *Amber* campaign can involve a dozen or more elder Amberites, each with their own ambitions and goals, all interacting with each other and with all the player characters, and all these significant events taking place without the player characters ever laying eyes on one of their elders.

The Chessboard that is Amber.

'Did you fear Martin's blood on your own?' Benedict asked.

'That bastard puppy!' Brand said. 'He was not truly one of us. He was only a tool.'

The Hand of Oberon

Cards in a deck, pieces on a board, tokens in a game. As a Game Master you have to see the board from the point of view of the elder Amberites. In their eyes the player characters are just pawns, or puppies too young to be taken seriously, but still capable of ripping down the drapes. Being young, immature, untried and untested simply means that they are easily manipulated. And, from the point of view of the elder Amberites, pawns are pretty expendable. They're never happy about losing a piece in the game, but sacrificing a pawn is never as painful as losing a more important piece.

Fathers & Mothers. The elder Amberite never take their own children lightly. They may manipulate or ignore their nieces and nephews, but it's hard for them not to have strong feelings about any of their own sons and daughters. Assigning a parent to a player character, and that parent's attitude to their child, is just as important to building a campaign as any villain or threat.

Villains and Bad Guys.

There's a tendency in some movies and television shows to present one-dimensional villains. You can spot them instantly, sometimes just because they wear black. They have nothing but evil in them, but with no real motivations for their acts.

Not so in *Amber*.

Sure, we're got our fair share of scoundrels, homicidal maniacs, and black-hearted rogues among the elder Amberites. It's just that you can never tell which characters are the rotten apples, which are truly innocent, and which ones just plain like looking evil. Even if the players do turn over a genuine dyed-in-the-wool archenemy, he (or she) is just as likely to be on side of goodness and fair play as not.

Chaos dwellers, for example, can be in opposition to the player characters, intent on the destruction of Amber, or on advancing some arcane cause, and all the while be very good-hearted people. After all, they have families, loves, and homes, just on the other side of the dividing line between the two realities.

Brand is definitely the villain of the piece, and by the end of the *Chronicles* he certainly seems to be pure evil. Yet there is more to him. He has a fondness for a relative or two, and even a favorite rug. Like any other Amberite he played the game open-ended, willing to change his plans if things worked out differently.

Ambiguous Bad Guys. The best *Amber* campaign is one where the villain isn't truly revealed until the final closure. The player characters should have a range of suspects, from dwellers in the Courts of Chaos, to their Aunts and Uncles of Amber, and even to other player characters.

Monsters, Monsters, Monsters...

The easy answer is just to build these creatures using the Artifact creation system.

Sometimes that's too easy. The fact is, there are no "conventional" monsters in Amber. Oh sure, there are plenty of big ugly suckers wandering around through shadow, through Arden, and they're thick as fleas out in Chaos. But these aren't really much of a challenge to Amberites.

No, the real monsters are those that are "constructed" as the creatures of other, elder, Amberites, or as the instruments of the Lords of Chaos.

As a Game Master feel free to ignore the whole point system when putting together a monster. For example, in the very first Amber campaign, an enormous black dragon ate Amber. When the characters next saw the creature, it had become larger than a continent and was rapidly consuming all the Shadows of Pattern.

The point is, the monster itself, though having unbeatable power, was not the real problem. Solving the player characters' dilemma involved tracking down the source of the dragon, and stopping the elder Amberite who has loosed it upon the universe.

Artifacts of Power.

Unlike monsters, it's always a good idea to figure out how many points go into each of the important artifacts of the campaign. It's not a question of limits. Feel free to thrown in as my points as you like. No, it's more a matter of ownership. The more points are involved in an item, the more strings are attached to it.

Zelazny's greatest artifact was the Jewel of Judgement, an item that changed hands at least a dozen times in the course of the story. Because it was worth so much, it was valued by many, and, being so powerful, it was impossible to conceal. It was sometimes even impossible to protect.

A campaign artifact in the hands of a player character becomes, in some ways, a ball of Bad Stuff. The character becomes "it," the one with the prize, and the object of a lot of unwelcome attention.

CHARACTER ADVANCEMENT

The old saying, "practice makes perfect," has to be wrong. In *Amber* it has to be "Practice **can't** make perfect."

Think about it.

Travel to a time-shifted Shadow, where the years fly by as seconds crawl in Amber, and there's no limit to how much practice you can get in any skill, power, and/or attribute. Anyone with Pattern, or Logrus, or other Powers, can figure out a way to use this trick.

Were it simply a matter of time and practice, any Amberite could improve. Worse, any Amberite could get *infinitely* good at anything.

So practice isn't enough.

True advancement, those things for which points are required, takes more than practice, more than time.

In *Amber*, advancement takes conflict.

In other words, without threat, there is no advancement. Likewise, without victory, there is no advancement.

Characters can only advance if they actually overcome some kind of obstacles.

Campaign Advancement.

Using Zelazny's *Chronicles* as a source, the idea here is to refrain from giving points until the players have achieved some major, group-wide goal. Then, when the obstacle has been overcome, or at least is no longer an obstacle, award all the players an identical number of points.

This isn't "fair" in the sense of better players being rewarded no more than weak ones. However, in Amber, better players are going to pound all over the weak ones anyway. The points won't make any difference.

Points, desirable as they may be, do not an Amberite make. After all, it's not a character's Attributes, Powers, and items that determines superiority. It is how they are used. Pit a skilled and experienced Amber player, with a 100 point character, against a raw beginner with 200 points, and the outcome is certain. To paraphrase an old adage, quite appropriate for *Amber*, "age and treachery beat youth and power every time."

How Many Points to Award.

As a rough guideline, it should be *possible* for players to gain a couple of points for every couple of sessions of play. There's no requirement that players ever be rewarded with advancement points.

Characters can spend a lot of time chatting with each other, exercising their role-playing muscles, and generally having a good time. There is absolutely nothing wrong with this. It's just that the player characters aren't facing any threats, and so they won't get any advancement points.

Major, universe-threatening, cataclysmic, really horrible-and-unbelievably-bad threats provide an opportunity for player characters to get ahead.

The bigger the threat, and the more obstacles to overcoming it, the more points. Assume that something is Amber-threatening, a nasty, rotten, and vile force on the verge of wiping out the Pattern. Only clever play, a concerted effort, and acts of courage and determination can save the day. That would be a threat worthy of the player character's efforts, and could be appropriately rewarded.

Take an overview of the Patternfall War, the cumulative events of the *Chronicles of Amber*. A major enemy, the Courts of Chaos, was threatening all of Amber. Inside the family itself was a vicious and powerful traitor. On the other hand, Oberon himself was manipulating things in a helpful manner, and certain forces of the enemy could be swayed into switching sides. All told, if a Game Master were to use points to evaluate the forces allied against Amber, the threat, including Brand, the Lords of Chaos, and the Black Road, would amount to a couple of thousand points. On the other hand, the defenders of Amber weren't exactly cream puffs. I'd estimate the whole affair at being worth somewhere between twenty-five and seventy-five advancement points. Call it fifty.

It's also possible for a threat to be facing just one player character. Solving the problem of a really vicious antagonist, out to make life utterly miserable (or just short) for the character, can also be worth Advancement Points.

I'm hesitant to give an exact formula for figuring out Advancement Points. Like everything in *Amber*, it depends too much on the individual players and Game Masters. Still, a rough estimate would be that about five percent (5%) of the points in the threat could be converted into Advancement Points.

Awarding Advancement Points.

Once the points are calculated, it's time for them to be spent.

Each player gets to choose how they would like to spend their points. They do that by writing down all the improvements and new things they want for their character. This can include Attributes, new Powers, Power advancements, Artifacts, Shadows and Allies. They should also write down how much Good Stuff or Bad Stuff they'd like.

However, there is a catch to this.

The players are not told how many points they're getting. They arrange their wish lists in priority order, starting with the thing they'd most like, and working down to the thing they want least. For each item they should specify whether or not they are willing to take Bad Stuff.

This contributes to the mystery of the game, and makes engaging in player versus player conflicts a tad more interesting, since the players can no longer be certain (at all!) of their own scores.

Describing the Wish Lists. Here's a sample conversation, showing how a Game Master might describe the advancement procedure to the players.

GM: What I need, in order to hand out your points, is a want list from each of you.

Peggy: How many points are we going to get?

GM: I'm afraid that's confidential.

Mick: You're not going to tell us?

GM: That's right.

Peggy: How can we possibly tell you what we want for our characters if we don't know how many points we have to spend?

GM: Just write down everything you want, in priority order. Then, next to each item, let me know whether or not you're willing to take Bad Stuff in order to get that item.

Ted: I don't get it.

GM: Okay, we'll use you and your character Ariel as an example.

Ted: Hey, don't pick on me! Why should you tell anyone else what I've got?

Cindy: Oh, I like this idea.

Willy: Yeah, as I recall Ted didn't bid on anything in the Attribute Auctions.

GM: This is exactly why I'm going to use Ted's character Ariel as the example. He didn't bid, so nobody can be sure of what his Attributes really are, and nobody will know how much of what I say is true.

Ted: I confess, I confess. All my Attributes are Human Rank.

Kevin: Really lame, Ted. We all know better than that.

GM: Okay, let's say, for the sake of argument, that Ted has a Human Rank in Psyche, a Chaos Rank in Strength, an Amber Rank in Endurance, and, ummm...

Beth: That he matched me in Warfare?

GM: Okay, what would that make him?

Beth: He'd have spent seventeen points and he be ranked just below me, under third place.

GM: Sounds perfect. So we've got ranks of Human, Chaos, Amber and three-point-five (3.5), or seventeen points. It's pretty obvious by now that Ted also has Pattern Imprint and Trump Artistry.

Alex: And about a thousand points of Good Stuff!

Ted: Only in comparison to Harick...

GM: Yes, we'll assume that he also has one point of Good Stuff. Now, Ted, how would you spend your advancement points?

Ted: Hypothetically, I'd want to, oh, buy a Shadow. I'd also like to improve my Trump Artistry and buy up some of my Attributes.

GM: Okay, let's put that in priority order. What's most important?

Ted: Human Rank Psyche is really awful, so I guess that would be first. I'd want to take that all the way up to Amber Rank.

GM: Nope, you can't do that.

Ted: Not enough points?

GM: No, I'm not going to tell you anything about how many points you've got to work with. It's just that you can't raise any Attribute more than one Rank. So you could boost your Human Psyche to Chaos, and your Chaos Strength to Amber. But you can't raise your Psyche from Human to Amber Rank. What next?

Ted: Since we're just making this up, I'd want the Shadow.

GM: Psyche improvement first, then Shadow. What next?

Ted: I'd say Trump, followed by, well, I'd want to get Strength up from Chaos Rank.

GM: Okay, sounds good for an example. Putting together a

priority list for Ariel we come up with, (1) Raise in Psyche, (2) New Shadow, (3) Improved Trump Artistry, and (4) Raise in Strength.

Cindy: So he's just get each of those things until the points run out?

GM: That's the idea. Except there's one more spin on it. For each of the items on Ariel's list it's up to Ted to decide if he'd like to go into Bad Stuff.

Kevin: Ich! Bad Stuff! No way!

Ted: Okay, so what if I say I don't want any Bad Stuff?

GM: Then I would just go down the list, give you each thing that you could afford, and stop when you run out of points. Any points you have left over I'd throw into Good Stuff.

Kevin: What's wrong with that?

GM: Well, let's say that I get down to number three on your list, improving Trump. Suppose you needed twenty points to get Advanced Trump, and you only had nineteen points available. If you said you didn't want any Bad Stuff, you wouldn't get Advanced Trump.

Kevin: Ah! Unless I say I want Bad Stuff you'll automatically cut me off, right?

GM: Bingo. So, what you want to do is tell me two more things. First, how much Bad Stuff you are willing to take. Second, for each item on your list, you should put down whether or not you're willing to go into Bad Stuff to get it.

Kevin: So if I say I'm willing to take, say, up to five points of Bad Stuff, you'd give me the Advanced Trump.

GM: It's two separate things. First, you tell me how many total points of Bad Stuff you're willing to take. Second, next to each item you write down whether or not you want it bad enough to take Bad Stuff.

Kevin: Could I write down that I'm willing to take five points of Bad Stuff to get Advanced Trump, but only three points to get a better Psyche.

GM: No. For each item you want, it's just a yes or no, do you want Bad Stuff or not.

Applying Advancement Points.

Once the wish lists are all complete, it's time for the Game Master to apply them to the characters. Working down through the list, each character should get whatever they can afford.

As soon as you get to an item that costs more than the number of points available to the character, stop. Did the player write down a willingness to take Bad Stuff for that item? If so, check to see if giving the item to the character would exceed the player's Bad Stuff limit. If it falls within the player's limits, then that item is added to the character.

The whole process is done item by item, according to the player's priority list. The first time an item can't be awarded, because of a lack of points, and an unwillingness to take Bad Stuff, then stop altogether. None of the items after that on the list should even be considered.

Attribute Rank Progress. Players make a list of Attributes they'd like to see improved. For each Attribute they should mark down whether or not they'd be willing to take Bad Stuff to advance. Advancement must be one Rank at time.

Example of Attribute Rank Progress. Let's take a look at the standings in Strength. From the start of the campaign Garvin has been Chaos Rank, Harick is ranked Amber, Dorell is ranked fourth (9 Points), Ariel bought up to fourth (also 9 Points), Farley is ranked third (12 Points), and Iresa is first (24 Points).

Garvin, for a price of ten points can go up from Chaos Rank to Amber Rank. He can't go any higher this time, but next time, if he likes, he can try to go higher.

Harick can go from Amber to fourth Rank, at a cost of nine (9) points. If Harick makes the advancement, he'll be Ranked "4.5," not fourth. That's Harick's limit for this one advancement.

Dorell and Ariel both have the option of buying up to third Rank, if they put in the extra three (3) points. In either case, they'll go up to "3.5," not third. If Dorell goes up, and Ariel does not, Ariel will still be ranked "4.5," and will not "move up" to third place. Only the original bidder can have the Rank of third.

Peggy's character Iresa can also advance, but she has no idea of where she's going. It's like climbing a ladder in the dark and having no idea of where the next rung is going to be. Here's another example of where choosing Bad Stuff can be really dangerous. If Iresa is only getting, say ten points, and the next Rank is forty points, that means she'd have to increase her current twenty-four (24) point rank by sixteen points, giving her six points of Bad Stuff.

Where the "rungs" are placed beyond the 1st Rank bidder is up to the Game Master. One possibility is to extend the player's Rank "ladder" up into the elder Amberites, so player characters will have to match their Aunts & Uncles.

Another opton is to let the player characters create their own "rungs," with the points they add on to their Attribute bid. This way the players have the choice of making small increases, so advancement is fast, but it's easy for others to follow. On the other hand, very large advancements might be expensive but would make it hard for others to catch up.

New or Advanced Powers.

Players can choose to spend their advancement points on upgrading or adding powers. However it is up to each Game Master to set the rules for how this is to be done. Many Game Masters require that advancement follow the specific rules, so gaining, for example, Trump Artistry, takes forty (40) points, and moving from regular Trump to Advanced Trump Artistry takes an additional twenty (20) points.

Game Masters, especially those who are creating new Powers and more exalted levels, sometimes allow for incremental advancement. In this case, the Game Master might allow the player on the way to Trump Artistry to gain some of the powers individually, for a fraction of the total points. Likewise, going up from basic to advanced could be done a few points at a time.

New or Improved Items.

New items can be bought, or any existing artifacts or creatures (found or Conjured) can be improved, using advancement points to add on additional Qualities or Powers, or to multiply their numbers. It is also possible to add points to a subset of the total number of items. Ffor example, a Power could be added to just one of a group of Named & Numbered creatures. Or a Quality could be improved on a Named & Numbered part of a Horde.

New or Modified Shadows.

Using advancement points, characters can add Shadows to their character's possessions. It's also possible, with the Game Master's consent, to pay to have a found Shadow turned into a personal one. Likewise, points can be used to "upgrade" a character's Shadow, adding barrier levels or the degree of control that one has over the Shadow.

Limitation to Shadow Change. It is impossible to change a Shadow's basic type. A Shadow purchased for one point as a "personal Shadow," can never become a two-point "Shadow of the Realm" or a four-point "Primal Shadow."

Allies & Such.

Players can use advancement points to gain new allies in Amber or Chaos, or even elsewhere. Turning a character that the player has encountered in the campaign into a "point" ally is subject to the Game Master's approval.

Family Friends and Devotees. The identities of a player characters Family Friends and Devotees are still left entirely up to the Game Master. In other words, the player can spend advancement points on a Friend or Devotee, but cannot specify who that character will be. Getting a Devotee with advancement points does *not* mean the player character becomes a blood relative and therefore gains potential admission to the Pattern or Logrus. If the player didn't make this arrangement at the time the character was created only the Game Master's mercy can change matters.

Eliminating Bad Stuff, or Adding Good Stuff.

Players always set their own limits on Good Stuff and Bad Stuff. It's impossible to "buy" both Good Stuff and Bad Stuff, since "Stuff" is a range from negative to positive.

Banking Points.

Players don't have to spend their advancement points. There are several ways they can "bank" points where they can get at them later.

Good Stuff. Here's the most obvious place to stick extra points. Until the next advancement period, the character is just lucky and destined to have good things happen.

Good Stuff for Artifacts, Creatures and Others. It's also possible for the player to stash points into Good Stuff dedicated to other things. Giving Good Stuff to an Artifact, Creature, Shadow or Ally, just makes everything rosy for awhile.

Carolan: An Example of Advancement.

Let's look at one example, Carolan, son of Random.

Carolan was first created by Don Woodward in 1985. Carolan was rewarded at the end of 1986, when the group had rescued Amber. Same thing happened at the end of 1987, when Carolan, along with the rest of the player characters, emerged victorious over another great threat posed to Amber. In both years Carolan was awarded the same number of points as the rest of the players.

By 1988, Carolan had managed to make his Uncle Caine into a dangerous and powerful enemy. In 1986, mistaking him for an enemy of Amber, Carolan had attacked, cutting off Caine's right hand. Caine's vengeance was forestalled through 1987 by virtue of Carolan's other family connections.

Carolan's next mistake caused him to be banished by the Court of Amber, his life forfeit.

Caine was free to pursue his revenge.

At which point, Caine became the major threat to Carolan. A threat, once resolved, that would be worth considerable points.

How many points? That would depend. How powerful is Caine? How fanatical in his pursuit of blood? For the sake of argument, if Caine is fantastically wily, and is a two hundred point character, then ten (10) points would be reasonable.

If Carolan dealt with the threat of Caine, solving his problem, then the character would gain 10 points. As it happened in the campaign, Carolan was not alone, and his companions were likewise in danger from Caine (though not as threatened), so they too might gain points when the Caine problem was resolved.

Resolution can take many forms. In this case, Carolan eventually managed to make peace with Caine (though not without the cost of a great deal of personal pain). If Caine were killed, that would take care of the problem. Except that it might also create a number of other problems, like eternal enmity of other elder Amberites.

DESIGNING SHORT TERM AMBER SESSIONS

Amber can be a lot of fun even in shorter sessions. Here are some guidelines for putting together more generic *Amber* role-playing adventures, not tailored around a specific group of player characters.

TOURNAMENT DESIGN

True *Amber* Tournaments, like any pure tournament form, has to be based on fairness to all the participants. As a result, you can't allow players to bring their campaign characters into a tournament.

Tournament Characters.

Players are not allowed to use their campaign characters in a tournament. Characters should either be pre-generated, written out and handed to the players, or else characters should be created at the beginning of the tournament.

Pre-Generated Characters. Simply put together a batch of characters, figuring out all the Attribute Ranks, Powers, and such in advance. Then, at the start of the session, hand out the character sheets.

One possibility is to use trimmed down, one hundred point, versions of Zelazny's characters. Another way to go is just copying down the sample characters from this book. Even better, the pre-generated characters can be designed specifically for the tournament, such that each character has elements that are crucial to solving the group's threat.

Tournament Attribute Auctions. Things start the same way as a regular campaign, with each player creating their own character. Some shortcuts include a shortened

Bidding War, where not all four of the Attributes are auctioned, suggested allocations of Powers ("we should have at least one Trump Artist, and no Advanced Pattern or Advanced Logrus will be allowed"), and limits to the number of Good Stuff or Bad Stuff available.

Tips for Fair Tournaments.

- If a tournament is based on group cooperation and group victory, then make that plain to the players.

- Each player, and each player character, should have an equal chance to participate. If there is going to be any kind of individual victory (as opposed to group victory) then the character abilities **and opportunities** must be equal.

- Each group should likewise be treated equally. For example, every critical choice presented to one group, should also be presented to every other group.

- Role-playing a character is what *Amber* is all about. If a tournament is going to be based on figuring out tricks, like problems that require a bit of out-of-character analysis and information sharing, then tell the players in advance. They should know that their primary goal is solving the problem, and that role-playing is secondary. That avoids penalizing good role players.

- As much as possible role-playing should be encouraged. Every aspect of a tournament should be analyzed to see whether or not role-playing is rewarded.

- Avoid the temptation of designing a tournament where the result hangs on a single inspired action, or where total failure is the result of a single fatal error. If the resolution hangs on either a single solution, or a single failure, then make sure that there are sufficient clues and signals to the players so they can make an informed choice. Remember that in any group there are bound to be *characters* inclined toward making hasty decisions; don't penalize the *player* for acting in character.

- In tournaments most errors should not be fatal (unless they are clearly forewarned). Errors should result in clear punishments but shouldn't wreck the whole session.

- Is the tournament winnable? If all things are done correctly, will the players win? This isn't quite as simple as it seems. Think through all the possible events of the role-playing adventure and picture what every possible character might do. Each and every *logical* choice, and even every illogical-but-in-character choice must be considered.

- Ideally, each problem in every tournament should be created with two possible solutions in mind. These can include the "right" decision based on the clues that have already been presented, and a "flailing about" solution for those who just try everything until something works. It's also a good idea to play-test any scenario to see if there are even more alternative solutions.

- Any really deadly situation must be avoidable. If the correct actions are taken there need *never* be a situation where a character's life is in danger.

CROSS-OVER SESSIONS

Cross-Overs are a particular kind of *Amber* institution. They are usually scenarios designed around a short, one-day, scenario. What makes them different from either tournaments, or campaigns, is that they allow player characters from different campaigns to role-play together.

The result is an adventure where players can role-play chatting with strangers from different Amber universes.

Cross-Over Mechanism.

The first thing to consider in designing a Cross-Over is how to bring characters in from different Amber universes. Ideally, the problem facing the player characters should be something threatening, so that the player characters will work together, and also somehow linked with the method of their coming together.

Here are a few of the ideas that have been used to justify Cross-Over sessions.

The Unicorn. Able to climb out of the Abyss, somehow connected with Primal Pattern itself, and blessed with supernatural abilities, there's no reason why the Unicorn could not lead characters from different versions of Amber to some common location. In fact, the Unicorn may be a singular entity, existing in all versions of Amber simultaneously.

Something Outside. An outside force or Power, threatening to consume or overwhelm all the versions of Amber, can be a powerful incentive for a Cross-Over. Sometimes this will be detected by powerful elder Amberites, who will then attempt to "save" the player characters by projecting them into the new reality. In other cases, the player characters are simply overwhelmed by the expanding force that is eating their Amber, and find themselves sucked into dealing with the problem.

Way off in the Future. Setting a Cross-Over in the far, far future of Amber is another way of getting various player characters together. This just assumes that some method of crossing-over has already been discovered.

The Jewel of Judgement. Since the Jewel contains a three-dimensional Pattern, it may serve as a gateway to other realms of Amber. In some Cross-Over sessions, King Random has managed to contact his counterparts in other realities, and player characters have been dispatched across the universes to help in emergencies.

Coming Together in Tir-na Nog'th. The dream city that is formed by moonlight over Mount Kolvir is plenty wierd enough for any Cross-Over. As a setting, where the characters simply climb up from their various separate universes, or as a jumping off point for another location, Tir-na Nog'th is a perfect vehicle for a Cross-Over.

The Caverns of Castle Amber. Another possible use for all those mysterious underground tunnels might be switch-overs to various Amber universes. Implementing this might be as simple as getting the player characters lost in the dungeons, and then having them meet each other in some other reality.

Is Chaos unique? There is a theory that there are many Ambers, because there can be many versions of the Pattern, but that there is only one Logrus. After all, the Logrus is not stable, it changes constantly, so how can there be more than one? Therefore, each version of the Courts of Chaos is dependent on the same, ever-changing, Logrus. That singular Logrus could form the bridge from one Amber universe to another.

Mixing Characters with Unknowable Attributes.

One of the problems with running player characters from different campaigns is that the Game Master has to figure out how each player character is put together. This isn't as easy as it may first appear. Since the specifics of advancement points are kept secret from the players, it's hard to figure out where the points belong.

The solution is to do a little questioning.

First, get whatever information the player has on the character. What were their final numbers when the character was first created? What Attributes are Amber or less, or have not changed?

Next, chat with the player about their character. Try to get enough of a verbal picture of the character to estimate which Attributes and Powers are most important. Since the character was created, what Attributes were used most? Least? Which Attributes and Powers did the player ask to have advanced?

Once you've got a rough estimate of what the character ought to be, get more specific in your questions. Here's a little sample:

CROSS-OVER GM: Let's see, you say that your character Garvin started out with first place in Psyche, spending fifty-two points.

Willy: Yeah, but I kept asking to improve my Psyche both of the times we got Advancement Points.

CROSS-OVER GM: Was Psyche your first priority both times?

Willy: Yes, absolutely. I didn't want Kevin's character going over me.

CROSS-OVER GM: Hmm. What about your other Attributes?

Willy: Endurance and Warfare both started at Amber Rank, and Strength at Chaos. I've asked to improve Endurance and Warfare, but never Strength.

CROSS-OVER GM: So Strength is still Chaos Rank?

Willy: Yeah. It's the other three Attributes that are confusing. I just have question marks on my character sheet.

CROSS-OVER GM: Between Endurance and Warfare, which was the highest priority? And did you ask for them both times?

Willy: Definitely Warfare. I know I asked for Warfare increases twice. I'm not sure about Endurance.

CROSS-OVER GM: If Garvin found himself in a hard-fought sword battle would he try to push his enemy with a quick-but-risky attack, or would he try to outlast his opponent?

Willy: That's a hard one. I guess it depends on the enemy...

CROSS-OVER GM: You know nothing about your foe, except you are well-matched and you have other problems.

Willy: I guess Garvin would go for the quick strike.

CROSS-OVER GM: How about if you were in a mind to mind battle, with someone who seemed to have Psyche nearly equal to yours. Would you push it with a possibly dangerous attack, or would you just wear them down?

Willy: Again, I think I'd shoot for the power attack.

CROSS-OVER GM: Thank you, that's all I need to know.

Once the questioning is over, it's up to the Cross-Over Game Master to create a new character summary table. If some things are less than certain, let the player's answers be the deciding factor. For example, Willy choose to keep his Endurance low, compared to his Psyche and Warfare.

Just because a character has a particular Attribute in one place doesn't mean it will necessarily translate into the next. The idea is to figure out how the character's points should be allocated, based on the character's personality. Being "right" is secondary.

Setting Cross-Over Point Limits. One way of making things fair for all the participants in a Cross-Over session is equalizing their points. All the player characters could be "built" from a base of one hundred and twenty-five (125) points. Starting characters can be boosted, and the advanced characters can be trimmed, creating a more balanced session.

GAME MASTER'S ELDER AMBERITE WORKSHEET

Good	Bad	Name	Psyche	Strength	Endurance	Warfare
		Benedict				
		Bleys				
		Brand				
		Caine				
		Corwin				
		Dara				
		Dierdre				
		Dworkin				
		Eric				
		Finndo				
		Fiona				
		Flora				
		Gérard				
		Julian				
		Llewella				
		Martin				
		Merlin				
		Oberon				
		Osric				
		Random				

OBERON

Oberon, Lord of Amber, stood before me in his green and his gold. High, wide, and thick, his beard black and shot with silver, his hair the same. Green rings in gold settings and a blade of golden color. It had once seemed to me that nothing could ever displace the immortal liege of Amber from his throne. What had happened? I still didn't know. But he was gone. How had my father met with his end?

Nine Princes in Amber

Oberon, who has been Lord of Amber for more millennia than anyone has counted, is not an old man. Middle-aged perhaps, with his thick black hair and beard streaked with silver, but not old. He is still powerful, he is still bigger than his children, still wields a golden blade, and his green-and-gold rings are not yet loose upon his fingers.

Chances are, if you're an Amberite, your grandfather will be Oberon. All of you Amberites have him in common. He is father to your parents' generation, all your powerful Aunts and Uncles. Any of them will tell you that Oberon was a very nasty father.

Well, do you think *your* Dad would kill your girlfriend? Oberon would do it. Not because he didn't like her, just because he thought you might be distracted from taking out the garbage, or some other household chore he thought important.

If he thought she was in the way? Bang. She'd be dead.

Oberon is also tough enough to beat up Gérard. Ruthless enough to scare Fiona, and Benedict, and Corwin. Oberon scares the pants off the whole lot. Kind of helps you understand how the relatives turned out the way they did, with a father like that.

They say that he's dead. Don't bet on it. After all, they buried Uncle Caine once. Maybe twice. You've met *him*, and he doesn't look dead.

OBERON'S IMPOSSIBLE DEEDS

Here's a guy who seems to be able to do impossible things. Not only is Corwin impressed, even Fiona can't figure out how he does this stuff.

One trick lets him send people *away*, using something that might feel like a Trump, but maybe not. Maybe it was a trick of the Jewel of Judgement, or maybe not. Another time, Oberon arranged things so that Trump stopped working, to the point where they didn't even feel cool. Maybe it was connected to the Primal Pattern, but maybe not.

Let's look at one series of events. When we first meet Benedict in the *Chronicles*, he is missing an arm, the result of his battle with the forces of the Black Road. Later on Corwin goes to Tir-na Nog'th where he is attacked by a dream-time ghost of Benedict that is wearing a mechanical arm. Corwin whacks off the arm and returns it to the "real" Benedict, who is delighted to add it to his body. Then, when things are really tense, the arm saves the day. Finally, in a mysterious episode, Benedict is attacked by a Shadowy figure who is wielding Corwin's sword, *Greyswandir*. Corwin himself, with *Greyswandir* at his side, actually watches the ghostly figure cut off Benedict's mechanical arm.

In other words, somehow Oberon warped time so that the past-time Corwin would steal the arm from the future-time Benedict, give it to the past-time Benedict, and then watch it being stolen again.

Sure makes my head spin. I mean, where did it come from? From Benedict. Where did it go? To Benedict. We can account for it from time A to time B and back again, but there's no gap for the thing to be created or destroyed. It's an eternal object caught in a time loop.

That's another of Oberon's impossible deeds.

OBERON - FATHER OF AMBER (400 Point Version)

...I do not really know what Dad was. He never encouraged intimacy, though he was not an unkind father. Whenever he took note of us, he was quite lavish with gifts and diversions. But he left our upbringing to various members of his court. He tolerated us, I feel, as occasionally inevitable consequences of passion. Actually, I am quite surprised that the family is not much larger. The thirteen of us, plus two brothers and a sister I knew who were now dead, represent close to fifteen hundred years of parental production. There had been a few others also, of whom I had heard, long before us, who had not survived. Not a tremendous batting average for so lusty a liege, but then none of us had proved excessively fertile either.

The Guns of Avalon

At every turn, according to Corwin, Oberon was superior to his children, beating each at his or her own game. Not just with power, though he wasn't reluctant to show off his power when he had to, but with a sense of guile.

Current Objectives. Well, assuming that he isn't dead, Oberon has definitely relinquished the crown of Amber, and he likely has other plans. Perhaps he's out having fun, playing at being something like Ganelon, out in some Viking Shadow.

ATTRIBUTES
PSYCHE - [55 Points]
STRENGTH - [80 Points]
ENDURANCE - [25 Points]
WARFARE - [35 Points]
POWERS
Advanced Pattern Imprint [75 Points]
Attuned to the Jewel of Judgement [10 Points]
Advanced Shape Shift [65 Points]
Sorcery [15 Points]
"Exalted" Conjuration [40 Points] - Somehow Oberon is able to bring extraordinary objects, things of total unreality, into being. From one point of view, this is nothing but a bookkeeping trick, since, after the conjured item has served its purpose, it is dispelled, and the universal balance is restored.

OBERON - REBEL SON OF CHAOS (500 Point Version)

'Oberon is a son of Chaos, a rebel son of a rebel father. But the power is still there..."

Dara, speaking of Oberon,
The Courts of Chaos

Assume that Oberon was born to Dworkin before the Pattern was created, or even in an intermediate time, before the two of them established Amber. In that case Oberon might very well have been a skilled Lord of Chaos, and that power will not have left him. After all, Dworkin describes Oberon as "...a lord of the living void, a king of chaos..."

Current Objectives. In this case Oberon, having overseen the successful prosecution of the Patternfall War, may well be back in his old haunts in the Courts of Chaos. His ancient enemies now in decline, he will once again seek to take control of that more antiquated throne. His strategy will be slow, involved and convoluted, and there may be many deaths, feuds and disappearances before he'll reveal his own hand in affairs.

ATTRIBUTES
PSYCHE - [15 Points]
STRENGTH - [30 Points]
ENDURANCE - [30 Points]
WARFARE - [145 Points]

POWERS
Advanced Pattern Imprint [75 Points]
Attuned to the Jewel of Judgement [10 Points]
Advanced Logrus Mastery [70 Points]
"Exalted" Shape Shift [85 Points] - Oberon has carried the ability of Shape Shifting to the point where he can be of

two minds, assuming the personality of who he would imitate, while keeping his own thoughts and secrets separate. He can perfectly imitate any Amberite, or any Lord of Chaos, so long as he has the memory of a Psychic contact with them.

Power Words [10 Points]
Sorcery [15 Points]
Conjuration [20 Points]

BAD STUFF
[+5 Points]

OBERON - WIELDER OF THE JEWEL OF JUDGEMENT (600 Point Version)

'You are not above murdering innocent people to manipulate me. Yet you would sacrifice your life for the kingdom.'

Corwin to Oberon,
The Courts of Chaos

A vicious and nasty player in the political game, this version of Oberon has no qualms about any action. For example, he was willing to set up Lorraine as a pawn, to kill her when she endangered his plans, and to stand by as Corwin hunted down and killed an innocent man in vengeance for her death.

Current Objectives. His plotting will not have stopped, and his death will be a ruse, designed to give his enemies a chance to expose themselves. In the Court he may be anyone, from Random to Corwin, freely using his power to root out any threat to Amber.

ATTRIBUTES
PSYCHE - [80 Points]
STRENGTH - [125 Points]
ENDURANCE - [50 Points]
WARFARE - [100 Points]

POWERS
Advanced Pattern Imprint [75 Points]
Master of the Jewel of Judgement [25 Points] - Oberon's "bauble" gives him the power to remake and mold the Pattern at will, editing the universe to suit his own design. That the Jewel may be attuned, or wielded, by someone else, no matter how far removed, is of no importance. Oberon's link with the Jewel is absolute, continuous and without limit.
Logrus Mastery [45 Points]
Advanced Shape Shifting [65 Points]
Sorcery [15 Points]
Conjuration [20 Points]

GM TIPS FOR PLAYING OBERON

Oberon is just about the pushiest character imaginable. When he reveals himself in his true identity he becomes a true King. Arrogant, demanding, and haughty. He'll order player characters around as if they were slaves, without regard to their wishes. Also, listening is not one of his strong points.

OBERON AS FATHER

To listen to his kids talk, especially Corwin, you'd think the guy was the nastiest, rottenest, most hard-hearted father ever.
Could he really be that bad?
He was probably worse. So inflicting him on a player character should probably only be done as punishment.

Good Guy Version: Sometimes parents are a lot better the second time around. Having been too harsh, or not attentive enough to the first batch of kids, Oberon could end up being a very fine father to the children he has in his dotage.

Distant, Remote, Unknown Version: Oberon, with his rather fickle marital relationships, can be a temperamental, and often absent, father. In this kind of situation the mother often turns her attention to the child, focusing her love and attention on a loyal son or daughter. The result may be that the player character has a rosy view of their mother, and an impression of Oberon wearing a demon's face.

Manipulator: It seems that Oberon is always looking at his children in terms to how they can best serve him. Like a doctor seeing another doctor in his son, so Oberon sees a potential monarch in any offspring, and he'll push, push, push, to get the kid to turn out that way. This is a good possibility for a player who describes their character as a real rebel.

Bad Guy Version: Oberon can be very cruel, turning away from a child, even denying the fact that he is the father. Even, in a mythical sense, you can imagine somebody like Oberon having all the babies of the right age killed, just to hunt down and destroy the woman and child that he sees as usurpers. In this version, Oberon could be the hunter, out to find and exterminate his young. The player character should have some other guardian, someone with Power, who can provide protection and give the character the opportunity to find their heritage.

OBERON AS A PLAYER IN THE GAMES OF THE AMBER COURT

If Oberon is still around somewhere, it's hard to imagine him resisting the temptation to meddle in the great game.
On the other hand, if he is alive, he went to a great deal of trouble to disappear. So, maybe, just maybe, Oberon is off on a beach somewhere, wearing nothing but flip-flops, a hawaiian shirt, and bermuda shorts, and grooming a beer belly. There he may sit right now, sipping Mai-Tai's under a palm tree, and writing his memoirs.
Why not?

Constant Allies & Enemies: As far as we can tell, Oberon's only ally was his Dad, Dworkin. He really doesn't have any allies among his children, but it seems likely that he would trust Corwin to help him if he needed it.

Oberon the Meddler: He's supposed to be gone. Oberon even said it himself, "No man can have everything he wants the way that he wants it."

However, the temptation may be just too much. Oberon may walk right back into the game, seizing pawns and playing for power and position.

Oberon as Guardian Angel: The last thing heard from Oberon were his parting words, spoken right before he set out to repair the Pattern. "My children, I cannot say that I am entirely pleased with you, but I suppose this works both ways. Let it be. I leave you with my blessing, which is more than a formality...." Aside from the general dispensing of Good Stuff on Amber, this may also mean that Oberon intends to act in Amber's best interest. In which case, he may appear from time to time, when things look particularly bad, to rescue the player characters.

GANELON: EITHER A SHADOW, OR OBERON IN DISGUISE

He wore a brown leather jacket over a black shirt, and his trousers were also black. They were bloused over the tops of his dark boots. He had about his waist a wide belt which held a hoof-hilted dagger. A short sword lay on the table before him. His hair and beard were red, with a sprinkling of white. His eyes were dark as ebony.

The Guns of Avalon

Ganelon alive?

How could that be? Oberon told Corwin of Ganelon's death. How could Oberon have lied, especially then, when he was just reconciling with his beloved son?

So he lied. He's Oberon, that's what he does.

After all, Ganelon is really useful to have around. While Oberon acted as Ganelon, the real Ganelon would have allowed him to be in two places at once.

Or perhaps just one Shadow of Ganelon is dead. In which case, in infinite Shadow, there are certainly more.

'How difficult was it for you, being Ganelon?'

He chuckled.

'Not hard at all,' he said. 'You may have had a glimpse of the real me.'

'I liked him. Or rather, you being him. I wonder whatever became of the real Ganelon?'

'Long dead, Corwin. I met him after you had exiled him from Avalon, long ago. He wasn't a bad chap. Wouldn't have trusted him worth a damn, but then I never trust anyone I don't have to.'

'It runs in the family.'

'I regretted having to kill him. Not that he gave me much choice. All this was very long ago, but I remembered him clearly, so he must have impressed me.'

The Courts of Chaos

Or maybe, just maybe, there never was a Ganelon. Maybe Oberon was playing at Ganelon all the time. Sure, it's twisted, but again, that's Oberon. He does that. He may even do it again.

ATTRIBUTES
PSYCHE - Chaos Rank
STRENGTH - Amber Rank
ENDURANCE - Chaos Rank
WARFARE - [5 Points]

POWERS
Shadow Path. [2 Points] - Capable of following others through Shadow, or of following a known pathway.

BAD STUFF [7 Points]

BENEDICT

Then there was Benedict, tall and dour, thin; thin of body, thin of face, wide of mind. He wore orange and yellow and brown and reminded me of haystacks and pumpkins and scarecrows and the Legend of Sleepy Hollow. He had a long, strong jaw and hazel eyes and brown hair that never curled. He stood beside a tan horse and leaned upon a lance about which was twined a rope of flowers. He seldom laughed.

Nine Princes in Amber

Benedict is a timeless warrior. He spends eons in Shadow, perfecting his craft, to the point where he is no longer challenged by his siblings. A lean scarecrow of a man, with few words, and fewer smiles, he does not give his trust lightly. Nor does he lightly regard the betrayal of Amber.

He has a reputation as the best warrior in the family. Not just with a sword. With anything. Give him any weapon, any group of fighting men, any tactical or strategic situation, anything.

He'll beat the snot out of you.

Let's say the guy is obsessed. For yuks Benedict fights wars. He'll find a place, somewhere out in Shadow where they're fighting a war. He joins up on the losing side, and fights until he wins. Then he finds a place, just like the first place, only where the losers have a smaller army, and he joins up again, wins again. And he'll repeat the process with a tactical disadvantage, bad terrain, different weather, varying weapons and technology. Benedict will try every possible combination.

Face him with a sword and he'll attack until you are beaten, giving no quarter, no edge, and never, ever making a single error. Take an army against him, and he'll appear at your weakest front, at your weakest moment, attacking relentlessly, until your soldiers scream and run, and are broken for all time.

He'll win every time.

Don't bother betting on whether Benedict will win or lose. The only interesting question, the only safe bet, is how fast.

BENEDICT - IDEAL WARRIOR (300 Point Version)

Benedict is devoted to the pursuit of his only true love, Warfare. Having virtually no other interests, Benedict has shaped himself into the most fearsome opponent possible.

Current Objectives. What he wants is straightforward and consistent with his past. He'll continue perfecting his every skill, trait and instinct.

Having sustained a less-than-perfect record against the forces of Chaos, he will be devoting the next phase of his life to learning their every strength and weakness. It's likely that he'll spend much of his time in far Shadows, those bordering the Courts of Chaos, where he can experiment with opponents similar to those he faced in the Patternfall War.

Like a good soldier, Benedict is loyal to his new King. If he is called, he will answer. Benedict, if he sees Amber in danger, will post himself wherever he feels he is most needed.

ATTRIBUTES
PSYCHE - [15 Points]
STRENGTH - [40 Points]
ENDURANCE - [45 Points]
WARFARE - [135 Points]

POWERS
Pattern Imprint [50 Points]
Power Words [20 Points]
> Has a total of fifteen Power Words, including Magic Negation, Chaos Negation, Psychic Defense, Psychic Disrupt, Neural Disrupt, Lifeforce, Resume True Form, Pain Attack, Trump Disrupt, Process Surge, Process Snuff, Shade, Light Strobe, Spark, and Weaken Structure.

BAD STUFF
[+5 Points]

BENEDICT - DEFENDER OF AMBER (400 Point Version)

In this version, it is assumed that Benedict is most obsessed with keeping Amber safe and strong. As always, he is personally fit, but he has also expended some of his energy gathering together a cadre of agents.

Current Objectives. Having sacrificed so much to the preservation of Amber, he'll not forsake her now. Standing by the new King, Benedict will take over responsibilities for coordinating Amber's defenses. Anyone threatening Amber will face Benedict's wrath directly. Benedict will also recruit even more agents, mobilizing them to patrol throughout Shadow...

ATTRIBUTES
PSYCHE - [35 Points]
STRENGTH - [15 Points]
ENDURANCE - [15 Points]
WARFARE - [150 Points]

POWERS
Advanced Pattern Imprint [75 Points]
Power Words [10 Points] - Magic Negation, Chaos Negation, Psychic Defense, Neural Disrupt, and Weaken Structure.
Sorcery [15 Points]
Conjuration [20 Points]

CREATURES AND ARTIFACTS OF POWER
See *"Agents of BENEDICT"* [65 Points]

BENEDICT - SON OF CHAOS (450 Point Version)

What if Benedict had actually been born in the Courts of Chaos, as part of Oberon's first family? If so, then Benedict, along with his full brothers Osric and Finndo, would be a true Lord of Chaos, as well as being a Prince of Amber.

Current Objectives. Now, with the ending of the Patternfall War, and an opening of diplomatic relations between the Courts of Amber and Chaos, Benedict may become Amber's foremost ambassador. He will set up an estate in the Courts of Chaos, and, from there, train and equip a new army. His goal will no more be the throne of Chaos than it ever was that of Amber, but he will prepare himself to defend his friends.

ATTRIBUTES
PSYCHE -   [15 Points]
STRENGTH - [10 Points]
ENDURANCE - [25 Points]
WARFARE - [200 Points]

POWERS
Pattern Imprint [50 Points]
Logrus Mastery [45 Points]
Shape Shift [35 Points]
Power Words [10 Points] - Magic Negation, Chaos Negation, Psychic Defense, Neural Disrupt, and Weaken Structure.
Sorcery [15 Points]

CREATURES AND ARTIFACTS OF POWER
Benedict's Lance [10 Points] - A creature of Chaos, molded by Benedict to take the form of various weapons. It's name is "Iisvigg" and it yearns to be set loose. However, Benedict will not part with it until he fulfills some ancient vow concerning a matter of honor in the Courts of Chaos.
> • Double Damage [2 Points]
> • Regeneration [4 Points]
> • Limited Shape Shift [4 Points] - Iisvigg is capable of taking the form of dozens of weapons. It is most effective, as either a lance, a sword, or a dagger.

Benedict's Tan Horse [10 Points] - Lately Benedict has been seen on a steed with black and red stripes, and with a red mane and tail. This is the same tan horse that appears on Benedict's Trump. In reality a creature of Chaos, the horse is capable of changing its appearance.
• Tireless, Supernatural Stamina [4 Points]
    • Resistant to Firearms [2 Points] - In addition, Benedict's horse is resistant to most physical distress, including heat, fire, cold and poison.
    • Psychic Resistance [1 Point]
• Shadow Path Movement [2 Points]
    • Alternate Shaping [1 Point] - While confined to a horse-like form, the creature can shift its appearance, including such things as hair color, hair pattern, length of hair, and the color of the eyes.

Game Master Note: Regardless of which version of Benedict you choose for your campaign, the Warfare of that version should be higher than that of any other Amberite. Either select versions of the other characters so they are consistently below Benedict, or modify Benedict's Warfare so it remains supreme.

GM TIPS FOR PLAYING BENEDICT

And Benedict, the gods know you grow wiser as time burns its way to entropy, yet you still neglect single examples of the species in your knowledge of people. Perhaps I'll see you smile now this battle's done. Rest, warrior.

The Courts of Chaos

The basic guideline for playing Benedict is simple.
Shut up.
When Benedict has nothing to contribute, he will say nothing. Even if he has something to say, he'll have to be coaxed into saying it.

His favorite answers are "perhaps," "no," and not answering at all. "Yes" is something that Benedict will say only with reluctance. As a Game Master, you might even want to cultivate a distinctive grunt.

When he's got something meatier to communicate, you can depend on him to say it in as few words as possible.

When Benedict does speak, he can really be annoying. It's not that he tries to be insulting, it's just that, from his point of view, everybody around him seems a little dense.

For example, Corwin once managed to get lost on Kolvir. Benedict, hearing this story, replied, "...You keep heading east, you know. That is the direction from which the sun has been known to take its course."

Benedict also can be a bit cranky about a couple of other things. Here are some examples.

Including Benedict in a Group. Saying things like "everybody" and "all Amberites" is irritating to Benedict. He doesn't like to be lumped in with anyone else. This might be part of his general contempt, or simply a desire, as the eldest in the family, to not be grouped in with the other kids.

Discussions of the Royal Succession.

'...the old arguments were resumed as to who was more legitimate. Of course, you and Eric are both my elders, but while Faiella, mother to Eric and myself, was his wife after the death of Clymnea, they-'

'Enough!' cried Benedict, slapping the table so hard that it cracked.

The lamp danced and sputtered, but by some small miracle was not upset. The tent's entrance flap was immediately pushed aside and a concerned guard peered in. Benedict glanced at him and he withdrew.

'I do not wish to sit in on our respective bastardy proceedings,' Benedict said softly. 'That obscene pastime was one of the reasons I initially absented myself from felicity. Please continue your story without the benefit of footnotes...'

The Guns of Avalon

Benedict has an aversion for discussions of his family's gossip, especially as it relates to his father's affairs. This can easily slide into violence.

Endangering Benedict's Retainers. Benedict does not view his soldiers, servants or Shadow friends as expendable. He will fiercely protect them from his relatives, and will kill to avenge any murder.

BENEDICT AS FATHER

Benedict as father tends toward the extremes. He can be very, very good, approaching the job with his typical single-minded dedication. On the other hand, Benedict isn't exactly a "people person" and he can easily be one of those fathers uncomfortable with all the strong emotions connected to the job.

Family-Oriented Benedict: Loyal, to the point of an absolute, Benedict will fiercely protect his children. He'll arrange that they be raised in a near-perfect environment. Probably a prosperous place in Shadow, in a strong manor house with Benedict as master. To a perceptive child, Benedict, for all his silence and outward calm, will obviously be loving and affectionate.

His daughters may grow up adoring Benedict, since he is willing to let them follow their own path. He'll lead his girls along the path to Warfare, if that is what they want. Or, if they desire another life, he'll arrange for another of his siblings to act as tutor.

To his sons he will endow a life-long training in the arts of combat. There will be no weapon, no tactic, no military force that they will not have experienced. Obviously, any player character assigned to this version of Benedict will have to have a high Warfare Attribute.

Distant, Remote, Menacing Benedict: Benedict is an absent father, and his business carries him away for long periods of time. In fact, if he does not expect a child (and what Amberite does?), he may be gone for many, if not all, of the formative years of the player character's life. In this case, the player character will know relatively little about Benedict and they will have a strained, cold relationship.

Benedict as Taskmaster: When you're the best there is, just how would you react to a child who isn't even in the running? Or a kid who has the potential, but who is too lazy to exploit it?

For boys having the perfect coach, the all-time world champion in his field, for a father can be less than pleasant. Benedict will push them to work harder, and to fight better. A particularly nasty touch would be to have a player character grow up, never having heard a single word of praise from Benedict. Player characters in this role should have relatively low Warfare Ranks.

Females with this version of Benedict as father may feel severely left out. Benedict not only emphasizes Warfare, but he deliberately leaves girls out, insisting that they pursue more traditional roles.

Quiet & Hostile Version: Cold as ice, silent as the grave, and an agent of death. Benedict could care less about anyone, or anything, because he's just so damn relentless. Nothing matters to him except his pursuit of the absolute, the excellence that is within him.

He didn't get to be the best warrior in the universe without selling a piece of his soul. Or, from the point of view of a child, maybe he sold the whole thing.

His children will see him as an ominous threat, the figure that scared their mother and every other living person around. As adults they'll be defiant, and perhaps even vengeful.

BENEDICT AS A PLAYER IN THE GAMES OF THE AMBER COURT

...Having served under Napoleon, Lee, and MacArthur, I appreciated the tactician as well as the strategist. Benedict was both, and he was the best I had ever known...

The Guns of Avalon

Benedict has absented himself from the Court of Amber for quite some time. Now that he's back, he's very hard to ignore. He doesn't seem to have any taste for the Throne, but he expects instant obedience from everyone around him. It's hard to see how he'll manage to remain civil towards young King Random, especially if Random fails to be polite.

Changing Benedict's opinion, or getting Benedict to make a snap decision, is just about impossible.

Sometimes a wonderful role for Benedict is as an obstacle. The player characters will know that Benedict can fix the whole situation, if only they can just get him moving.

Since Benedict is, tactically and strategically, about ten moves ahead of everybody, he tends to fall into his "officer" and "leader of men" role. Which means, in effect, "never explain spit," and "bash the hell out of anybody who gets out of line."

And, Benedict always knows what constitutes an emergency. There's no point in arguing with him.

"Take this sword." Benedict might command. "Go to the north road. Kill everyone who tries to pass."

See, it's quite clear. He wouldn't expect any questions, and if there were any exceptions, he would have said something else.

Constant Allies & Enemies: Benedict doesn't seem to have any particular enemies among his siblings. Fear, rather than hatred, seems to be the way he is usually regarded. His main contact, his first friend, would seem to be Gérard. Julian and Llewella always stay on Benedict's side of things. Corwin is a great admirer of Benedict, but not necessarily his ally.

Benedict as Victim: A handy role for Benedict is that of a victim. Not that he's particularly vulnerable, but any true enemy of Amber will likely want to neutralize the threat that Benedict poses. Take the throne of Amber if you can. If you hold it unjustly, Benedict has the ability to raise an army, invade, and wrest Amber back from any usurper.

Bad Benedict: To quote Corwin, "Benedict could be as tough and mean and nasty as any of us when he wanted to be. Tougher, even."

Once Benedict has determined you to be an enemy, pretty much all that's left is for you to pick out your grave site. Funeral arrangements will nearly always be open coffin, since Benedict leaves a very neat corpse.

Agents of BENEDICT

Identified only by their secret symbol, Benedict's yellow-and-brown flower, his agents are found throughout infinite Shadow. Each agent is "built" of twenty points based on the following possibilities:

ATTRIBUTES

Agent Psyche
Human Rank [0 Points]
Chaos Rank [4 Points]
Amber Rank [8 Points]

Agent Strength
Chaos Rank [0 Points]
Amber Rank [4 Points]

Agent Stamina
Chaos Rank [0 Points]
Amber Rank [4 Points]

Agent Warfare
Chaos Rank [0 Points]
Amber Rank [4 Points]

Low Ranked [8 Points] - The Agent will be ranked somewhere between the player characters' second and third places. A highly trained warrior, though limited in the range of weapons that can be used with this degree of skill.

Mid-Ranked [12 Points] - Equally matched, or superior, to the greatest of the player characters. Each Agent of this Warfare is regularly and personally trained by Benedict himself. Agents at this level are master strategists and will be difficult to outwit.

Top-Ranked [16 Points] - One of Benedict's most favored Agents. The Agent is close to a match for Benedict himself, can hold his own with Amberites like Bleys and Corwin. Will likely be clearly superior to player characters. There are no more than two Agents of this caliber. As with Benedict, these prime Agents are usually a step or two ahead of any competition.

POWERS & ABILITIES

NOTE: In order to use any of these Powers, the Agent will have to possess an Attribute Rank of Chaos or better in Psyche.

Agent Shadow Walking [3 Points] - Allows one of Benedict's Agents to walk through Shadow, towards an objective. The Agent is also very adept at following anyone else moving through Shadow.

Agent Shadow Manipulation [3 Points] - The Agent can shift aspects of the Shadow, both large and small. Shifting tiny details to make passage easier can be done quickly, in a few minutes. Shifting living things and the Shadow environment can take hours.

Trump Facility [2 Points] - An Agent must have a Psyche of Chaos Rank or better in order to use Trump Facility. With this skill the Agent can passively commune with a Trump, listening in on the subject's Trump calls and even some spoken conversation. It also allows the Agent to sense the proximity and location of the subject of the card.

Agent Power Words [1 Point each] - Most agents have at least one Power Word, which will be identical to one of Benedict's Power Words. Magic Negation, Chaos Negation, Psychic Defense, Neural Disrupt, and Weaken Structure.

Limited Sorcery [6 Points] - They have no way of creating new spells, and can only cast those provided to them by Benedict. Likewise they cannot store or rack spells, and are limited to casting spells the long way.

AGENT ARTIFACTS & CREATURES

Suit of Armor [0 Points] - Any of the Agents will have a standard suit of armor. Includes breastplate, arm and leg pieces, undersuit of chainmail, gauntlets, heavy boots, and helmet.

Activated Mail Armor [2 Points] - The armor is impervious to non-magical edged and pointed weapons. Purely magical attacks will be deflected and blunted. Bullets and other high-impact attacks will be resisted. Appears as an armored suit with both plate overpieces and chain undergarments.

Technical Bullet-Proof Armor [3 Points] - A fabric suit lined with bullet-proof material. Looks like a padded uniform, with a bulky vest and a helmet with a clear plastic pull-down facepiece. Colored either dark brown or with camouflage pattern.

High-Tech Energized Armor [4 Points] - The basic garment looks like a suit of long underwear, with attached gloves and socks, made of a silky material. When attached to a power source, the armor will deflect most any energy or high speed projectile attack. Typically worn under other garments or armor.

Extra Hard Edged Weapon [0 Points] - All of Benedict's Agents will be armed with a sword, or comparable weapon, of very high quality.

Double Damage Weapon [1 Point] - Usually a sword, but can be either another kind of hand to hand style weapon (axe, hammer, etc.), or a selection (Named & Numbered) of missiles for a bow or crossbow.

Deadly Damage Weapon [2 Points] - Generally a sword or other edged weapon that can be used for penetrating magical armor (all the way through "Invulnerable to Conventional Weapons"). The damage inflicted will be severe.

Destructive Force Weapon [3 Points] - The Agent has a special, custom-built weapon, capable of delivering damage equivalent to modern weapons or magical forces. Very often this takes the form of a modern sniper rifle, specially designed to work with a variety of ammunition, in a variety of Shadow environments.

Sample Agent Team. The following three Agents have been assembled into a powerful hunter-killer team by Benedict:

"BULL"

The leader of this three-man team, Bull's specialty is pure Warfare and combat. His Warfare Attribute will match or better that of most player characters. Bull is the contact man, and generally carries a Trump for Benedict. He also carries Trump of any surveillance targets the team is assigned to follow.

ATTRIBUTES
Chaos Rank Psyche [4 Points]
Chaos Rank Strength [0 Points]
Chaos Rank Stamina [0 Points]
Mid-Ranked Warfare [12 Points]

POWERS & ABILITIES
Trump Facility [2 Points]
Agent Power Word [1 Point] - "Chaos Negation"

AGENT ARTIFACTS & CREATURES
Double Damage Weapon - Sword [1 Point]

"LORAK"

This agent works as part of the team, but is very often the focus of their work. Lorak is a trained assassin and sniper. He can use either modern rifles, or a finely crafted crossbow, to deliver a deadly blow from some distance. Lorak also provides the team with transportation through Shadow.

ATTRIBUTES
Amber Rank Psyche [8 Points]
Chaos Rank Strength [0 Points]
Chaos Rank Stamina [0 Points]
Chaos Rank Warfare [0 Points]

POWERS & ABILITIES
Agent Shadow Walking [3 Points]
Agent Shadow Manipulation [3 Points]

AGENT ARTIFACTS & CREATURES
Technical Bullet-Proof Armor [3 Points]
Extra-Hard Crossbow Bolts [0 Points]
Sniper Rifle [3 Points]

"HANK"

Hank is the muscle of the group, but also their main liaison throughout Shadow. He has a natural affinity for people, and can make himself welcome just about anywhere. He is also very good in picking up new languages.

ATTRIBUTES
Human Rank Psyche [0 Points]
Amber Rank Strength [4 Points]
Amber Rank Stamina [4 Points]
Low Ranked Warfare [8 Points]

AGENT ARTIFACTS & CREATURES
Activated Mail Armor [2 Points]
Deadly Damage Sword [2 Points]

BLEYS

Then came a fiery bearded, flame-crowned man, dressed all in red and orange, mainly of silk stuff, and he held a sword in his right hand and a glass of wine in his left, and the devil himself danced behind his eyes, as blue as Flora's, or Eric's. His chin was slight, but the beard covered it. His sword was inlaid with an elaborate filigree of a golden color. He wore two huge rings on his right hand and one on his left: an emerald, a ruby, and a sapphire, respectively.

Nine Princes in Amber

Called valiant, exuberant and rash by his brother Corwin, who knows a thing or two. It's impossible to put a label on Bleys, he's larger than life, with flaming red hair and beard, with a demon flash in those blue eyes.

He is the master swordsman, who, but for the suicide leap of a dead man, and an ill-placed pool of blood, might have fought his way up the stairs of Kolvir into Amber itself. As it was, he killed hundreds of warriors, single-handed, in a fight that has yet to be equalled.

Don't let those blue eyes blind you. Don't be distracted by the glitter of his blade. Bleys spent time at the knee of Dworkin, wielding his famous charm to gain the secrets that shape Pattern, Trump, and all other things of importance and power.

Wizard? Warrior? Master Manipulator? Tactician? Bleys may be all those things. He will charm you, beguile you, convince you to help him in grand pursuits, persuade you of the rightness of his cause, and give you the crown on a platter.

But, just before you go reaching for that bejeweled cap, be sure you see exactly where Bleys might be keeping his other hand.

In fact, you should suspect even being in the presence of Bleys. Do you think you meet by chance? It's unlikely, since mere chance is too small a tool for Bleys. He'd rather manipulate Shadow, drawing in the royalty of Amber to his whim, at his convenience, to serve his own purpose.

The man is also a galling egotist. For example, he treats his older brother Corwin as if Corwin were the younger, were the child. He never shows a flicker of doubt of his own superiority. Yet he is unflinching in his praise. He simply does not believe that anyone is truly his better, so he has envy for no one, and a glad-hearted, wide smile for all.

A sweeping bow for the ladies, a shared joke with the men. When need be, a master of sword, or spell. It might be that Bleys is simply second best at everything.

BLEYS -
THE CORWIN VERSION
(300 Point Version)

He was good, even better than I remembered him to be. He advanced like a whirlwind, and his blade was alive with light. They fell before it—how they fell, my friend! Whatever else you might say of Bleys, on that day he acquitted himself as became his rank. I wondered how long he could keep going.

He'd a dagger in his left hand, which he used with brutal efficiency whenever he could manage corps à corps. He left it in the throat of his eleventh victim.

I could see no end to the column which opposed us. I decided that it must stretch all the way to the landing at the top. I hoped my turn wouldn't come. I almost believed it.

Three more men plummeted past me and we came to a small landing and a turn. He cleared the landing and began the ascent. For half an hour I watched him, and they died and they died. I could hear the murmurs of awe from the men behind me. I almost thought he could make it to the top.

He used every trick available. He baffled blades and eyes with his cloak. He tripped the warriors. He seized wrists and twisted, with his full strength.

We made it to another landing. There was some blood on his sleeve by then, but he smiled constantly, and the warriors behind the warriors he killed were ashen. This helped him, too. And perhaps the fact that I stood ready to fill the gap also contributed to their fears and so slowed them, worked on their nerves. They'd heard of the naval engagement, I later learned.

Bleys worked his way to the next landing, cleared it, turned again, began to ascend. I hadn't thought he could make it that far, then. I didn't think I could make it as far as he had. It was the most phenomenal display of swordsmanship and endurance I'd seen since Benedict had held the pass above Arden against the Moonriders out of Ghenesh.

Nine Princes in Amber

Corwin sees Bleys as talented, vibrant, and hot-tempered, but also brilliant. Definitely not an inferior in any way, but not necessarily superior. In Corwin's eyes, Bleys is the closest to himself. Yet, as the story progresses, Corwin must come to realize that Bleys has powers, ability and knowledge beyond his own.

Current Objectives. Bleys will be pretty bored by the current situation in the Court of Amber. He'll likely seek entertainment elsewhere, out in Shadow. He'll start a new plot, raise a new army, and start his arcane research anew.

ATTRIBUTES
PSYCHE -   [15 Points]
STRENGTH - [25 Points]
ENDURANCE - [25 Points]
WARFARE - [87 Points]

POWERS
Advanced Pattern Imprint [75 Points]
Trump Artistry [40 Points]
Power Words [10 Points] - Resume True Form, Defensive Luck, Pattern Negation, Trump Disrupt and Burst of Magic.
Sorcery [15 Points]

CREATURES AND ARTIFACTS OF POWER
Bleys' Golden Pattern Sword [8 Points] - The next step above a Deadly Damage Weapon, incorporating a bit of the Pattern into the blade so that it becomes destructive to creatures of Chaos.
• Destructive Force [8 Points]

BLEYS -
MASTER MANIPULATOR
(400 Point Version)

...as a tactician, I had always thought him brilliant; and when he laid out the maps of Amber and the outlying country which he himself had drawn and when he explained the tactics to be employed therein, I knew that he was a prince of Amber, almost matchless in his guile.

Nine Princes in Amber

There's no doubt, in light of later events, that Corwin was manipulated shamelessly by Bleys. And, as Bleys fit neatly into a cabal with Fiona and Brand, all students of Dworkin, it would seem he acquired a great many arcane talents and tricks.

Current Objectives. Having been far too long away from Amber, Bleys may seek a high position in Random's new Court of Amber. How long this would last is another question. Bleys is easily bored.

ATTRIBUTES
PSYCHE -   [45 Points]
STRENGTH - [15 Points]
ENDURANCE - [10 Points]
WARFARE - [115 Points]

POWERS

Advanced Pattern Imprint [75 Points]
Advanced Trump Artistry [60 Points]
Power Words [10 Points] - Chaos Negation, Lifeforce, Defensive Luck, Shade, and Spark.
Sorcery [15 Points]
Conjuration [20 Points]

CREATURES AND ARTIFACTS OF POWER

Bleys' Golden Pattern Sword [16 Points] - Like its companion blade, *Greyswandir*, it is like a piece of the Pattern made into metal. A touch of contact will set the blood of Chaos dwellers on fire. It is also deadly against supernatural creatures and inflicts wounds that may not heal with Regeneration or Shape Shifting.
• Primal Damage to Creatures of Chaos [16 Points]

The Psychic Rings of Bleys [6 Points] - All the rings are specially "linked" so that they are in continual Psychic communication with each other, no matter how distant. Only a Shadow Barrier against Psyche can sever a ring's connection with the others. Fiona keeps one of the rings. Others may be given out by Bleys as gifts or tools, either for communication, or as spy devices.
• "Linked" Remote Psychic Sensitivity [3 Points]
• Named & Numbered [*2 Points]

Bleys' Emerald Ring [4 Points] - On his right forefinger Bleys wears the primary ring of the "link." This one is intelligent and is capable of monitoring the activities and sensations of all the other rings.
• Able to Speak in Tongues and Voices [4 Points]

Bleys' Ruby Ring [7 Points] Bleys wears this one on his right ring finger. It is also part of the "link" but also confers a magical protection on the wearer.
• Resistant to Firearms [2 Points]
• Confer Quality [5 Points]

Bleys' Sapphire Ring [2 Points] - On Bleys' left pinky finger is another of his "links." This one can also be used to Sense Danger.
• Sensitivity to Danger [2 Points]

BLEYS - DWORKIN'S APPRENTICE (450 Point Version)

Discovering that Dworkin is the key to the ultimate domination of Amber, Bleys has enrolled as a permanent student. He is continually looking for ways of expanding his knowledge.

Current Objectives. Once again, Bleys will seek Dworkin's advice and counsel. Working with Fiona he will again seek to stretch his knowledge of the universe and its workings.

ATTRIBUTES

PSYCHE -   [35 Points]
STRENGTH - [35 Points]
ENDURANCE - [35 Points]
WARFARE - [95 Points]

POWERS

Advanced Pattern Imprint [75 Points]
Advanced Trump Artistry [60 Points]
Shape Shift [35 Points]
Sorcery [15 Points]
Conjuration [20 Points]

CREATURES AND ARTIFACTS OF POWER

Bleys' Golden Pattern Sword [16 Points] - Like its companion blade, *Greyswandir*, it is like a piece of the Pattern made into metal. A touch of contact will set the blood of Chaos dwellers on fire. It is also deadly against supernatural creatures and inflicts wounds that may not heal with Regeneration or Shape Shifting.
• Primal Damage to Creatures of Chaos [16 Points]

Bleys' Emerald Ring [14 Points] - On the right forefinger Bleys wears a ring that allows him to magically regenerate from any of his wounds.
• Regeneration [4 Points]
• Confer Power [10 Points]

Bleys' Ruby Ring [3 Points] - On his right ring finger he wears a ring that he can use to inflict extra damage in a punch, or if he has need of a weapon, can be shape shifted into a dagger form.
• Double Damage [2 Points]
• Alternate Form [1 Point] - The blade will be golden, with a black handle and the ruby will become inset in the pommel.

Bleys' Sapphire Ring [9 Points] - On Bleys' left pinky finger is the ring he uses to block Psychic contact. This makes it easy for him to avoid unwelcome Trump calls.
• Psychic Barrier [4 Points]
• Confer Quality [5 Points]

BLEYS' SHADOW AVERNUS

So I walked the land called Avernus and considered its foggy valleys and chasms, its smoking craters, its bright, bright sun against its crazy sky, its icy nights and too hot days, its many rocks and carloads of dark sand, its tiny, though vicious and poisonous beasts, and its big purple plants, like spineless cacti; and on the afternoon of the second day, as I stood on a cliff overlooking the sea, beneath a tower of massed vermilion clouds, I decided that I rather liked the place for all that, and if its sons would perish in the wars of the gods, I would immortalize them one day in song if I were able.

Nine Princes in Amber

The land was known as Avernus, and the assembled troops were not quite men. I reviewed them the following morning, walking behind Bleys. They were all of them around seven feet in height, had very red skins and little hair, cat-like eyes, and six-digited hands and feet. They wore garments that looked as light as silk, but were woven of something else and were mainly gray or blue in color. Each bore two short blades, hooked at the end. Their ears were pointed and their many fingers clawed.

Nine Princes in Amber

Bleys' Shadow Avernus [6 Points] - Bleys uses this place as a base of operations, and also as a source for loyal and effective troops in any army he might assemble.
- Personal Shadow [1 Point]
- Communication Barrier [1 Point]
- Control of Shadow's Destiny [4 Points]

BAD STUFF
[+3 Points]

GM TIPS FOR PLAYING BLEYS

Bleys, you are still a figure clad in light to me - valiant, exuberant and rash. For the first, my respect, for the second, my smile. And the last seems to have at least been tempered in recent times. Good. Stay away from conspiracies in the future. They do not suit you well.

The Courts of Chaos

Bleys is the most exuberant of the Princes of Amber. The one most willing to laugh, or love, or rush into battle. He has a wicked sense of humor and a wonderful over-confidence in his own abilities.

When faced with a challenge, Bleys will often go out of his way to perform in a difficult and spectacular way. For example, if he needs to make a rifle shot, he'd do it blindfolded, or backwards over his shoulder while sighting through a mirror. If he knew that an assassin's bolt was coming, he'd rather spill something on the floor, and reach to pick it up at just the right moment, pretending not to notice the near miss.

Merely telling a young Amberite of a bit of danger would cramp Bleys' style. He'd much prefer a flashy demonstration, or a bit of deception designed to conceal more than it reveals.

He is generally arrogant. With younger Amberites he will likely act in an even more condescending and superior way. Yet he will be unflinching in his praise.

He is capable of remaining in Shadow, hidden, through any crisis. Then, at that point when things are at their most desperate, he will appear with a grand entrance, arriving just in time to prevent disaster.

BLEYS AS FATHER

Bleys is a father that no child can ever live up to. No matter how good they are, at no matter what specialty, it seems that Bleys is always ten times better. He's not evil, or cruel, he just doesn't know how to be humble.

Bleys, The Golden God Parent: Not just likable, Bleys is charismatic! He's the life of the party, the natural leader of any group. Regardless of his aspect, Bleys is always radiant, shining, larger than life, and dazzling.

Still, having the life of the party appear when you're looking for a quiet heart-to-heart talk can be pretty irritating.

Bleys, the Missing Father: Bleys is a busy man. What with the shifting of time through Shadow, years can pass between his infrequent visits. For a child, this can be pretty devastating.

What makes it worse, in some ways, is that Bleys can always win over a shy, reluctant, or angry child.

A child can spend years cursing him out for not being there when he is needed, building up a festering resentment. Then, when Bleys does show up, he'll win the kid over with wonder and merriment and sheer charm. A few weeks later Bleys will be gone again, probably without even saying goodbye. The child goes through the whole thing all over again, Eventually swearing never to forgive Bleys, and promising to hate him forever.

Bleys shows up again, when the character has grown into a bitter teenager. It doesn't matter. Bleys is just too charming, too entertaining, too wonderful. He'll break down the teenager's barriers. He'll make the character love him all over again.

And then Bleys will disappear again.

Quite, quite infuriating.

BLEYS AS A PLAYER IN THE GAMES OF THE AMBER COURT

From a Bleys point of view, why do anything simply? With a bit of effort, and imagination, any action can be turned into a grand gesture.

In the Court he will rush on the scene with the grand entrance. A sweeping bow for the ladies, a shared joke with the men. While he is there, he will dominate center stage. He departs with a flourish, long before his welcome is worn thin.

Constant Allies & Enemies: Wherever you find Bleys, whatever his plans, you can be sure that Fiona is not far behind. They rarely appear together, these two fiery redheads, but their minds are always close. Bleys is such a lovable guy that nobody seems to really hate him. On the other hand, nobody extends him trust.

Bleys & Corwin. Bleys must remain eternally confused by Corwin's great sacrifice. When Bleys fell from Kolvir, Corwin selflessly tossed Bleys his only Trump, his only possible escape from his own trap.

It's possible that Bleys may still feel some sense of obligation, or even guilt, about putting Corwin in such a situation.

On the other hand, Bleys being a true son of Oberon, maybe he's still smarting over Corwin's calling him a fool as the Trump were thrown.

BRAND

Then there was a figure both like Bleys and myself. My features, though smaller, my eyes, Bleys' hair, beardless. He wore a riding suit of green and sat atop a white horse, heading toward the dexter side of the card. There was a quality of both strength and weakness, questing and abandonment about him. I both approved and disapproved, liked and was repelled by, this one.

Nine Princes in Amber

"Things are never what they seem," says Brand, green eyes aglow, "your friend today is your enemy tomorrow."

His words fit the speaker. Brand is the most changeable of them all. His temper is unpredictable, his depressions black, his charm enormous, and his ambitions boundless. Of all Amberites, he is the most dangerous, for he has dreamed the most evil of Amber dreams.

Brand is perhaps Dworkin's greatest student in sorcery. A man capable, in both capacity and will, of destroying Amber, the Pattern, and just about everything. He is frightening in the scope of his power. Do not think that there is anything that Brand cannot do, for he is as accomplished a Trump Artist as any of his generation. And, it is said, he can sit in a chair, anywhere, and reach out with the power of his mind alone, as if he were a living Trump.

Brand has conceived of the complete destruction of the Amber universe, of the complete erasure of the Pattern. In its place, Brand would see a new Pattern, one more to his own liking. It would reflect a new universe where there would be but one god. One god named Brand.

Or so Corwin would have us believe. But Corwin won that fight, and wrote the history as a victor justifying his own actions. From another point of view, who can say but that Corwin might look just as evil...

"In its place," Brand will tell you, putting off your question, revealing only what he wants to reveal. And then he will tell you nothing that you do not already suspect, and he will wring information from you, just in simple conversation, as if he were squeezing the juice from a ripe fruit. "Sequence and order," he will say, as you press him further, "time and stress - they are most important in this matter..."

And, if you do happen to meet Brand, and leave him safely, remember that you may never again be at peace.

Watch your back. Always.

BRAND - MAD VISIONARY OF AMBER
(325 Point Version)

Brand has seen a vision of the universe, of something mad and wonderful. It is the universe redesigned in Brand's own image. And he found that it wanted it. All that it required was removing the old order, removing the Pattern. No murder was too brutal, no victim too innocent, to stop Brand's plan to erase the Primal Pattern of Amber.

Current Objectives. If Brand lives, and should he return, or even if he finds a way to extend his influence beyond the eternal trap of the Abyss, he is unlikely to abandon his old dream.

ATTRIBUTES
PSYCHE -   [135 Points]
STRENGTH - Amber
ENDURANCE - Amber
WARFARE - Amber

POWERS
"Exalted" Pattern Imprint [100 Points] - Brand can do things that seem quite impossible. He can find anyone, anywhere in Shadow, and use the Pattern to transport himself there and back. To quote Brand, "...I learned the last of Dworkin's lore. I have gone on since then and paid dark prices for greater insight into the workings of the universe..."

Advanced Trump Artistry [60 Points]
Sorcery [15 Points]
Conjuration [20 Points]

BAD STUFF
[+5 Points]

BRAND - THE LIVING TRUMP
(425 Point Version)

'I do not understand the power that he possesses,' he said, 'but it is considerable. I know that he can travel through Shadow with his mind, that he can sit in a chair, locate what he seeks in Shadow, and then bring it to him by an act of will without moving from the chair; and he can travel through Shadow physically in a somewhat similar fashion. He lays his mind upon the place he would visit, forms a kind of mental doorway, and simply steps through. For that matter, I believe he can sometimes tell what people are thinking. It is almost as if he has himself become some sort of living Trump. I know these things because I have seen him do them. Near the end, when we had him under surveillance in the palace he had eluded us once in this fashion. This was the time he traveled to the shadow Earth and had you placed in Bedlam. After his recapture, one of us remained with him at all times. We did not yet know that he could summon things through Shadow, however. When he became aware that you had escaped your confinement, he summoned a horrid beast which attacked Caine, who was then his bodyguard. Then he went to you once again. Bleys and Fiona apparently got hold of him shortly after that, before we could, and I did not see him again until that night in the library when we brought him back. I fear him because he has deadly powers which I do not understand.'

Julian, speaking of Brand,
The Hand of Oberon

Taking his lessons from Dworkin, and surpassing them, Brand acquired the frightening ability to turn himself into a living Trump.

Current Objectives. Perhaps the Abyss holds no great menace for Brand. Falling eternally merely gives him the time to somehow deal with his wounds. Then, Caine's arrows removed, and his wounds healed, he can Trump himself anywhere.

If free upon the universe, Brand's plans may change. He had thought that the Primal Pattern of Amber must be destroyed before his new image could be inscribed with the Jewel of Judgement. Since then he has discovered that Corwin's Pattern could co-exist in the universe with that of Amber.

Brand's new plan may have him creating a new Pattern. He'll wait until his new Pattern is a secure base, and then he will take whatever moves are necessary to destroy the existing Patterns. First, finding Corwin, or a child of Corwin, to erase Corwin's Pattern. Then moving on to eliminate the Primal Pattern.

ATTRIBUTES
PSYCHE -   [165 Points]
STRENGTH - Amber
ENDURANCE - [5 Points]
WARFARE - Amber

POWERS
Advanced Pattern Imprint [75 Points]
"Exalted" Trump Artistry [140 Points] - Brand has turned
himself into a Trump, into a living artifact. Becoming part
of the cosmic Trump deck, he is able to read the subjects of
any of the cards, locating them anywhere in Shadow,
spying upon their movements and their Trump
conversations, and even looking into their thoughts.
Shape Shift [35 Points]
Sorcery [15 Points]

BAD STUFF
[+10 Points]

BRAND - HUSBAND OF CHAOS (400 Point Version)

Dara admitted that Brand had a alliance with the Courts
of Chaos. This version assumes that Brand has entered into the
most traditional of treaties among royals; a marriage contract.

Having a wife and family in the Court of Chaos, how far
from rescue could Brand be? The Abyss is no great mystery to
the folk of the Courts of Chaos. Since they have the means to
recover him, his family will not be long in reclaiming Brand.

Current Objectives. After recuperating, Brand will
build up a new base of operations near the Courts of Chaos.
When he finally returns to Amber it will be with an entirely
new identity, courtesy of his newly acquired Advanced Shape
Shifting. Who knows? He could even mask himself as one of
those brash members of the new generation!

ATTRIBUTES
PSYCHE -   [85 Points]
STRENGTH - Amber Rank
ENDURANCE - [25 Points]
WARFARE - [25 Points]

POWERS
Advanced Pattern Imprint [75 Points]
Logrus Mastery [45 Points]
Advanced Trump Artistry [60 Points]
Advanced Shape Shift [65 Points]
Sorcery [15 Points]
Conjuration [20 Points]

BAD STUFF
[+5 Points]

GM TIPS FOR PLAYING BRAND

*And you Brand... With bitterness do I regard your
memory, mad brother. You almost destroyed us. You
nearly toppled Amber from her lofty perch on the
breast of Kolvir. You would have shattered all of
Shadow. You almost broke the Pattern and redesigned
the universe in your own image. You were mad and
evil, and you came so close to realizing your desires
that I tremble even now. I am glad that you are gone,
that the arrow and the abyss have claimed you, that
you sully no more the places of men with your
presence nor walk in the sweet airs of Amber. I wish
that you had never been born and, failing that, that
you had died sooner. Enough! It diminishes me to
reflect so. Be dead and trouble my thinking no more.*

The Courts of Chaos

Brand is one slippery customer. He's got a knack for
asking the right questions, and giving away no answers that
have any value.

He's also amazingly sneaky.

When he was looking for information on Martin, he didn't
ask any questions of Random directly. Instead, he made a rude
remark about "bastards" in general. Then, when Random got
upset, Brand apologized (kind of unusual for an Amberite), and
got Random talking about Martin.

See? Brand didn't bring up Martin. Instead he got Random
to mention him first, and then, just out of "politeness" had a
chance to satisfy his curiosity.

BRAND AS FATHER

Assigning Brand as a father is likely one of the nastiest
things a Game Master can do to a player. This is definitely an
opportunity to use up some of a player's Bad Stuff.

It's important to set the player character up with the right
background. Since the character has no direct evidence of
Brand's wrong-doings, they may believe that Brand was
framed, and that the whole thing was a scheme put on by
Fiona, or Caine, or that blabbermouth, Corwin.

Brand, Misunderstood Hero: Nurturing, caring,
and placing the child's welfare above all else. This, in some
cases, may mean growing up in a Shadow where Brand is
absolute lord and god, and where the character was treated as
the perfect "god-child." Pampered and spoiled, that may have
been the golden period of a player character's life.

Brand as Manic Depressive: A likely version of
Brand-as-Father is that of the unpredictable manic depressive.
One week he's brooding, driving everyone from his presence,
and scaring the heck out of the player character as a little kid.
Then, the next thing you know, he appears, and pulls the
character out of a classroom, making some excuse, or telling
some lie. Then, when the child asks the reason, Brand will say,

"because it's a beautiful day," and will gift the player character, at maybe age ten, with a beautiful horse, and an adventure of a lifetime.

Brand can be just great. Until the pendulum of his mood swings back again.

Brand the Psycho: Finally, there's the possibility that Brand was a complete psychopath of a father. Killing family pets, committing outrageous abuses on both the child and the mother (or even being responsible for the mother's death), and generally burning with hatred, contempt and venom.

AS A PLAYER IN THE GAMES OF THE AMBER COURT

Lord knows what kind of awesome power that Brand may be concealing. Here's a guy who just doesn't know when to quit, doesn't know the concept of limits, or proportion.

Remember, just because Brand was the villain of the *Chronicles*, it doesn't mean that he'll be instantly killed, or even hunted, if he reappears. After all, King Random is establishing a new order. Also, if there's anybody who can talk his way out of a corner, it's Brand.

Good Guy: Here's a wierd possibility. What if a "reformed" Brand showed up in Amber? He shows, up, a healthy Dierdre in tow, smiling and happy that everything is finally over and resolved. This Brand, asking to let bygones be bygones, to forgive and forget, would definitely make some people rather uneasy.

Another possibility is that Brand could show up just at the right time, just as some horrible threat is about to swallow up Amber. Brand could swoop in, save the day, and come out smelling like a hero. Of course, the suspicious among the player characters might consider the possibility that Brand staged the whole thing.

A Neutral Player: Brand may decide to take a back seat in the affairs of Amber, at least for some time to come. His opponents proved too vicious, his enemies too powerful. Why risk his life again, when everyone stands so strong against him? Brand could retire for centuries, study quietly and basically just wait out any hostility that people may feel for him. In the meantime he could offer to do little "favors" for anyone who asks.

CAINE

Then came the swarthy, dark-eyed countenance of Caine, dressed all in satin that was black and green, wearing a dark three-cornered hat set at a rakish angle, a green plume of feathers trailing down the back. He was standing in profile, one arm akimbo, and the toes of his boots curled upwards, and he wore an emerald-studded dagger at his belt.

Nine Princes in Amber

There's one in every bunch. A guy who won't face you with sword in hand, who'd rather push a dagger in your back. In this family, the guy's name is Caine.

Swarthy, with the darkest eyes in Amber, you can't call him cold-blooded, just coldly calculating. After all, it's not everybody who can slit the throat of their own twin, just for a positional advantage in the family game.

Hiding in a corner, dressed in his dark satins, fingering one of his emerald studded daggers, Caine is the most menacing of Amberites. Plus, he'll throw a deal, turn on his own word, or bald-faced lie, if it suits him.

"Never trust a brother," that's the Amber family motto. Remember it, and remember that it might as well have been written with Caine in mind. He won't mind, because for damn sure he has no intention of trusting you!

Caine is an Amberite's Amberite, he plays the games so well, he'd be hard put to tell you exactly where the truth starts and where his lies stop.

So why can't we just call Caine a villain?

Because, in Corwin's story of Amber, Caine is one of the good guys. He wasn't wearing a white hat at the time, but he did show up in the nick of time, did dispatch the bad guy when everything looked pretty hopeless, and had a good excuse for every scummy thing he did.

In other words, you can't fault Caine's motives; he's always on the side of the angels.

He ought to be the hero. Whoever kills the villain ought to be the hero, but Caine is no hero...

CAINE - PROTECTOR OF AMBER (200 Point Version)

...I cast the cards several times, and the same thing came up on each occasion.

His name was Caine.

He wore satin that was green and black, and a dark three-cornered hat with a green plume of feathers trailing down behind. At his belt there was an emerald-studded dagger. He was dark.

Nine Princes in Amber

Caine, regardless of his methods, and his numerous flaws, has always put the security and safety of Amber before anything else. Of course, Caine being Caine, that means engaging in whatever sneaky, underhanded and low-down tricks that he can use against the enemies of Amber.

Current Objectives. In the continuation of the story of Amber, Caine might return to his old habits, of drinking and womanizing. Or, he might turn over a new leaf and become more of a responsible citizen.

ATTRIBUTES
PSYCHE -   [43 Points]
STRENGTH - [10 Points]
ENDURANCE - [25 Points]
WARFARE - [40 Points]

POWERS
Pattern Imprint [50 Points]
Advanced Trump Artistry [60 Points]
Power Words [10 Points] - Magic Negation, Defensive Luck, Trump Disrupt, Process Surge and Process Snuff.

CREATURES AND ARTIFACTS OF POWER
Caine's Emerald-Studded Daggers [12 Points]
   • Deadly Damage [4 Points]
   • Alternate Named & Numbered Shapes [2 Points] - Aside from their shape as daggers, they can be shifted into arrows (perhaps the ones used by Caine at the end of the *Chronicles*), bolts, or even small creatures.
   • Named & Numbered [*2 Points]

CAINE - MASTER MANIPULATOR (350 Point Version)

'...Caine had ambitions of his own—long-term ones—but ambitions nevertheless. He was in no position to pursue them, however. So he decided that if his lot was to be a lesser one, he would rather serve it under Eric than under Bleys. I can see his point, too.'

Fiona, speaking of Caine,
The Hand of Oberon

It has suited Caine's plans to aid Corwin, and to dispatch Brand. Yet it means nothing, for Caine continues to play Amber's favorite game. The outcome of the Patternfall War, and the loss of certain of his competitors is a satisfactory score for Caine. He continues to play his moves, and each "piece" he can remove from the board is yet another victory.

Current Objectives. Caine, in this case, is likely to stay close to the throne. He has always been an effective counselor and advisor to the Court of Amber, and there is no reason for him to stop doing so under King Random.

ATTRIBUTES
PSYCHE -   [145 Points]
STRENGTH - Amber
ENDURANCE - Amber
WARFARE - [30 Points]

POWERS
Advanced Pattern Imprint [75 Points]
Trump Artistry [40 Points]
Shape Shift [35 Points]

CREATURES AND ARTIFACTS OF POWER
Caine's Emerald-Studded Daggers [28 Points] - The daggers are "linked" Psychically to each other, so it's as if they were all one item, each able to sense the environment of all the others. Caine can leave one concealed in, say, the Throne Room of Amber, concentrate on another, and learn of all the events that are taking place in the room. Caine's "gift" of a dagger would effectively allow him to spy upon a character.
   • Double Damage [2 Points]
   • Extraordinary Psychic Sense [4 Points]
   • Psychic Neutral [2 Points]
   • Shape Shift [1 Point] - Capable of shifting to invisible form.
   • Confer Quality on Owner [5 Points] - This is used with the daggers' Extraordinary Psychic Sense, allowing Caine to directly spy through any of the daggers.
   • Named & Numbered [*2 Points]

BAD STUFF
[+3 Points]

CAINE - AMBERITE OF A THOUSAND FACES (400 Point Version)

This is a picture of Caine as a Shape Shifter, as the lurker in Shadow, as the unseen hand of fate. He has a thousand hiding places, and a million plans for retreat, confusion and mystery. Most important of all, Caine trusts no one. Not with the usual family paranoia, but with a suspicion so deep that Caine is totally incapable of ruling out anyone as his potential enemy.

Current Objectives. As usual, Caine will hit the ground running every time trouble is brewing. He's not above using his old trick of finding a Shadow of himself, and using it as a decoy. Although this time he's more likely to use the clone in its living form, letting this other "Caine" fill his shoes while he roams through Shadow and spies on his siblings.

ATTRIBUTES
PSYCHE - [71 Points]
STRENGTH - [25 Points]

ENDURANCE - [35 Points]
WARFARE - Amber Rank

POWERS
Pattern Imprint [50 Points]
Advanced Trump Artistry [60 Points]
Advanced Shape Shift [65 Points]
Sorcery [15 Points]
Conjuration [20 Points]

CREATURES AND ARTIFACTS OF POWER
Caine's Emerald-Studded Daggers [26 Points] - Trump
Artifacts of unusual power and depth. One of their special
abilities is that of passively receiving Trump, listening in to
Trump conversations and sometimes hearing even the
surface thoughts of the subject of a Trump. The tip of the
dagger must be in contact with the particular Trump for
this to work properly.
- Double Damage [2 Points]
- Psychic Sensitivity [1 Point]
- Psychic Neutral [2 Points]
- Able to "Mold" Shadow Creatures [2 Points]
- Capable of Racking Named & Numbered Spells [2
  Points]
- Powered by Trump [4 Points]
- Named & Numbered [*2 Points]

Caine's Shadow Servants [33 Points] - These Shadow
humanoids have the capacity to Shape Shift in several
powerful ways. Most useful is their ability to take on the
identity of humans and Amberites, making them perfect
spies for Caine. Several of them are also equipped with
Caine's Emerald-Studded Daggers.
- Tireless, Supernatural Stamina [4 Points]
- Psychic Sensitivity [1 Point]
- Shadow Path [2 Points]
- Limited Shape Shift [4 Points]
- Horde [*3 Points]

GM TIPS FOR PLAYING CAINE

*Caine, I never liked you and I still do not trust
you. You have insulted me, betrayed me and even
stabbed me. Forget that. I do not like your methods,
though I cannot fault your loyalty this time around.
Peace, then. Let the new reign begin with a clean slate
between us.*

The Courts of Chaos

Caine believes in the old fiction writer's adage, "don't tell,
show!" Which means he would not explain anything. Instead
he will arrange for demonstrations. The players will find a
dead body, a smoking gun, or a mystic pentacle still vibrating
with the traces of a transportation spell.

CAINE AS FATHER

If someone told me Caine was my father, my first
thoughts would be skepticism. Can I believe it? Caine is such a
slippery character, I wouldn't put it past him to pose as
someone's father, for years on end, all as part of some arcane
stratagem.

Caine, The Confidant: Caine could be a terrific
father. He could view his children as his only true allies, and
will likely devote himself to protecting them. One possibility is
that Caine might see that his child could be the only ally he
could trust.

Caine, The Hidden One: Imagine a father who
only appears as if out of a dream. He comes out of the
shadows, or late at night as the child lays sleeping. The player
character will be placed with a "normal" family out in Shadow,
but will experience a few rare visits from Caine. As the hidden
one, he'll watch, and protect, but do little to interfere or help the
characters development.

Caine, Neglectful Father: Caine the carousing
sailor would make a poor father. He would have a rough time
settling down with a wife, and an even rougher time trying to
relate to a child. There would be a few golden memories, but
mostly the player character will remember the fights and the
long periods when Caine was simply gone.

CAINE AS A PLAYER IN THE GAMES OF THE AMBER COURT

What a snake!
Let's not call him cold-blooded, but rather calculating. He
kills his own Shadow, attempts to kill two of his own brothers
with a knife in the back, and then shows up with the right
force, at the right time, in the right place, to deliver the final
blow to the main enemy of Amber.

Constant Allies & Enemies: Caine is loyal to his
brother Julian, the closest thing he has to a true friend.
Likewise, his relationship with Gérard has always been
cordial, if not warm. As to the others, he suspects everyone. He
has a particular dislike of Random, something that may be
hard for him to put aside in the new order.

Caine as Betrayer: On more than one occasion
Caine has entertained offers to change sides. Corwin asked
him to betray Eric, and the coalition of Bleys and Fiona also
courted his allegiance. In both cases Caine appeared to be
changing sides, but eventually sided with Eric and Julian. This
can easily happen again, and the player characters could easily
misinterpret Caine's "betrayal" of Amber for the real thing.

CORWIN

Green eyes, black hair, dressed in black and silver, yes. I had on a cloak and it was slightly furled as by a wind. I had on black boots, like Eric's, and I too bore a blade, only mine was heavier, though not quite as long as his. I had my gloves on and they were silver and scaled. The clasp at my neck was cast in the form of a silver rose.

Me, Corwin.

Nine Princes in Amber

He describes himself as tall, dark, wearing silver and black, and clasping his cloak with his trademark, a silver rose. With his blade *Greyswandir* at his hip, Corwin soldiered and travelled through Shadow, driven by vengeance and pride, battling brother Eric for the Throne. And, though he saved Amber from destruction, he remains a tragic figure.

Corwin is especially important to us because it's through him that we know all about Amber. His voice recounts the events of the books we know as *The Chronicles of Amber*. We see Amber first through Corwin's eyes, learn its powers, weaknesses, secrets and intrigues, and we grow to love, or hate, the members of his family through his experiences.

At one point a couple of his brothers got together and decided to put Corwin on ice for awhile. Just to keep him out of harm's way. So they stuck him in the deepest, nastiest dungeon cell they could find. In solitary. Bread and water diet. No talking. No walks. Straw, stone walls, four steps one way, five steps the other, a slot to shove in the bread and water, and a small hole for a toilet. And, just to make sure Corwin wouldn't try anything, they put his eyes out. With white-hot pokers.

Still, Corwin had plenty of chances to escape. They opened his cell exactly once a year. Cleaned him up, gave him dinner, and threw him back in for another year.

That tells you something about his brothers. Your relatives.

Corwin got out. Grew his eyes back.

Which tells you something about Corwin. Beaten, battered, drained by everything a variety of universes could throw at him, Corwin just keeps going.

Just for laughs, Corwin and Random once held a fencing contest just to see who would get winded first and give up the fight. Hard fencing, like running, exhausts most folks in a couple of minutes, ten minutes if you've trained like crazy for months. Maybe the best fencer in the world could go for half an hour.

Twenty-six *hours* later the brothers stopped. They could have gone on, but Corwin gave up. It seems he had a hot date.

We could also tell you about the time that Corwin carried a big guy on a stretcher. Corwin did it with his arms straight out, for fifteen *miles*. And that was when he was weak and pale. But we think you get the point by now. Corwin never stops. Ever.

CORWIN - CHAMPION OF AMBER (300 Point Version)

And the man clad in black and silver with a silver rose upon him? He would like to think that he has learned something of trust, that he has washed his eyes in some clear spring, that he has polished an ideal or two. Never mind. He may still be only a smart-mouthed meddler, skilled mainly in the minor art of survival, blind as ever the dungeons knew him to the finer shades of irony. Never mind, let it go, let it be. I may never be pleased with him.

The Courts of Chaos

Here is Corwin as he has described himself.

Current Objectives. What he will do after the Patternfall War is unclear. He has promised to explore his newly made Pattern. Perhaps he will wander through its Shadows for a time. He also has unfinished business with his son Merlin. Then there is the fact that he will feel somewhat responsible for Random's new position on the Throne of Amber. Corwin is not the Amberite that he was before his time on Shadow Earth, nor is he yet the Prince who sought the throne afterwards. Perhaps Corwin will merely be looking for his new self.

ATTRIBUTES

PSYCHE - [21 Points]
STRENGTH - [16 Points]
ENDURANCE - [81 Points]
WARFARE - [85 Points]

POWERS

Pattern Imprint [50 Points]
Initiate of the Jewel of Judgement [10 Points]
Power Words [15 Points] - Magic Negation, Chaos Negation, Psychic Disrupt, Pain Attack (customized as a hot foot), Process Surge, Process Snuff, Spark, Burst of Magic, Weaken Structure, and Thunder.

CREATURES AND ARTIFACTS OF POWER

Corwin's Sword, Greyswandir [16 Points] - The blade is forged of a piece of the Pattern itself, somehow cast from the moonlight on the steps of Tir-na Nog'th. A touch of the blade against the blood of any creature of Chaos, including even the High Lords of Chaos, is usually fatal. Their blood will flame up, and shoot out of their bodies. Greyswandir is also effective against supernatural creatures, like Weir, who otherwise regenerate from wounds.
 • Primal Damage to Creatures of Chaos [16 Points]

Corwin's Silver Scale Gloves [4 Points]
 • Resistant to Normal Weapons [1 Point]
 • Extra Hard [1 Point]
 • Named & Numbered [*2 Points] - There are, of course, two gloves in Corwin's pair.

PERSONAL SHADOWS

I walked among Shadows, and found a race of furry creatures, dark and clawed and fanged, reasonably man-like, and about as intelligent as a freshman in the high school of your choice - sorry, kids, but what I mean is they were loyal, devoted, honest, and too easily screwed by bastards like me and my brother. I felt like the dee-jay of your choice.

Nine Princes in Amber

Corwin's Shadow Ri'ik [2 Points] - Used as a recruiting ground by Corwin on at least two occasions. His furry soldiers are good fighters, and adapt easily to new technology, including modern weapons.
 • Personal Shadow [1 Point]
 • Control of Contents [1 Points]

ALLIES

Corwin's Shadow Earth Ally [1 Point] - Bill Roth, an old friend, is loyal enough to stand by Corwin and support his claim to sanity, even when every piece of rational evidence says otherwise. The fact that Bill is a lawyer makes him doubly useful to Corwin on Shadow Earth.

BAD STUFF

[+1 Point]

CORWIN - PATTERN MASTER (400 Point Version)

...there is a journey that I must make. I must ride to the place where I planted the limb of old Ygg, visit the tree it has grown to. I must see what has become of the Pattern I drew to the sound of pigeons on the Champs-Élysées. If it leads me to another universe, as I now believe it will, I must go there, to see how I have wrought.

The Courts of Chaos

Now that Corwin has created a Pattern of his very own, he has changed into an entity of great importance. Just as Dworkin and Oberon gain power from their relationships to Pattern and Amber, so Corwin will have become more than he was by being a part of the new Pattern.

Current Objectives. Returning to the Pattern of his making, Corwin may well find the desire to fill it with native defenders. He might recruit other Amberites, but there are few that he would trust, and his best friend, Random, is somewhat preoccupied. A better choice will be to father a new dynasty, a brood of children with the blood of the new Pattern.

ATTRIBUTES
PSYCHE - [35 Points]
STRENGTH - [30 Points]
ENDURANCE - [75 Points]
WARFARE - [100 Points]

POWERS
Advanced Pattern Imprint [75 Points]
Initiate of the Jewel of Judgement [10 Points]
Sorcery [15 Points]

CREATURES AND ARTIFACTS OF POWER
Corwin's Sword, Greyswandir [16 Points] - as above.
 • Primal Damage to Creatures of Chaos [16 Points]
Corwin's Silver Scale Gloves [4 Points]
 • Resistant to Normal Weapons [1 Point]
 • Extra Hard [1 Point]
 • Named & Numbered [*2 Points]

PERSONAL PATTERN
Corwin's Pattern [40 Points] - Rather than a simple Shadow, Corwin has become the personification of a whole new Pattern. This is a rather simple explanation of how the points might be organized.
 • Primal Plane [4 Points]
 • Infinite Quantity [*10 Points] - Represents the infinite number of Shadows connected with a Pattern, a multiplier a couple of steps up from "Ubiquitous."

CORWIN - SORCERER OF AVALON (500 Point Version)

*'You make me think of that line from the Holy Book—**The Archangel Corwin shall pass before the storm, lightning upon his breast...** You would not be named Corwin, would you?'*

'How does the rest of it go?'

*'...**When asked where he travels, he shall say, 'To the ends of the Earth,' where he goes not knowing what enemy will aid him against another enemy, nor whom the Horn will touch.'***

The Courts of Chaos

What if Corwin is more, much more, than he admits? We know that he was feared as a sorcerer in Avalon, and perhaps elsewhere. Perhaps his lack of great power was simply a side-effect of his amnesia, and that by the end of his saga he has regained his legacy? After all, his reputation in the Courts of Chaos seems very large.

Current Objectives. Nowadays, in the aftermath of the Patternfall War, Corwin will have the time to return to his old locus, to restore the fallen towers and...

ATTRIBUTES
PSYCHE - [55 Points]
STRENGTH - [15 Points]
ENDURANCE - [125 Points]
WARFARE - [135 Points]

POWERS
Pattern Imprint [50 Points]
Initiate of the Jewel of Judgement [10 Points]
Sorcery [15 Points]
Conjuration [20 Points]

CREATURES AND ARTIFACTS OF POWER
Corwin's Sword, Greyswandir [16 Points] - as above.
 • Primal Damage to Creatures of Chaos [16 Points]
Corwin's Silver Rose Clasp [4 Points] - Perhaps Corwin's personal symbol is also the container for his magic, and a convenient place for him to hang his spells.
 • Psychic Resistance [1 Points]
 • Self-Healing [1 Point]
 • Capable of Racking Named & Numbered Spells [2 Points]
Corwin's Silver Scale Gloves [12 Points]
 • Invulnerable to Conventional Weapons [4 Point]
 • Double Damage [2 Point]
 • Named & Numbered [*2 Points] - There are, of course,

two gloves in Corwin's pair.

PERSONAL SHADOWS

'Yes, I remember Avalon,' he said, 'a place of silver and shade and cool waters, where the stars shone like bonfires at night and the green of day was always the green of spring. Youth, love, beauty - I knew them all in Avalon. Proud steeds, bright metal, soft lips, dark ale. Honor...'

Ganelon, on Shadow Avalon,
The Guns of Avalon

Corwin's Shadow Avalon [6 Points] - It is the home of Sir Lancelot du Lac, and perhaps others of the Knights of the Round Table. Corwin claimed that it was destroyed, but he had no qualms about seeking it out. It's also worth noting that in that place Corwin had a reputation as a Sorcerer King. Avalon seems to cast Shadows of its own, so the place that Benedict found, and came to rule, had many of the same features, right down to the Field of Thorns. Still, the fact that Benedict could not reach Corwin's Avalon indicates that it is powerfully barred.
• Primal Plane [4 Points]
• Restricted Access [2 Points]

PERSONAL PATTERN
Corwin's Pattern [40 Points] - as above.
Primal Plane [4 Points]
Infinite Quantity [*10 Points] - as above.

BAD STUFF
[+3 Points]

GM TIPS FOR PLAYING CORWIN

'Yes,' he said. 'But I wonder... I've a peculiar feeling that I may never see you again. It is as if I were one of those minor characters in a melodrama who gets shuffled offstage without ever learning how things turn out.'

'I can appreciate the feeling,' I said. 'My own role sometimes makes me want to strangle the author. But look at it this way: inside stories seldom live up to one's expectations. Usually they are grubby little things, reducing down to the basest of motives when all is known. Conjectures and illusions are often the better possessions.'

Bill Roth and Corwin,
Sign of the Unicorn

In action, Corwin is superb. Not that he plays fair. No, he'll take whatever advantage is offered. If and when he comes to the rescue of player characters he should always deal with the enemy in a blunt fashion.

Out of combat, Corwin is the moodiest of Amberites. Even in his best moods he's somewhat intense. When disappointed he'll retreat to his room, or the library, locking himself in for hours or days. Typically, at one point he says, "I just noticed that I am morbid and drunk and bitter..."

It's easy to forget that Corwin the warrior is also Corwin the balladeer. In years past he was a composer for many of the songs and tunes still heard on the streets of Amber, and throughout Shadow.

CORWIN AS FATHER

Corwin, as a father, expects little, demands nothing, and offers much. The big question is whether or not Corwin could settle down, turning his back on all his intrigues and adventures, for the long years needed to be a good father.

Don't forget that there are really two Corwins. The Corwin we see in the *Chronicles*, has been somehow softened by his centuries of human life here on Shadow Earth. As Fiona says, half seriously, "I have just noticed that this is not really Corwin! It has to be one of his shadows! It has just announced a belief in friendship, dignity, nobility of spirit, and those other things which figure prominently in popular romances!" This recent Corwin would make a caring father.

The old Corwin was none of those things. We see a bit of that old character creep out in every one of Corwin's wicked deeds. The old Corwin was far more brutal, and far less compassionate. Not a good situation for any child born to him in those days.

Carl Corey, or Cordell Fenneval as Father: Corwin had four hundred years here on Shadow Earth to father children. His kids could have come from anywhere from Elizabethan England, to Revolutionary France, where he was known as Cordell Fenneval, to the United States of the Twentieth Century, where he is still called Carl Corey.

If he had known of a child, it's entirely possible that Corwin might have settled down to a solid twenty or thirty years of domestic life, becoming the ideal father. Then, over the years, as it became obvious that the child was no more likely to grow old than the father, he might have filled the kid in on what he knew of his immortality. Which wouldn't have been a whole lot, just what he would remember since recovering back in the 1600s from the Black Death.

Corwin the Traveler: It's possible that a child could have been the result of some casual fling (Corwin's morals have never been particularly high), at some distant port (Corwin also traveled a lot). The character could be from Shadow, or even City Amber itself.

King Corwin: For a long time, probably for centuries, Corwin was King of Shadow Avalon. For all we know, he started an entire line of succession in that distant land. Cut off from the rest of Shadow by Corwin's magical barriers, one, two, or an entire group of player characters could have been raised in the Arthurian-style Court of Avalon. Now that Corwin has the time, he could return for his offspring and introduce them to their Amberite heritage.

Corwin, The New Pattern's Oberon:
Traveling to the Pattern that he created, Corwin might there
found a new dynasty. In this case the Game Master could set
up a whole series of wives and liasons, depending on the kind
of structure Corwin decided to impose on the place. With time
as wierd as always, the kids from Corwin's Pattern might
already be hundreds of years old, even though only a few
years may have passed in Amber after the Patternfall War.

CORWIN AS A PLAYER IN THE GAMES OF THE AMBER COURT

*I am part of the evil which exists to oppose other
evils...and on that Great Day of which prophets speak
but in which they do not truly believe, on that day
when the world is completely cleansed of evil, then I,
too, will go down into darkness, swallowing curses.
Perhaps even sooner than that, I now judge. But
whatever... Until that time, I shall not wash my
hands nor let them hang useless.*

The Guns of Avalon

Constant Allies and Enemies. Corwin's closest
and longest-running alliance is with his full sister Dierdre. In
recent years Corwin has been able to count on Random's
whole-hearted support as well. In the aftermath of the
Patternfall War, Corwin is at peace with most of the rest of his
siblings. He has had a falling out with Dara, the mother of his
son Merlin.

DIERDRE

...and then there was a black-haired girl with the same blue eyes, and her hair hung long and she was dressed all in black, with a girdle of silver about her waist. My eyes filled with tears, why I don't know. Her name was Dierdre.

Nine Princes in Amber

Enter Dierdre, framed by lights and bright shapes, in perfect silhouette. She is mistress of the dramatic entrance, skilled in catching the eye, taking the center of attention at any gathering. She dresses to match her pale skin and black hair, in constant garb of black and silver. Her only concession to color are her stunning blue eyes.

If you find yourself in Dierdre's company, don't let your attention wander. She is accustomed to the limelight, and easily annoyed when she isn't the center of attention.

A sexist in Amber, one of those fools who thinks a woman less combat-ready than a man, would be wise to keep very quiet in Dierdre's presence. For Dierdre has as much genius for combat as any of her brothers, and is just as likely to charge into the fray as any of her male relatives.

Like all Amberites, she has awesome strength, enough to pick up a snarling werewolf, and to snap its back like a toothpick.

In battle, while her sisters remain among the archers, you'll find Dierdre charging in at the front lines, clad in black and silver armor, wielding a battleaxe with deadly accuracy and force.

Dierdre matches her brothers in the family game as well. In fact, she is often the only one to question the purity of even the most innocent of Amberites. She trusts no one, not even Corwin, though he can hardly think of Dierdre without a tear coming to his eye. And, while Corwin discounts his other sisters as having any chance at the throne, he carefully avoids mentioning Dierdre's place in the line of kings.

Let's give her our highest complement. You can trust her like a brother.

DIERDRE - CORWIN'S PARAGON (250 Point Version)

Anything is possible. It's even possible that Corwin, totally smitten by Dierdre, may actually present her accurately. In this case Dierdre is good of heart and intention, one of the best of the Amberites. Unlike those who dabble in the darker arts, Dierdre is more of a purest, intent on honing her military abilities.

Current Objectives. Once recovered from the Abyss, Dierdre will certainly go back to Amber. There she will either become part of Random's Court, or join up with Corwin, or simply return to her own Shadow realms.

ATTRIBUTES
PSYCHE -   [25 Points]
STRENGTH - [35 Points]
ENDURANCE - [25 Points]
WARFARE - [90 Points]

POWERS
Pattern Imprint [50 Points]
Power Words [12 Points] - Has a total of seven Power Words. Magic Negation, Psychic Defense, Psychic Disrupt, Pain Attack, Process Surge, Process Snuff and Weaken Structure.

ARTIFACTS OF POWER
Dierdre's Black & Silver Armor [10 points] - Dierdre's fighting gear is quite durable. She is also quite adept at using the armor's points and edges as weapons in any close combat.
  • Invulnerable to all Conventional Weapons [4 Points]
  • Double Damage Edges and Points [2 Points]
  • Sensitivity to Danger [2 Points]
  • Rapid Healing [2 Points]
Dierdre's Battle Axe [4 Points]
  • Deadly Damage [4 Points]

BAD STUFF
[1 Point]

DIERDRE - VALKIRIE OF AMBER (350 Point Version)

Among all the Princesses of Amber, Dierdre seems the most suited to pursuing her potential in the martial arts.

Current Objectives. Once rescued from the Abyss, Dierdre will go back to her warrior ways. She will seek out battle wherever she can, from the Court of Amber, through Shadow, and all the way to the Courts of Chaos.

ATTRIBUTES
PSYCHE - Amber
STRENGTH - [55 Points]
ENDURANCE - [55 Points]
WARFARE - [186 Points]

POWERS
Pattern Imprint [50 Points]

ARTIFACTS OF POWER
Dierdre's Black & Silver Armor [4 points]
  • Invulnerable to all Conventional Weapons [4 Points]

DIERDRE - FUTURE QUEEN OF AMBER
(400 Point Version)

'Clarissa? What became of your mother?'

'She died in childbirth. Dierdre was the child. Dad did not remarry for many years after mother's death. When he did, it was a redheaded wench from a far southern shadow. I never liked her. He began feeling the same way after a time and started fooling around again. They had one reconciliation after Llewella's birth in Rebma, and Brand was the result. When they were finally divorced, he recognized Llewella to spite Clarissa. At least, that is what I think happened.'

'So you are not counting the ladies in the succession?'

'No. They are neither interested nor fit. If I were, though, Fiona would precede Bleys and Llewella would follow him. After Clarissa's crowd, it would swing over to Julian, Gérard, and Random, in that order. excuse me—count Flora before Julian. The marriage data is even more involved, but no one will dispute the final order. Let it go at that.'

Ganelon's questions to Corwin,
Sign of the Unicorn

What is interesting about the above passage is what is left out. Corwin claims that he and Eric have the strongest claims to the throne. Yet, when questioned, he fails to mention his full sister Dierdre as a candidate.

Why shouldn't Dierdre be next in line for the crown?

Which leaves us with the conclusion that Dierdre may very well have her eyes upon the throne.

Current Objectives: Returning from the Abyss, Dierdre may be quite unhappy with Random on the throne. Since she, unlike the others, never swore him fealty, there is nothing to stop her from heading out into Shadow, seeking the means to conquer Amber.

ATTRIBUTES
PSYCHE -   [35 Points]
STRENGTH - [55 Points]
ENDURANCE - [35 Points]
WARFARE - [128 Points]

POWERS
Advanced Pattern Imprint [75 Points]
Sorcery [15 Points]
Conjuration [20 Points]

CREATURES AND ARTIFACTS OF POWER
Dierdre's Girdle of Silver [21 Points] - This belt allows Dierdre to pass through Shadow unnoticed, both cloaking her mind, as well as allowing her to take on a variety of forms.
- Psychic Neutral [2 Points]
- Confer Quality on Wearer [5 Points]
- Limited Shape Shift [4 Points] - Includes the forms of invisibility, a range of identities (including that of a male form of Dierdre), and at least one demonic form suitable for a disguise while in the Courts of Chaos.
- Confer Power on Wearer [10 Points]

Dierdre's Black & Silver Armor [10 points] - as above.
- Invulnerable to all Conventional Weapons [4 Points]
- Double Damage Edges and Points [2 Points]
- Sensitivity to Danger [2 Points]
- Rapid Healing [2 Points]

Dierdre's Battle Axe [6 Points]
- Deadly Damage [4 Points]
- Capable of Storing Named & Numbered Spells [2 Points]

GM TIPS FOR PLAYING DIERDRE

...Dierdre did spot an interesting possibility, however. Namely, that Gérard could have done the stabbing himself while we were all crowded around, and that his heroic efforts were not prompted by any desire to save Brand's neck, but rather to achieve a position where he could stop his tongue - in which case Brand would never make it through the night...

Sign of the Unicorn

An "interesting" possibility? Corwin is, as usual, far too kind to his bloody-minded sister. Had Julian, or Fiona, made the same remark, Corwin would surely have reported it in quite a different way.

In any case, this is a great clue as to how to play Dierdre. She will be suspicious, paranoid, and as untrusting as any Amberite.

DIERDRE AS MOTHER

Dierdre, portrayed by Corwin as the real martyr of the Patternfall War, offers a great opportunity as a player character's mother.

Mother Amber: Dierdre has the potential to be the best mother of all the Amberites. She is independent of Amber and so is likely to raise her child out in Shadow. With her various skills she will see to it that her children receive everything they need, including love, affection, training, and discipline.

Child of Brand's Victim: Regardless of how the player character was reared, there's plenty of role-playing potential in a son or daughter who knows of Dierdre's fate. The picture of Dierdre being pulled into the Abyss by Brand should be haunting.

Dierdre, Holder of the Strings. The fact that Dierdre shamelessly manipulates Corwin suggests another version of motherhood, one where Dierdre seeks to totally control her offspring. She would raise a child for the advantages the adult might offer, thinking to groom a King or Queen of Amber. From the point of view of a player character this version of mother Dierdre could be hellish.

DIERDRE AS A PLAYER IN THE GAMES OF THE AMBER COURT

'I was prisoner myself, till I made it out one of the secret ways two days ago. I thought I could walk in Shadows till all things were done, but it is not easy to begin this close to the real place. So his troops found me this morning...'

Dierdre's tale in
Nine Princes in Amber

If she were prisoner in Amber, why would she choose to walk Shadow to get away? There are several decks of Trump in the library, and the Pattern in the basement, each offering far safer and speedier transport.

It seems possible that Dierdre might have been working with Eric. Offering her assistance, Eric could easily arrange for her to be encountered by Corwin. Not that she necessarily intended to betray Corwin. By putting herself in Corwin's way, she could discover his intentions, and then choose whether to support Eric or Corwin.

Constant Allies & Enemies: Dierdre has always been able to count on Corwin's complete attention and support. None of her siblings seem to have gained her undivided trust or support.

Or Perhaps Dierdre is Brand's Partner: Here's a bloody awful thought. What if Dierdre was really working with Brand all along? She might have staged that whole drama at the edge of the Abyss as a way of buying Brand some time. Certainly it's odd that she didn't contribute anything to the conversation.

Need to explain her biting Brand's hand and trying to break away at the last minute? Well, look at it from her point of view. Caine has just appeared, and is pointing a silver-tipped arrow right at you. Do you think Caine would hesitate to shot *through* Dierdre to kill Brand?

This also explains why she was still close enough to Brand for him to grab her as he toppled backwards.

There are a lot of possibilities to be derived from this conspiracy theory. First, Dierdre must have been privy to a lot of Brand's secrets, even up to sharing his limited control over the Jewel of Judgement. Once in the Abyss her plans might have been foiled by the Unicorn's recovery of the Jewel of Judgement.

So, either Brand is dead, and Dierdre is the inheritor of his powers, or Brand is alive, saved by Dierdre. Dierdre is either trapped in the Abyss, awaiting rescue, or has, over time, managed to free herself. If Dierdre is out, and the villainess portrayed here, then she will be hatching all kinds of nasty schemes a Game Master can incorporate into a campaign.

ERIC

'Then there was Eric. Handsome by anyone's standards, his hair was so dark as to be almost blue. His beard curled around the mouth that always smiled, and he was dressed simply in a leather jacket and leggings, a plain cloak, high black boots, and he wore a red sword belt bearing a long silvery saber and clasped with a ruby, and his high cloak collar round his head was lined with red and the trimmings of his sleeves matched it. His hands, thumbs hooked behind his belt, were terribly strong and prominent. A pair of black gloves jutted from the belt near his right hip.'
Nine Princes in Amber

Eric, smiling.

Devilishly handsome, telling all who can see, by his smile, that he is master, he is in control, and he is the wall that separates lesser Amberites from their ambitions.

He's the best politician in the family. Where every brother and sister has their own personal vendetta, Eric pushes for the coalition. He's the organizer, the doer. The man of action.

Of all the Princes of Amber, Eric may be the one most suited to be a King. Certainly, with his sharp get-ups, always black with flashes of red and silver, he looks good as a king. Eric is also one of the only family members who is really practical. He'll put aside a personal grudge if there's something more important cooking. He never lets his personal feelings get in the way of getting a job done, although he *is* occasionally a little too paranoid for his own good.

Quick, too. When King Oberon disappeared mysteriously, Eric was first to the throne, and gathered enough support to make his claim unshakable. True, Corwin doesn't care much for Eric, but that's probably because they are just too much alike. Most of the rest of the family supports Eric. Most of the time.

Eric's suit is strong as a politician, but let's not forget that he's one of Amber's best swords. You can't meet the man without noticing his hands, strong, powerful, quick.

One last thing. If Eric has a single weakness, it is that of mercy. He takes care to hide it. And you shouldn't count on it in mid-battle. But, when an opponent is down and out, Eric nearly always shows compassion. Even when it isn't in his own best interest.

A practical man, not a vengeful one, Eric.

Just 'cause he's dead, don't discount the power of Eric. His image is strong in Amber, and his death not yet accepted...

ERIC - RIGHTFUL KING OF AMBER (250 Point Version)

'...Damn Eric, anyway! ...He once accused me of cheating at cards, did you know that? And that's about the only thing I wouldn't cheat at. I take my card playing seriously. I'm good and I'm also lucky. Eric was neither. The trouble with him was that he was good at so many things he wouldn't admit even to himself that there were some things other people could do better. If you keep beating him at anything you had to be cheating.'

Random's comments on Eric,
Sign of the Unicorn

Eric, give him his due, made a good king of Amber. He tried to mend fences with just about everybody, forgiving all the petty slights that kept the family apart for so long. He also was great at manipulating everybody and playing them against each other.

Current Objectives. Should Eric be recovered somehow, brought back from the dead, or from distant Shadow, he'll still consider himself to be King of Amber. He could make the case that the Unicorn, who surely must have known of Eric's inevitable return, selected Random, not as King, but only as the temporary guardian of the Throne. Once reinstated on the Throne of Amber, Eric will resume his just rule, and will make peace with as many of his relatives as possible.

ATTRIBUTES
PSYCHE - [40 Points]
STRENGTH - [15 Points]
ENDURANCE - [15 Points]
WARFARE - [75 Points]

POWERS
Advanced Pattern Imprint [75 Points]
Initiate of the Jewel of Judgement [10 Points]
Power Words [10 Points] - Magic Negation, Chaos Negation, Psychic Defense, Pattern Negation and Weaken Structure.

CREATURES AND ARTIFACTS OF POWER
Eric's Silvery Saber [8 Points] - Although it does not contain the Pattern, Eric's blade is even more destructive to a variety of supernatural and unnatural creatures.
  • Destructive Force Damage [8 Points]
Eric's Black Gloves [2 Points]
  • Resistant to Normal Weapons [1 Point]
  • Named & Numbered [*2 Points] - There are two gloves in Eric's pair.

ERIC - EXPLOITER & OPPORTUNIST (400 Point Version)

Eric ended up on the throne because he was in the right place, at the right time. Not because he was lucky, but because he made his own luck by being able to spot the right time and the right place.

Current Objectives. If Corwin turned down the job of King, why should Eric want it any more? It was a piece in the game they played, and, from Eric's point of view, he won it, fair and square. Now there are other games to be played. For example, since Corwin now has a Pattern all his own, it's possible that Eric will feel the need to compete. In that case, his plan might involve getting the Jewel of Judgement (to which he is, of course, already attuned), finding an appropriate place on the fringes of reality, and then inscribing his own Pattern.

ATTRIBUTES
PSYCHE - [55 Points]
STRENGTH - [15 Points]
ENDURANCE - [15 Points]
WARFARE - [175 Points]

POWERS
Advanced Pattern Imprint [75 Points]
Initiate of the Jewel of Judgement [10 Points]
Power Words [10 Points] - Magic Negation, Chaos Negation, Psychic Defense, Pattern Negation and Weaken Structure.
Sorcery [15 Points]
Conjuration [20 Points]

CREATURES AND ARTIFACTS OF POWER
Eric's Silvery Saber [8 Points] - as above.
  • Destructive Force Damage [8 Points]
Eric's Right Glove [1 Point] - Worn on the right hand and used to protect him from the scrapes and scratches of an enemy's blade.
  • Resistant to Normal Weapons [1 Point]
Eric's Left Glove [1 Point] - Worn on the left hand and used for punching and the like.
  • Extra Hard [1 Point]

ERIC - THE MAGE OF AMBER (475 Point Version)

Eric's mystical powers seem to be somewhat overlooked. Caine and Julian felt that Eric was capable of standing against the cabal of Brand, Bleys and Fiona. Therefore Eric must have had some kind of power to counter theirs. His command over the *Weir* indicates his control over supernatural forces, and he may have had other resources.

Current Objectives. Death provides a convenient cover for Eric. He has no reason to inform anyone of his return to the living, especially since he can freely go about his new business unmolested by his pesky brothers and sisters.

ATTRIBUTES

PSYCHE -   [35 Points]
STRENGTH - [15 Points]
ENDURANCE - [15 Points]
WARFARE - [175 Points]

POWERS

Advanced Pattern Imprint [75 Points]
Initiate & Wielder of the Jewel of Judgement [15 Points] - In addition to the normal uses of the Jewel of Judgement, Eric has also discovered how to tap it for magical energy, and how to "rack" spells within the Jewel.
"Pattern" Sorcery [30 Points] - Eric has found the secret of creating spells based on Pattern energy.
"Pattern" Conjuration [40 Points] - Eric, just before his death, stumbled on a method of producing Pattern objects. At this point he can create things like Corwin's Pattern sword with just a few days of work.

CREATURES AND ARTIFACTS OF POWER

Eric's Silvery Saber [8 Points] - as above.
  • Destructive Force Damage [8 Points]
Eric's Red Sword Belt [14 Points] - The belt allows Eric to take on different forms, and to spy without being seen.
  • Limited Shape Shift [4 Points] - Includes invisibility, a Weir shape, and also a form suitable for the Courts of Chaos.
  • Confer Power on Wearer [10 Points]
Eric's Black Glove [13 Points] - One of Eric's gloves is used to increase a wearer's Psychic power. Touching a Trump, or a person, with the glove, gives the wearer an increased "boost" of Psyche. Note that this is strictly offensive and that the glove does nothing to help a defensive Psyche battle.
  • "Augmented" Psychic Force [8 Points]
  • Confer Quality on Wearer [5 Points]
Eric's "Weir" Servants [45 Points] - Early on, in *Nine Princes in Amber*, Corwin, Dierdre and Random were confronted with a trio of things they called "Weir." These Shape Shifters were no match for an equal number of Amberites, but they would have posed a threat if their numbers had been larger. Even though one had its back broken, and the other had lost its head, they remained a threat until Corwin pierced them with his silvery Pattern blade, *Greyswandir*.
  • Tireless, Supernatural Stamina [4 Points]
  • Combat Reflexes [2 Points]
  • Shadow Seek [4 Points]
  • Regeneration [4 Points] - The Weir can even regenerate after fatal damage, or even dismemberment, unless destroyed by some destructive force based on real Power.
  • Alternate Shape [1 Point] - Able to shift from human to werewolf form.
  • Horde [*3 Points]

GM TIPS FOR PLAYING ERIC

And good-bye, Eric. After all this time I say it, in this way. Had you lived so long, it would have been over between us. We might even one day have become friends, all our causes for strife passed. Of them all, you and I were more alike than any other pair within the family...

The Courts of Chaos

Eric has a nobility that most of his siblings lack. This boils down to one thing. He has the capacity to forgive.

However, in day to day life, Eric is just as much an arrogant, pushy, self-righteous and conceited fiend as any Amberite. He just smiles more.

ERIC AS FATHER

Eric would likely have been an excellent father. He understood the importance of family, and of sharing with his relatives. That would make it all the more painful for a player character, knowing that their father died unjustly.

Eric of Shadow Earth: Since Eric spent a fair amount of time on Shadow Earth, he might very well have fathered a child there. In fact, since it was Eric who took Corwin to Shadow Earth, it's possible that a child of Eric could have been there years before.

Heirs to Eric's Crown: Since Eric was officially King of Amber before Random, then Eric's heirs should, technically, have precedence over the children of Random. This makes for some rather interesting political situations.

ERIC AS A PLAYER IN THE GAMES OF THE AMBER COURT

Eric proved the winner in the great game of Amber. He took the throne, crowned himself, and ruled as King. It was something he could not do alone, and it is the way he arranged for support among his brothers that marks him as a true genius in the game.

Constant Allies & Enemies: While he was King, Eric had the active support of Julian, Caine, Gérard, and Florimel. He eventually gained support from Random and Benedict, and possibly Fiona. Which is about as big a coalition of Amberites as anyone has ever assembled.

Eric also has a nasty habit of bad-mouthing just about everybody. According to Corwin (not the best or most unbiased of witnesses), Eric said, "...Julian I spit upon. Caine is a coward. Gérard is strong, but stupid..." This makes it hard to figure out who Eric actually likes. There is the suspicion that all of Eric's true allies hid themselves away, the better to ferret out Eric's true enemies.

REINCARNATING ERIC

'What does it mean,' I said, 'when you are wearing the Jewel and everything begins to slow down about you? Fiona warned me that this was dangerous, but she was not certain why.'

'It means that you have reached the bounds of your own existence, that your energies will shortly be exhausted, that you will die unless you do something quickly.'

'What is that?'

'Begin to draw power from the Pattern itself - the primal Pattern within the Jewel.'

'How is this achieved?'

'You must surrender to it, release yourself, blot out your identity, erase the bounds which separate you from everything else.'

'It sounds easier said than done.'

'But it can be done, and it is the only way.'

The Hand of Oberon

Eric died with the Jewel of Judgement around his neck, throbbing and leaching his life force away. As Fiona points out, his wounds didn't seem to be enough to kill him.

'Think again of Eric's death, Corwin. I was not there when it occurred, but I came in early for the funeral. I was present when his body was bathed, shaved, dressed—and I examined his wounds. I do not believe that any of them were fatal, in themselves. There were three chest wounds, but only one looked as if it might have run into the mediastinal area...'

'One's enough, if -'

'Wait,' she said. 'It was difficult, but I tried judging the angle of the puncture with a thin glass rod. I wanted to make an incision, but Caine would not permit it. Still I do not believe that his heart or arteries were damaged. It is still not too late to order an autopsy, if you would like me to check further on this. I am certain that his injuries and the general stress contributed to his death, but I believe it was the jewel that made the difference...'

Fiona's observations on Eric's death,
Sign of the Unicorn

Put all these things together and you end up with a lot of little oddities. Here are a few possible conspiracy theories for bringing Eric back from the dead.

1. Caine's Plot. Why wouldn't Caine permit an autopsy? Amberites aren't known for their fussiness about the dead. Could it be that Caine, who staged his own death shortly after, was not working alone?

2. Amberites Don't Just Heal, They Regenerate. As Fiona observed, none of Eric's wounds were necessarily fatal. Whether Caine was aware of it or not, Eric might have simply regenerated and come back to life. Seeing the possible advantages of being "dead," it would be no great matter to find (or Conjure) a suitable corpse and depart.

3. Storage in the Jewel of Judgement. If the Jewel of Judgement were capable of automatically rescuing Corwin when he was stabbed, why didn't it do something similar for Eric? After all, the artifact is capable of inscribing a Pattern and thereby creating an infinite number of universes. Rescuing a single individual seems a fairly small matter in comparison. Perhaps, when Eric died with the pulsing of the Jewel of Judgement, Eric's "Pattern" was contained within the Jewel. In this case Eric might have been released later, or he could still be waiting for release.

FIONA

Fiona—five-two, perhaps, in height—green eyes fixed on Flora's own blue as they spoke, there beside the fireplace, hair more than compensating for the vacant hearth, smoldering, reminded me, as always, of something from which the artists had just drawn back, setting aside his tools, questions slowly forming behind his smile. The place at the base of her throat where his thumb had notched the collarbone always drew my eyes as the mark of a master craftsman, especially when she raised her head, quizzical or imperious, to regard us taller others. She smiled faintly, just then, doubtless aware of my gaze, an almost clairvoyant faculty the acceptance of which has never deprived of its ability to disconcert.

Sign of the Unicorn

A tiny, sharp-tongued, green-eyed, red-head. Corwin calls her a sorceress. But then, she always seems to be one step ahead of Corwin.

A bit of secret lore, an edge of perspective, and a spell or conjuring that puts her ahead of the pack. Fiona has the knack of making her larger brothers seem clumsy and oafish.

Perhaps they are.

At the climax of the great battle of Patternfall, when the family's worst enemy held the Jewel of Judgement and mocked them all, it was Fiona's power that saved the day.

Any Amberite can engage in mind to mind combat under special circumstances. A Trump contact, where the minds actually connect, allows for the battle of wills, to impose paralysis, or even injury, on an opponent. Fiona's special trick allows her to reach minds without the restriction of Trump, using her magic, and imposing her will even through the defenses of other Amberites.

Fiona knows all too well that any power, including that of the great Pattern, is limited in its range, force, and subtlety, by the mind of the initiate. Her other powers, Trump, and magic, are likewise powered by her psychic will. She knows this, she keeps her mind sharp, and none of her relatives are eager to contest her in the mystic arts.

In moving through Shadow, especially those hard-to-forge pathways through Amber, movement usually involves the subtle manipulation of Pattern for hours, even days. But when there was an emergency, and when Corwin needed to get to a place days away, Fiona found a short-cut, a way of mere minutes. It is proof of her ability to manipulate the stuff of Shadow virtually without contest, and also, as Corwin put it, an endorsement for higher education, the advantage of a lifetime of study.

One other thing. Don't irritate Fiona. Her temper is considerable. And, on the sly, she has slipped a dagger between the ribs of a helpless relative.

FIONA - SORCERESS OF AMBER (300 Point Version)

'...I have always been very fond of Fiona. She is certainly the loveliest, most civilized of us all.'

Julian's observation,
The Hand of Oberon

Fiona, though originally allied with Brand and Bleys, seems to have outmatched both of her brothers in her mastery of arcane Powers. Her objective has always been to bring together Dworkin's knowledge and influence in the affairs of state.

Current Objectives. Having sacrificed much to foil Brand, she will expect a place close to King Random. Using her Powers to survey and control Shadow, she will extend Amber's reach and authority to unprecedented heights.

ATTRIBUTES

PSYCHE -   [122 Points]
STRENGTH - Amber Rank
ENDURANCE - Amber Rank
WARFARE - Amber Rank

POWERS

Advanced Pattern Imprint [75 Points]
Trump Artistry [40 Points]
Power Words [10 Points] - Chaos Negation, Pattern Negation, Trump Disrupt, Burst of Magic and Weaken Structure.
Sorcery [15 Points]
Conjuration [20 Points]

CREATURES AND ARTIFACTS OF POWER

For one thing, all six had uniformly bloodshot eyes. Very, very bloodshot eyes. With them, though, the condition seemed normal.

For another, all had an extra joint to each finger and thumb, and sharp, forward-curving spurs on the backs of their hands.

All of them had prominent jaws, and when I forced one open, I counted forty-five teeth, most of them longer than human teeth, and several looking to be much sharper. Their flesh was grayish and hard and shiny.

Nine Princes in Amber

Fiona's Shadow Creatures [15 Points] - These are the creatures that Fiona used to imprison Brand, when he was locked away in the Shadow tower. In chasing Random, after he failed in rescuing Brand, the creatures showed themselves able to follow him in spite of his best Shadow Shifting efforts, and to adapt with appropriate clothing and weapons in every different Shadow. For example, in the California of Shadow Earth, they ambushed him in a Men's Rest Room, rushing out of six different stalls, each wearing grey overcoats and brandishing flat-looking pistols.
• Combat Reflexes [2 Points]
• Shadow Trail [1 Point]
• Able to "Mold" Shadow Creatures [2 Points]
• Horde [*3 Points]

PERSONAL SHADOW

Then I saw the landscape—over his shoulder, out a window, over a battlement. I can't be sure. It was far from Amber, somewhere where the shadows go mad. Farther than I like to go. Stark, with shifting colors. Fiery. Day without a sun in the sky. Rocks that glided like sailboats across the land. Brand there in some sort of tower—a small point of stability in that flowing scene. I remembered it, all right. And I remembered the presence coiled about the base of that tower. Brilliant. Prismatic. Some sort of watch-thing, it seemed—too bright for me to make out its outline, to guess at its proper size...

Random, on Brand's Prison,
Sign of the Unicorn

Fiona's Prismatic Shadow [3 Points] - Fiona seems to have achieved a Shadow well outside of the normal boundaries of Amber, to a place very near to the Courts of Chaos. Only occasionally, and with great effort could Brand pierce the barrier against Trump contact, perhaps only by the use of some other Power.
• Personal Shadow [1 Point]
• Communication Barrier [1 Point]
• Control of Contents [1 Point]

FIONA - MISTRESS OF THE PATTERN (350 Point Version)

'Turn here.'

I entered a cleft in a hillside. The way was narrow and very dark, with only a small band of stars above us. Fiona had been manipulating Shadow while we had talked, leading us from Ed's field downward, into a misty, moorlike place, then up again, to a clear and rocky trail among mountains. Now, as we moved through the dark defile, I felt her working with Shadow again. The air was cool but not cold. The blackness to our left and our right was absolute, giving the illusion of enormous depths, rather than nearby rock cloaked in shadow. This impression was reinforced, I suddenly realized, by the fact that Drum's hoofbeats were not producing any echoes, aftersounds, overtones...

The band of stars had narrowed, and it finally vanished above us. We advanced through what seemed a totally black tunnel now, with perhaps the tiniest flickering of light a great distance ahead of us...

The light ahead grew larger, brighter, but there were no drafts, sounds, or smells from that direction...

...The light had grown large, become a circular opening. It had approached at a rate out of proportion to our advance, as though the tunnel itself were contracting. It seemed to be daylight that was rushing in through what I chose to regard as the cave mouth.

> Fiona's "shortcut" from
> Shadow Earth to Primal Pattern
> *The Hand of Oberon*

Perhaps Fiona's connection with the Pattern is even greater than anyone realized. She may have learned more from Dworkin than simple tricks, she may have achieved a rapport with the Pattern itself.

Current Objectives. Her curiosity about Corwin's Pattern will be intense. She will pester him to allow her entry. She may even become a co-ruler with Corwin in that place, founding a line of offspring parallel to Corwin's own.

ATTRIBUTES
PSYCHE -   [125 Points]
STRENGTH - Amber Rank
ENDURANCE - [20 Points]
WARFARE - [5 Points]

POWERS
"Exalted" Pattern Imprint [100 Points] - For Fiona the Pattern has become a living thing. Not conscious, in any sense that Amberites understand, but with a purpose and a pulsating interaction with the universe of Shadow. Using this knowledge, and her affinity for its movements, she is able to cut across the usual boundaries of Shadow, travelling across the waves of reality that generate Shadow.

Trump Artistry [40 Points]
Shape Shifting [35 Points]
Sorcery [15 Points]
Conjuration [20 Points]

FIONA - TRUMP TRICKSTER (450 Point Version)

His image faded as I released the contact, but a strange thing happened then. The sense of contact, the path, remained with me, objectless, open, like a switched on radio not tuned to anything.

Bill was looking at me peculiarly.

'Carl, what is happening?'

'I don't know, Wait a minute.'

Suddenly, there was contact again, though not with Gérard. She must have been trying to reach me while my attention was diverted.

'Corwin, it is important...'

'Go ahead, Fi.'

> *The Hand of Oberon*

Fiona's ability to control Brand, even at the height of his greatest Power, seemed to allow her to push through his defenses and hold him in place, unable to exercise his greatest Powers.

Current Objectives. Having won the battle of the Patternfall War, Fiona will once again seek out her old mentor, Dworkin, and continue her studies. She will have a renewed interest in Pattern, and the Jewel of Judgement, and will be attempting to find the link between the Power of Pattern and the Power of Trump.

ATTRIBUTES
PSYCHE -   [200 Points]
STRENGTH - Amber Rank
ENDURANCE - [15 Points]
WARFARE - Amber Rank

POWERS
Advanced Pattern Imprint [75 Points]

"Exalted" Trump Artistry [125 Points] - Fiona seems to have gained a level of control over Trump that involves more subtlety than raw Power. She can open up contacts without the subject being aware of them, or engage in Psychic battle without the victim having the opportunity to block the contact.

Sorcery [15 Points]
Conjuration [20 Points]

GM TIPS FOR PLAYING FIONA

Then there was Fiona, with hair like Bleys or Brand, my eyes, and a complexion like mother of pearl. I hated her the second I turned over the card.
Nine Princes in Amber

There's something about Fiona that isn't quite grown up.

She's perpetually a little red-headed nine-year-old, with freckles and a really snotty attitude. No matter that she's the ageless sorceress of Amber, in my mind's eye I always see her, taking the arm of some massive soldier of Amber, giving him an encouraging, approving look, and then glancing back over her shoulder, and sticking her tongue out at whoever she has out-maneuvered.

FIONA AS MOTHER

Fiona can be quite a Mom, devoted and fiercely loyal to her children. On the other hand, she doesn't suffer fools. Her disapproval for any offspring who doesn't meet her exacting standards would be loud, cruel, sharp and torturous.

Instructor Mother: For player characters with an emphasis on the advanced Powers, and with the Good Stuff to pay for it, Fiona can be the perfect Mother. She is capable of showing her child the mysteries of any of Amber's Powers, and she always has the freedom to introduce her children to the life of Castle Amber.

FIONA AS A PLAYER IN THE GAMES OF THE AMBER COURT

'...and I, of course, am innocent of all but malice.'

Fiona, describing herself,
Sign of the Unicorn

Fiona doesn't play *in* the game, she plays *with* the game. Her manipulations go to the extent of questioning the rules, the goals and even making fun of moves made by others. She treats her bigger brothers as if they were over-sized tokens, brutes who occasionally have their uses, but who fail to grasp the finer points.

For example, when, in the Library, sides were chosen between Corwin's faction and that of Benedict, Fiona waited until she would be the deciding vote. This she did, knowing full well that it would make her position all the stronger in whichever side she chose.

Constant Allies & Enemies: Fiona's full brother Bleys is her closest ally and confidant. The two of them are never far apart in their plans and desires. Fiona has many enemies in the family, and is disliked by all her sisters and most of her brothers. Her temper, and her biting wit, have kept most of them from getting too close. Even though she currently has a friendship with Corwin, it's unlikely that it will last long. They are simply of different temperaments.

FLORA

The woman behind the desk wore a wide-collared, V-necked dress of blue-green, had long hair and low bangs, all of a cross between sunset clouds and the outer edge of a candle flame in an otherwise dark room, and natural I somehow knew, and her eyes behind glasses I didn't think she needed were as blue as Lake Erie at three o'clock on a cloudless summer afternoon; and the color of her compressed smile matched her hair.

Nine Princes in Amber

In every family, somebody gets the title for best looking. In this family, her name is Flora.

She looks good. Movie-star good, with perfect hair, long, with bangs, and perfect blue eyes, and always attired perfectly in a low-cut dress, in colors to match.

How should Amber's best-looking woman deal with her rambunctious brothers?

Florimel, Flora to those her know her well, knows she's not cut out for the throne. Not that she doesn't want it, she just doesn't have any particular edge over her brothers. And, save for Llewella, she doesn't get along all that well with her sisters.

So she has come up with an alternate position. One right next to the throne. And, regardless of who ends up in power, there's little Flora whispering in their ear. It's her great strength to be a political survivor in a changing game.

Flora avoids direct combat. On Shadow Earth, where she has spent a lot of time, she keeps her house filled with an assortment of trained Irish Wolfhounds, big enough, and numerous enough, to give an Amberite serious thought before interfering in her ways.

If she comes across as the weak sister, don't buy it. She's as cunning as any of them. She's just figured out that it doesn't pay to fight with her siblings. And one thing all of her brothers and sisters want is allies.

So Flora is the perfect ally.

FLORA -
FAITHFUL SERVANT OF
THE CROWN
(150 Point Version)

I raised my arm and pointed at her and the lightning flashed at my back, just outside the window. I felt a tingle, a mild jolt. The thunderclap was also impressive.

'You sin by omission,' I tried.

She covered her face with her hands and began to weep.

'I don't know what you mean!' she said. 'I answered all your questions! What do you want? I don't know where you were going or who shot at you or what time it occurred! I just know the facts I've given you, damn it!'

Sign of the Unicorn

Here we see Florimel just as she appears to Corwin. A little flighty and a lightweight on the Amber scene. She knows her weaknesses and tries to keep herself allied with someone strong enough to protect her, and with enough power to keep her door to Amber open.

Current Objectives. Likely as not Flora is now firmly settled in the Court of Amber. She will be a significant part of Random's court, and will use her early connection with the Random-Corwin alliance to gain influence and political power.

ATTRIBUTES
PSYCHE - [28 Points]
STRENGTH - [20 Points]
ENDURANCE - [20 Points]
WARFARE - [20 Points]

POWERS
Pattern Imprint [50 Points]
Power Words [10 Points]
   Psychic Defense, Pattern Negation, Trump Disrupt, Light Strobe and Weaken Structure.

CREATURES AND ARTIFACTS OF POWER

At that point an enormous dog entered the room—an Irish wolfhound—and it curled up in front of the desk. Another followed and circled the globe twice before lying down...

'This is an ultrasonic dog whistle. Donner and Blitzen here have four brothers, and they're all trained to take care of nasty people and they all respond to my whistle.'

Nine Princes in Amber

Flora's Irish Wolfhounds [2 Points]
   • Extra Hard Teeth [1 Point]
   • Named & Numbered [*2 Points]

FLORA - PROPRIETOR OF
SHADOW EARTH
(200 Point Version)

'The beginning...' she said. 'Yes... It was in Paris, a party, at a certain Monsieur Focault's. This was about three years before the Terror—'

'Stop,' I said. 'What were you doing there?'

'I had been in that general area of Shadow for approximately five of their years,' she said. 'I had been wandering, looking for something novel, something that suited my fancy. I came upon that place at that time in the same way we find anything. I let my desires lead me and I followed my instincts.'

'A peculiar coincidence.'

'Not in light of all the time involved - and considering the amount of travel in which we indulge. It was, if you like, my Avalon, my Amber surrogate, my home away from home. Call it what you will, I was there...

Sign of the Unicorn

Flora spent an awful lot of time here on Shadow Earth. Perhaps she had already been in residence long before the French Revolutionary period she mentions. Two possibilities come to mind. One, that Eric led her there to keep track of Corwin. Two, that she suggested it to Eric in the first place.

Current Objectives. She has too many irons on the fire on Shadow Earth for her to neglect the place for long. Flora will continue to make regular trips, and to keep up the maintenance of any traps, forces or barriers that she may have erected.

ATTRIBUTES
PSYCHE - [55 Points]
STRENGTH - [20 Points]
ENDURANCE - [20 Points]
WARFARE - [20 Points]

POWERS
Pattern Imprint [50 Points]
Sorcery [15 Points]

CREATURES AND ARTIFACTS OF POWER
Flora's Irish Wolfhounds [10 Points] - As above, but with a
few additions.
  • Combat Reflexes [2 Points]
  • Double Damage Teeth [2 Points]
  • Shadow Trail [1 Point]
  • Named & Numbered [*2 Points]

PERSONAL SHADOW
Flora's Shadow Earth [10 Points] - After hanging out on
Shadow Earth for a couple of hundred years, not disabled
like Corwin, but in full command of a range of powers,
she's gained a huge amount of control over its access and
destiny.

For example, when Corwin first showed up, Flora
supposedly attempted to return to Amber from Shadow
Earth. She returned, claiming that the path had been
blocked and blaming Corwin. A short time later, when
Corwin and Random started their travel from Shadow
Earth to Amber, they ran into a number of difficulties.

Could their problems have been caused by Eric? It
seems unlikely. At the time Eric had his hands full, and, in
any case, had he been interested in blocking Corwin's
escape he could have done so easily with the Jewel of
Judgement.

So we come around to Flora. Perhaps her attempted
departure was merely a ruse. It seems very likely that she
was the agent responsible when Corwin and Random ran
into trouble.
  • Primal Plane [4 Points]
  • Restricted Access [2 Points]
  • Control of Shadow Destiny [4 Points]

FLORA -
POWER IN WAITING
(350 Point Version)

That she would seem so helpless is suspicious in and of
itself. Flora's meek exterior may be concealing something more
sinister. Something more powerful and in keeping with the rest
of her family.

Current Objectives. With Random on the throne
and her place on the winning team firmly established, Florimel
may attempt to exploit resources that she just couldn't get to
before. The family library and Dworkin's Cave being two

examples.

ATTRIBUTES
PSYCHE - [63 Points]
STRENGTH - [45 Points]
ENDURANCE - [50 Points]
WARFARE - [35 Points]

POWERS
Pattern Imprint [50 Points]
Shape Shifting [35 Points]
Sorcery [15 Points]
Conjuration [20 Points]

CREATURES AND ARTIFACTS OF POWER
Flora's Irish Wolfhounds [2 Points] - as above.
  • Extra Hard Teeth [1 Point]
  • Named & Numbered [*2 Points]

GOOD STUFF
[35 Points] - This the "reserve" that Flora will tap when she
gains access to some greater Power.

GM TIPS FOR PLAYING FLORA

*Flora... Charity, they say, begins at home. You
seem no worse now than when I knew you long ago...*

The Courts of Chaos

Florimel has no particular Power or Attribute that sets her
apart from her relatives.

All Flora has going for her is that she's intensely
interesting. Here's a woman who tosses back full glasses of
whiskey like water, keeps a hand grenade in her purse, and can
lie her head off while weeping great big tears. There's no way
to know how much of her patter is foolish rambling, and how
much is calculated deception.

Either way, a Flora that is truly flaky, or a Flora just
pretending to be a twit, she provides an opportunity for some
great role-playing.

Flora can be played as an all-too-human member of the
Amber Court. She will be concerned with appearances, will
worry about protocol, and can attend to all the little details of
Courtly life that are likely to drive King Random crazy. In fact,
since she is so good at the social aspects of the royal life, it's
even possible that Random will name her regent in his
occasional absences.

While that may seem preposterous, remember that neither
Benedict, Julian, Gérard or Caine are particularly diplomatic.
They each, in their own way, are irritating as all get out. If
you'd like to keep people from being alienated, Flora is an
excellent choice.

It's also interesting having Flora be a very female
influence in an otherwise male Court. She may insist on
redecorating, especially clearing out some of Oberon's archaic
fashions. Even her sisters don't care all that much about the
feminine touches.

Flora in Conversation. In interacting with the players, the Game Master can have Flora inserting all kinds of down-to-earth comments and observations. Right in the middle of some discussion of great Power, or of a threat to Amber, she can say the most mundane (and annoying) things. This is especially effective with her own children.

- "I really think you should give a little more thought to your wardrobe. Did you even look at yourself in the mirror this morning?"

- "Well, I know you've had a hard day, but I don't think that excuses you from a little basic hygiene. There are showers here, you know."

- (to a female character) "Dear, with a little blush, and a touch of mascara, I think you'd be far more ready for what you have in mind."

- (on learning of an invasion/return of some enemy) "Have you even considered how we are going to arrange the entertainment? I really think we should plan a menu and notify the kitchen staff."

- "Really darling, have you looked at this place? It's absolutely Medieval! I think we're going to have to go with something more colorful, and perhaps tear out a wall or two."

- "Precious lamb, I know you are new here in Amber, but there are, after all, proper ways to approach your problems. It wouldn't kill you to study a little of the Court Protocol."

FLORIMEL AS MOTHER

Flora can make a great mother! She can be loving, or annoying, or nagging, or prim and proper. In other words, she can be just like anybody's mother.

Born on Shadow Earth. Player characters starting out on Shadow Earth could have been raised by Mrs. Evelyn Flaumel (i.e Flora), and her maid Carmella. Chances are their upbringing would be based in the rich society circles of the Eastern United States, with frequent worldwide travel.

Flora: Flaky, but Lovable. You can easily picture Flora as one of those eccentric, slightly distracted mothers. Someone too rich for her own good, and heavily reliant on the hired help to care for her children's messy or tedious problems. There's also the possibility that Flora just wasn't around most of the time.

Flora, User of Children. Flora could see in any child a chance to cultivate the perfect proxy for her long-term ambitions. By pushing, shoving, and cajoling, she would get the kid to take up all the courtly arts, and become, in her mind, a fit candidate for the throne. This is most likely if she were to have a son, especially one who seemed to have the potential to stand up against Eric, Corwin and Bleys. In that case, Flora might have pushed her son's martial arts training to the point of abusiveness.

FLORIMEL AS A PLAYER IN THE GAMES OF THE AMBER COURT

'Yes. I'd made no secret of my whereabouts. In fact, all of them came around to visit me at one time or another.'

'That includes Random?'

She curled her lip.

'Yes, several times,' she said.

'Why the sneer?'

'It is too late to start pretending I like him,' she said. 'You know. I just don't like the people he associates with—assorted criminals, jazz musicians... I had to show him family courtesy when he was visiting my shadow, but he put a big strain on my nerves, bringing those people around at all hours - jam sessions, poker parties. The place usually reeked for weeks afterward and I was always glad to see him go. Sorry, I know you like him, but you wanted the truth.'

'He offended your delicate sensibilities...'

Flora's chat with Corwin,
Sign of the Unicorn

Is this Flora's real problem with Random? A resentment because he took advantage of her hospitality? Or could it be a calculated pettiness, designed to avoid the discussion of real issues?

Probably a little of both.

Flora keeps her fingers in the pot constantly. There are no plots, cabals or schemes that she isn't interested in pursuing. It's interesting that she always lands on her feet, and ends up allied with whatever coalition is currently in power.

Constant Allies & Enemies: Flora seems only to have friends of the moment. Her enemies seem few, probably because she doesn't seem to pose a threat to anyone.

GÉRARD

And a big, powerful man regarded me from the next card. He resembled me quite strongly, save that his jaw was heavier, and I knew he was bigger than I, though slower. His strength was a thing out of legend. He wore a dressing gown of blue and gray clasped around the middle with a wide, black belt, and he stood laughing. About his neck, on a heavy cord, there hung a silver hunting horn. He wore a fringe beard and a light mustache. In his right hand he held a goblet of wine.

Nine Princes in Amber

The strongest guy in the universe.

The average Amberite, with not much work, is pretty strong. Listen to Corwin: "Suddenly, and without thinking, I picked up a huge overstuffed chair and hurled it perhaps thirty feet across the room. It broke the back of the man it struck."

Or how about after Corwin gets free of his dungeon cell, still not completely recovered, and starts hefting rocks, big rocks. "One of them must have weighed around four hundred pounds, and I did not roll it. I hefted it and set it in place."

So here's Corwin, the guy who lifts cars (yeah, he can do that too), throws chairs, and lifts boulders. Gérard is so strong Corwin is scared of him.

It's not just that Gérard can lift heavier things, although he can certainly handle double the load of his weaker brothers, it's that, in hand to hand combat, he is unbeatable.

Imagine a muscular world-class fighter, boxer or wrestler, your choice. They're built for dishing out damage, yeah, but they can also take it. How many punches will it take for a weaker fighter to do any real damage to that kind of guy?

Against Corwin, Gérard took half a dozen punches to the face, neck and body, and at least one solid kick to the stomach, all blows that would have killed a normal man. And Corwin did not win that fight.

Aside from his great Strength, Gérard is also known for his dogged loyalty to Amber. His instant, unthinking response will be to rescue, or defend, any relative he sees in danger. And woe be unto anyone who gets in the way of Gérard when he's on a mission of mercy! He, alone among his siblings, is willing to protect an innocent, to give the benefit of the doubt.

The flip side of this is that he'll give you absolute Hell if he thinks you're up to no good. A "lesson" from Gérard usually means having to engage in a little friendly wrestling. If you're an Amberite with no particular weaknesses, you'll be out of traction within the week.

GÉRARD - STRONGMAN OF AMBER (250 Point Version)

If I had had to choose a place to fight with Gérard, this would not have been it. He, of course, was aware of this. If I had to fight with Gérard at all, I would not have chosen to do so with my hands. I am better than Gérard with a blade or a quarterstaff. Anything that involved speed and strategy and gave me a chance to hit him occasionally while keeping him at bay would permit me to wear him down eventually and provide openings for heavier and heavier assaults. He, of course, was aware of this also. That is why he had trapped me as he had. I understood Gérard, though, and I had to play by his rules now.

Sign of the Unicorn

Gérard stands to protect Amber. His will is as strong as his body, and he will not yield to temptation or game playing. Nothing is as important as his family.

Current Objectives. He will seek to keep Amber safe. To do this he will remain by Random's side, as guardian and protector. He will have a concern and caring for any young newcomers to Amber and will likely provide them with a warm welcome and any assistance he can provide.

ATTRIBUTES
PSYCHE - Amber Rank
STRENGTH - [120 Points]
ENDURANCE - [40 Points]
WARFARE - [35 Points]

POWERS
Pattern Imprint [50 Points]

GOOD STUFF
[5 Points]

GÉRARD - MASTER OF THE SHIPS OF AMBER (350 Point Version)

Currently master of the fleet of Amber, Gérard wields a great deal of influence in affairs of commerce and of City Amber.

Current Objectives. The events of the Patternfall War have done damage to the fleet, the business of sea trade, and the reputation of Amber. Gérard will seek to remedy all of these things. The fleet will be expanded, and possibly modernized, and he will open new Shadows to trade.

ATTRIBUTES
PSYCHE - [10 Points]
STRENGTH - [176 Points]
ENDURANCE - [50 Points]
WARFARE - [55 Points]

POWERS
Pattern Imprint [50 Points]

CREATURES AND ARTIFACTS OF POWER
Gérard's Big Sword [9 Points] - Gérard's favorite weapon is a huge sword, so big that it is unwieldy for even six-footers like Corwin. The sword is tough enough to sustain or deflect most any attack, and also tough enough to withstand the power of Gérard's swings. No extra damage is really necessary, since a swipe by Gérard tends to pass through even the toughest armor.
• Impervious to Damage [8 Points] - No natural force or material can damage the sword, nor can it be affected by Magic or any lesser Power.
• Extra Hard [1 Point]

GÉRARD - DEFENDER OF AMBER (500 Point Version)

...I had a clear shot at his groin with my right, but I restrained myself. It is not that I have any qualms about hitting a man below his belt. I knew that if I did it to Gérard just then his reflexes would probably cause him to break my shoulder...

Sign of the Unicorn

Perhaps Gérard knows more of Amber's secrets than anyone suspects. After all, Oberon kept Gérard in Amber, when he ordered everyone else out to deal with the Courts of Chaos. Gérard may be the one Amberite who really knows all the secrets of Castle Amber, and of the dungeons below.

Current Objectives. His place, as always, is in Amber and the surrounding seas. He will stand by King Random, supporting the realm, and governing in Random's absence.

ATTRIBUTES
PSYCHE - [35 Points]
STRENGTH - [253 Points]
ENDURANCE - [60 Points]
WARFARE - [60 Points]

POWERS

Pattern Imprint [50 Points]

Power Words [30 Points] - Magic Negation, Chaos Negation, Psychic Defense, Psychic Disrupt, Neural Disrupt, Lifeforce, Resume True Form, Defensive Luck, Pattern Negation, Pain Attack, Trump Disrupt, Process Surge, Process Snuff, Shade, Light Strobe, Spark, Burst of Magic, Weaken Structure, Thunder, and Burst of Psyche.

CREATURES AND ARTIFACTS OF POWER

Gérard's Silver Hunting Horn [8 Points] - Used for creating the cross-Shadow seaways that link the port of Amber with other destinations.

• Create Shadow Path [8 Points] - Actually rips a pathway through Shadow, that can later be used by someone with the "Shadow Path" Power, or anyone who has memorized the particular course. Gérard can also use the Horn to close down Shadow pathways.

Gérard's Big Sword [9 Points] - as above.

• Impervious to Damage [8 Points] - No natural force or material can damage the sword, nor can it be affected by Magic or any lesser Power.

• Extra Hard [1 Point]

GM TIPS FOR PLAYING GÉRARD

Gérard, slow, faithful brother, perhaps we have not all changed. You stood rock-like and held to what you believed. May you be less easily gulled. May I never wrestle you again. Go down to your sea in your ships and breathe the clean salt air.

The Courts of Chaos

Big and intimidating, Gérard is the only Amberite who admits to being lost and outsmarted. Not that he feels very good about it. He can get downright violent with anyone who tries to make fun of him. And there will be trouble for any player character who deliberately confuses Gérard with some convoluted line of logic.

Game Master Note: No matter which version of Gérard you choose for your campaign, the Strength of that version should be the higher than that of any other Amberite. Either select versions of the other characters so they are consistently below Gérard, or modify Gérard's Strength so it remains supreme.

GÉRARD AS FATHER

Gérard loves every member of his family. He even cares about Shadow folk, and will defend them if they are in need. This makes him about the most compassionate of all the Amberites, and the best father for those players characters blessed with a lot of Good Stuff.

On the other hand, Gérard is not an indulgent man. If he sets rules, he expects his children to live by them. He's also not one to avoid inflicting a bit of corporal punishment if he feels it is necessary.

GÉRARD AS A PLAYER IN THE GAMES OF THE AMBER COURT

I was hanging high in the air. By turning my head slightly I could see for a very great distance, down.

I felt a set of powerful clamps affixed to my body - shoulder and thigh. When I turned to look at them, I saw that they were hands. Twisting my neck even farther, I saw that they were Gérard's hands. He was holding me at full arm's length above his head. He stood at the very edge of the trail, and I could see Garnath and the terminus of the black road far below. If he let go, part of me might join the bird droppings that smeared the cliff face and the rest would come to resemble washed-up jellyfish I had known on beaches past.

'I am not a clever man,' he said. 'But I had a thought—a terrible thought. This is the only way that I know to do something about it. My thought was that you had been away from Amber for an awfully long while. I have no way of knowing whether the story about your losing your memory is entirely true. You have come back and you have taken charge of things, but you do not yet truly rule here. I was troubled by the deaths of Benedict's servants, as I am troubled now by the death of Caine. But Eric has died recently also, and Benedict is maimed. It is not so easy to blame you for this part of things, but it has occurred to me that it might be possible—if it should be that you are secretly allied with our enemies of the black road.'

'I am not,' I said.

'It does not matter, for what I have to say,' he said. 'Just hear me out. Things will go the way they will go. If, during your long absence, you arranged for this state of affairs—possibly even removing Dad and Brand as part of your design—then I see you as out to destroy all family resistance to your usurpation.'

'Would I have delivered myself to Eric to be blinded and imprisoned if this were the case?'

'Hear me out!' he repeated. 'You could easily have made mistakes that led to that. It does not matter now. You may be as innocent as you say or as guilty as possible. Look down, Corwin. That is all. Look down at the black road. Death is the limit of the distance you travel if that is your doing. I have shown you my strength once again, lest you have forgotten. I can kill you, Corwin. Do not even be certain that your

blade will protect you, if I can get my hands on you but once. And I will, to keep my promise. My promise is only that if you are guilty I will kill you the moment I learn of it. Know also that my life is insured, Corwin, for it is linked now to your own.'

'What do you mean?'

'All of the others are with us at this moment, via my Trump, watching, listening. You cannot arrange my removal now without revealing your intentions to the entire family. that way, if I die forsworn, my promise can still be kept.'

'I get the point,' I said. 'And if someone else kills you? They remove me, also. that leaves Julian, Benedict, Random, and the girls to man the barricades. Better and better—for whoever it is. Whose idea was this, really?'

'Mine! Mine alone!' he said, and I felt his grip tighten, his arms bend and grow tense. 'You are just trying to confuse things! Like you always do!' he groaned. 'Things didn't go bad till you came back! Damn it, Corwin! I think it's your fault!'

Then he hurled me in the air.

Sign of the Unicorn

Slow, steady and strong, Gérard will stop at nothing to protect Amber and those he loves. Suspecting evil deeds, he wasn't above teaching a "lesson" that involved dangling Corwin up-side-down at the top of a cliff

By the same logic, he'll be willing to teach the young pups of Amber similar lessons. If angered, he'll leave a player character bruised and bloody.

Constant Allies & Enemies: Gérard has no fierce enemies. Each Amberite knows that Gérard is a safe haven, a man who will always offer a place of safety when the others turn traitor. He backed Eric, but was willing to make a deal with Corwin. He stood with Julian, Benedict and Llewella, before Corwin managed to get a majority behind his regency.

Neutral Gérard: Gérard has straddled more than one fence in his time. Always sympathetic to the underdog, he won't condemn anyone without sure evidence, and even then he is unlikely to vote for a death penalty. He protected Brand, when Brand was hurt and helpless, and he's likely to protect others.

Gérard as Bully: Gérard, in the opinion of some, is an over-sized jock. A muscle-bound bully who gets his way by pushing people around, squeezing them, and bashing their heads. There's certainly something to be said for that opinion.

JULIAN

Next, there was the passive countenance of Julian, dark hair hanging long, blue eyes containing neither passion nor compassion. He was dressed completely in scaled white armor, not silver or metallic-colored, but looking as if it had been enameled.

Nine Princes in Amber

If you want to approach Amber by land, that means travelling through Forest Arden. And that means getting by Prince Julian, the appointed defender of Arden.

Not as strong as some, nor as skilled in warfare by Amber standards, Julian is a second rater. But how are you going to get at this guy?

First, he knows the depths of Forest Arden, not just the vast dominion seen from Kolvir, but the thousands of Shadows of Arden that form the true land border to Amber. He patrols the Shadow ways through Arden wearing invincible scale armor, surrounded by hand-picked rangers, Hellhounds, hawks, and who knows what else.

And, he rides a horse, if you can call Morgenstern a horse. As Corwin says, "He did create Morgenstern, out of Shadows, fusing into the beast the strength and speed of a hurricane and a pile driver." What do you call a creature who looks like a horse, but who can't be hurt by ordinary bullets, runs up to a hundred miles an hour, and can leap its own height over a wall from a standing start? Something more than just a horse.

Which brings us to the Hellhounds. Hellhounds are pretty fast too, and they don't just chase cars, they *eat* them. Julian has *lots* of Hellhounds.

And we haven't even mentioned Julian's hawks, or his handpicked forest guard...

Julian is also a guy with an attitude. Don't expect him to explain his reasons, his motivations, or his methods. If he likes you, fine; he'll stand by you (he may never bother to <u>tell</u> you, but he will stand by you). Confront him, and, unless you're holding a dagger to his throat, don't expect him to justify himself. Julian is a proud man, and, from his point of view, if you don't believe him, or trust him, you can always start a new career as Hellhound chow.

JULIAN -
THE CORWIN VERSION
(250 Point Version)

I am certain the dogs would have finished it, but at that moment the riders topped the hill and descended. There were five of them, Julian in the lead. He had on his scaled white armor and his hunting horn hung about his neck. He rode his gigantic steed Morgenstern, a beast which has always hated me. He raised the long lance that he bore and saluted with it in my direction. Then he lowered it and shouted orders to the dogs. Grudgingly they dropped away from the prey. Even the dog on the manticora's back loosened its grip and leaped to the ground. All of them drew back as Julian couched the lance and touched his spurs to Morgenstern's sides.

The beast turned toward him, gave a final cry of defiance, and leaped ahead, fangs bared. They came together, and for a moment my view was blocked by Morgenstern's shoulder. Another moment, however, and I knew from the horse's behavior that the blow had been a true one.

A turning, and I saw the beast stretched out, great gouts of blood upon its breast, flowering about the dark stem of the lance.

Julian dismounted. He said something to the other riders which I did not overhear. They remained mounted. He regarded the still-twitching manticora, then looked at me and smiled. He crossed and placed his foot upon the beast, seized the lance with one hand, and wrenched it from the carcass. Then he drove it into the ground and tethered Morgenstern to its shaft. He reached up and patted the horse's shoulder, looked back at me, turned, and headed in my direction.

The Hand of Oberon

Julian is not as skilled, or as strong, or as well versed in Powers, as his brothers. However, when it comes to his particular task, that of protecting the road to Amber that passes through Forest Arden, he is superb.

Current Objectives. Now that the troubles are over, Julian will return to Amber, expecting to regain his old title and position, as advisor to the King. The question is just how well he will be treated by Random.

ATTRIBUTES
PSYCHE -   [5 Points]
STRENGTH - [35 Points]
ENDURANCE - [15 Points]
WARFARE - [45 Points]

POWERS
Pattern Imprint [50 Points]
Power Words [10 Points] - Lifeforce, Trump Disrupt, Process Snuff, Weaken Structure, and Thunder.

CREATURES AND ARTIFACTS OF POWER
Julian's White Scale Armor [6 Points] - Dressed in his armor, Julian is immune from most weapons, especially from bullets, bolts and arrows. The Extra Hard quality adds a little extra to Julian's powerful kicks and punches.
• Invulnerable to all Conventional Weapons [4 points]
• Extra Hard [1 Point]
• Self Healing [1 Point]

Morgenstern was six hands higher than any other horse I'd ever seen, and his eyes were the dead color of Weimaraner dog's and his coat was all gray and his hooves looked like polished steel. He raced along like the wind, pacing the car, and Julian was crouched in his saddle—the Julian of the playing card, long black hair and bright blue eyes, and he had on his scaled white armor.

Nine Princes in Amber

Julian's Horse Morgenstern [21 Points] - Nearly a primal force, Morgenstern is possibly the most powerful animal presented by Zelazny throughout the *Chronicles.*
• Immense Vitality [4 Points]
• Engine Speed [4 Points]
• Amber Stamina [2 Points]
• Combat Reflexes [2 Points]
• Resistance to Firearms [2 Points]
• Extra Hard Hooves and Teeth [1 Point]
• Psychic Sensitivity [1 Point]
• Psychic Resistance [1 Point]
• Shadow Seek [4 Points]
Julian's Hellhounds [39 Points] - Anywhere within the limits of Forest Arden Julian can call on virtually limitless numbers of Hellhounds. While not as fast as Morgenstern, they are capable of running down a speeding car, and eating it.
• Double Vitality [2 points]
• Engine Speed [4 points]
• Amber Stamina [2 points]
• Combat Training [1 Point]
• Double Damage Teeth [2 points]
• Shadow Path [2 Points]
• Horde Quantity [*3 Points]
Julian's Hawks [26 Points] - Julian uses his hawks as scouts in Forest Arden. Whatever they see, he can also see through their eyes.
• Double Speed [2 Points] - Twice as fast as a normal hawk. That means a maximum speed of better than one hundred miles per hour.
• Amber Stamina [2 Points]

- Combat Training [1 Point]
- Extraordinary Psychic Sense [4 Points] - Each of the hawks is capable of communicating with Psyche over a range of many miles, or across Shadow.
- Shadow Seek [4 Points]
- Named & Numbered [*2 Points]

BAD STUFF
[+2 Points]

JULIAN - ARDEN'S CHAMPION (350 Point Version)

'Julian has no friends,' she said. 'That icy personality of his is thawed only by thoughts of himself. Oh, in recent years he seemed closer to Caine than to anyone else. But even that... even that could have been a part of it. Shamming a friendship long enough to make it seem believable, so that he would not be suspect at this time. I can believe Julian capable of that because I cannot believe him capable of strong emotional attachments.'

Fiona, on Julian,
Sign of the Unicorn

Julian's real allegiance may be to Arden itself. This most beautiful of forests may be what softened Julian's heart and what gives him a purpose.

Current Objectives. When the Patternfall War is over Julian will return to his old haunts in the forest. He has sworn fealty to King Random, but he is no man's slave. He'll serve Amber as he sees fit.

ATTRIBUTES
PSYCHE -   [40 Points]
STRENGTH - [25 Points]
ENDURANCE - [25 Points]
WARFARE - [87 Points]

POWERS
Pattern Imprint [50 Points]
Sorcery [15 Points]

CREATURES AND ARTIFACTS OF POWER
Julian's White Scale Armor [6 Points] - as above.
Julian's Horse Morgenstern [21 Points] - as above.
Julian's Hellhounds [39 Points] - as above.
Julian's Hawks [26 Points] - as above.
Julian's Forest Arden Guard [12 Points] - Specially trained and equipped by Julian, each man is trained in horsemanship, woodcraft, and combat.
- Double Normal Stamina [1 Point]

- Combat Reflexes [2 Points]
- Shadow Trail [1 Point]
- Horde [*3 Points]

PERSONAL SHADOWS
Julian's Shadow Arden [4 Points]
  Shadow of the Realm [2 Points]
  Guarded [2 Points]

JULIAN - MASTER CONJURER (450 Point Version)

All of Julian's "investment" in the guardians of Forest Arden may be items that he has created through a unique form of Conjuration.

Current Objectives. Having learned something of the Chaos and its creatures in the recent war, Julian may return to Amber set on performing some new experiments. He may try creating a number of new guardians for the forest, and he will certainly be implementing measures to detect any Logrus users or other Chaos dwellers who attempt its passage.

ATTRIBUTES
PSYCHE -   [15 Points]
STRENGTH - [25 Points]
ENDURANCE - [25 Points]
WARFARE - [87 Points]

POWERS
Advanced Pattern Imprint [75 Points]
Sorcery [15 Points]
"Pattern" Conjuration [30 Points] - Julian is capable of Conjuring creatures and then endowing them with a bit of Pattern reality. He will have all of the creatures described above, but as items that he created rather than "bought."

CREATURES AND ARTIFACTS OF POWER
Julian's White Scale Armor [6 Points] - as above.

GM TIPS FOR PLAYING JULIAN

Julian, Julian, Julian... Is it that I never really knew you? No. Arden's green magic must have softened that old vanity during my long absence, leaving a juster pride and something I would fain call fairness—a thing apart from mercy, to be sure, but an addition to your armory of traits I'll not disparage.

The Courts of Chaos

Let's look at one of the least complicated characters in terms of Power, but, from the point of role-playing games, one of the most interesting.

The Master of the Forest of Arden is a guy who rides around on a creature of legend, Morgenstern, a creature who looks like a horse, but who can't be hurt by ordinary bullets, is capable of out-running a car, and can leap over a car from a standing start. Julian also leads countless numbers of Hellhounds. And he sees through the eyes of a hawk. And he leads a mounted patrol straight out of legend.

Yet, in the books, our first impression of the guy tells us he's a wimp. After all, Corwin manages to totally defeat him in one-on-one combat.

Or does he?

Let's look at this logically. Here's Julian, armored in invincible scale armor, mounted on Morgenstern the Monster, surrounded by Hellhounds. Hellhounds who, unlike ordinary dogs, don't just chase cars, they <u>eat</u> them.

Perhaps Julian, like most Amberites, had a reason for losing to Corwin...

From the point of view of the Game Master, Julian is a harsh character. He'll be pretty rough with the player characters, seeing them as brash youngsters who need an occasional firm hand. When asked a direction, rather than just saying and pointing, he's likely to take a shoulder and turn the character in the right direction.

Any player character who mouths off to Julian is, sooner or later, going to be taught a lesson. This will usually come when the character ends up in Forest Arden. Julian can arrange for someone to be lost and hungry for a few days, or threatened by nasty critters, or just chased by Hellhounds. He won't do them (much) damage, and he'll personally supervise their rescue, but not without a caustic comment or two.

JULIAN AS FATHER

Julian could be a wonderful father, the best of the bunch. His home, Arden, could be a wonderland for any growing player character. It's possible that its hunting lodges (and even tree houses), could be home to a loving family. Julian would be there often, always a present in hand, a mystery to be discovered, or another gem of Arden to be explored, and with the time to spend as a father.

It's also likely that Julian will gift any son or daughter with their own creatures. A horse related to Morgenstern, a Hellhound or two, or even a hawk. They'll be part of Julian's points, and subject to Julian's orders, but devoted to protecting the player character.

Julian the Hunter. For other player characters there might be a problem with Julian's sports. Opposed to Julian's blood sports, a player character of delicate sensibilities might have a more distant relationship, punctuated with frequent arguments.

Julian of the Hounds. On the other hand, Julian is so tough, and so cold, that he could inspire hatred in anyone. A player character may say, "Damn him! He thinks he can me control me the way he shapes his hounds! I'll never let him tell me what to do..."

JULIAN AS A PLAYER IN THE GAMES OF THE AMBER COURT

'Your words are ill-considered,' I said.

'Not so. I considered every one of them,' he answered. 'We spend so much time lying to one another that I decided it might be amusing to say what I really felt. Just to see whether anyone noticed.'

Corwin & Julian,
Sign of the Unicorn

Julian was one of Eric's strongest allies, yet Eric treated him with contempt. Julian stood by Corwin, who showed him no trust. Fact is, Julian seems more devoted to Amber and Arden, than to any ruler. He sees the maneuvering of his siblings and he does not approve.

Constant Allies & Enemies: Julian's closest friend is no doubt Caine and the two of them will likely be allies, conspirators, and possibly even partners. He has a lot of respect and love for Fiona, in spite of the fact that she probably despises him. Julian is not popular with Random or Corwin, and he is distrusted by most of the others.

LLEWELLA

Next was Llewella, whose hair matched her jade-colored eyes, dressed in shimmering gray and green with a lavender belt, and looking moist and sad. For some reason, I knew she was not like the rest of us. But she, too, was my sister.

Nine Princes in Amber

Not every Amberite is a player in the game of power.

One who sits out the game is Llewella, the green-haired Amberite maiden.

If there was an invitation to join the family of Oberon, Llewella must have accepted with reluctance. Or, as Corwin put it, hers was not a presence often felt in Amber.

Llewella prefers the sea, the underwater city of Rebma, an exact reflection of Amber under the water.

Were it not that her personal retreat, the undersea realm of Rebma, is so close to Amber, she might never have appeared in Corwin's tale. And, like the other absent ones, who Corwin numbers but never describes, Llewella might have remained, as she wishes, unknown and untroubled.

In Rebma, things are different. There, in the watery city, Llewella is sister to the Queen, Moire, and wields considerable power. Enough to offer sanctuary if she wishes. Or punishment...

And so, in Amber, you'll find her reluctant to participate in the games. She'll neither criticize, nor praise. Although she may, if you perform particularly well, give you the occasional kiss on the cheek.

Llewella, off in a corner, pretending to study a book, had her back to the rest of us, her green tresses bobbed a couple of inches above her dark collar. Whether her withdrawal involved animus, self-consciousness in her alienation, or simple caution, I could never be certain. Probably something of all these. Hers was not that familiar a presence in Amber.

Sign of the Unicorn

LLEWELLA - RELUCTANT PRINCESS
(200 Point Version)

Corwin always describes Llewella with her back to the rest of the family. A sister, but a reluctant sister, as if their insistence on her involvement were an unpleasant imposition.

Current Objectives. Llewella will continue her previous policy. She will avoid any entanglements, give help if asked, and otherwise stay uninvolved.

ATTRIBUTES
PSYCHE -   [45 Points]
STRENGTH - [35 Points]
ENDURANCE - [15 Points]
WARFARE - [35 Points]

POWERS
Pattern Imprint [50 Points]
Power Words [15 Points] - Chaos Negation, Lifeforce, Trump Disrupt, Process Surge, Process Snuff, Shade, Light Strobe, Spark, and Weaken Structure.

GOOD STUFF
[5 Points]

LLEWELLA - DEEP PLAYER IN AMBER'S FAVORITE GAME
(300 Point Version)

'Rebma is the ghost city,' he told me. 'It is the reflection of Amber within the sea. In it, everything in Amber is duplicated, as in a mirror. Llewella's people live there, and dwell as though in Amber.'

Nine Princes in Amber

We were taken to the palace in the center of the city, and I knew it as my hand knew the glove in my belt. It was an image of the palace of Amber, obscured only by the green and confused by the many strangely placed mirrors which had been set within its walls, inside and out...

Nine Princes in Amber

Consider also the description of Rebma's palace. Corwin speaks of strangely placed mirrors. Could it be that they have some function other than decoration? Perhaps they could be used to spy upon the real Amber?

Occasionally, throughout the tale, we hear that events, and even objects (such as the Jewel of Judgement), have their reflection in Rebma.

Yet it is said that Llewella "retired" to Rebma. It seems strange that someone so disinterested would find a home in a place so perfectly suited for keeping tabs on events in Amber.

The fact that Llewella spends all of her time in Rebma may have more importance than Corwin suspects. She may, in fact have discovered a means of keeping track of all the various events in Amber.

Current Objectives. Her base in Rebma will remain her constant home and seat of power. She will return there as soon as possible and will be reluctant to leave.

ATTRIBUTES
PSYCHE -   [85 Points]
STRENGTH - [30 Points]
ENDURANCE - [40 Points]
WARFARE - [25 Points]

POWERS
Advanced Pattern Imprint [75 Points]
Sorcery [15 Points]
Modified Conjuration [25 Points] - Llewella is the creator of the various mirrors that hang in strange places throughout the Castle in Rebma. Each is empowered to look into the related position in Castle Amber.

GOOD STUFF
[5 Points]

LLEWELLA - QUIETLY COMPETENT AMBERITE
(350 Point Version)

No more and no less than she appears, Llewella is simply a capable Princess of Amber. While waiting her turn in the succession to the throne, she hasn't wasted her time like her more frivolous siblings. After all, she will either be Queen or she will not, and if it is to be, then she will be the best Queen possible.

Current Objectives. Continuing to hone her many talents, it is possible that Llewella will seek to be named as one of the emissaries to the Courts of Chaos. Another possibility is that she will seek out Corwin and persuade him to take her to the new Pattern.

ATTRIBUTES
PSYCHE - [50 Points]
STRENGTH - [50 Points]
ENDURANCE - [50 Points]
WARFARE - [50 Points]

POWERS

Pattern Imprint [50 Points]
Trump Artist [40 Points]
Shape Shifting [35 Points]
Sorcery [15 Points]

GOOD STUFF

[10 Points]

GM TIPS FOR PLAYING LLEWELLA

Llewella, you possess reserves of character the recent situation did not call upon you to exercise. For this, I am grateful. It is sometimes pleasant to emerge from a conflict untested.

The Courts of Chaos

Corwin thinks Llewella to be quiet and demure, the most withdrawn of Amberites. That leaves the door open for the Game Master to visualize any number of possible interior Llewellas.

LLEWELLA AS MOTHER

Here's a wide open set of possibilities. Llewella could be quiet and loving, or quiet, withdrawn and unknowable, or quiet and hostile. She could fulfil a variety of roles as parent of a player character.

An upbringing in Rebma would probably be a positive experience, with the player character raised as one of the royal family. On the other hand, most player characters will obviously be strangers in Rebma, with quite a different appearance. That means they may have some problems growing up.

LLEWELLA AS A PLAYER IN THE GAMES OF THE AMBER COURT

Llewella returned, bearing a tray containing slabs of meat, half a loaf of bread, a bottle of wine, and a goblet. I cleared a small table and set it beside Gérard's chair. As Llewella deposited the tray, she asked, 'But why? That leaves only us. Why would one of us want to do it?'

I sighed.

'Whose prisoner do you think he might have been?' I asked.

'One of us?'

'If he possessed knowledge which someone was willing to go to this length to suppress, what do you think? The same reason also served to put him where he was and keep him there.'

Her brows tightened.

'That does not make sense either. Why didn't they just kill him and be done with it?'

I Shrugged.

'Must have had some use for him,' I said. 'But there is really only one person who can answer that question adequately. When you find him, ask him.'

'Or her,' Julian said. 'Sister, you seem possessed of a superabundance of naïveté, suddenly.'

Her gaze locked with Julian's own, a pair of icebergs reflecting frigid infinities.

'As I recall,' she said, 'you rose from your seat when they came through, turned to the left, rounded the desk, and stood slightly to Gérard's right. You leaned pretty far forward. I believe your hands were out of sight, below.'

'I would have had to do it with my left hand - and I am right-handed.'

'Perhaps he owes his life to that fact.'

'You seem awfully anxious, Julian, to find that it was someone else.'

'All right,' I said. 'All right! You know this is self-defeating. Only one of us did it, and this is not the way to smoke him out.'

'Or her,' Julian added.

Sign of the Unicorn

It seems that Llewella could care less about the politics of Amber. Her home is elsewhere, and the doings of her bickering brothers and sisters is merely annoying. She seems to be sitting out this particular dance.

But wait! Llewella hasn't retreated to some quiet Shadow. No, she lives in another Court, the Court of Rebma where Queen Moire reigns. A court that is described as the perfect mirror image of Amber. She may serve as informer and spy, and play a full part in the game of the Amber Court.

Constant Allies & Enemies: She seems something of an enigma, and has neither enemies nor close friends among her siblings. However, when everyone was pushed into taking sides her choice was interesting. It happened that fateful day in the Library of Castle Amber, just after Caine's murder. At one end of the room Dierdre and Random clustered around Corwin. At the other end, Julian and Gérard greeted Benedict. Llewella choose Benedict's party.

RANDOM

...a wily-looking little man, with a sharp nose and a laughing mouth and a shock of straw-colored hair. He was dressed in something like a Renaissance costume of orange, red, and brown. He wore long hose and a tight-fitting embroidered doublet. And I knew him. His name was Random.

Nine Princes in Amber

It's hard to talk about one without the other. Random would not have become king without Vialle, that blind maiden who he was forced to marry. He would probably not have found the greatness inside himself until he found her love. She brought out a fierce gentleness in Random, drawn out of the hurt of his inner self.

Think about Random's upbringing. His older brothers were these guys, Benedict, Bleys, the whole lot of angry shouting gods. How would you like to be little brother to these guys? Worse, what would it be like to be the shortest guy in the family of Amber?

Random ended up with something of an attitude problem. It got him in big trouble, that attitude...

In Rebma, he pushed a little too far. Ran off with Morganthë the Princess Royal. But couldn't, or wouldn't keep her. She ended up pregnant, and, after baby Martin was born, Morganthë committed suicide.

His escape was out into Shadow, to hang-gliders, late-night jazz clubs (he's got a second career as a drummer), later-night high-stakes poker, and good-looking women. As Random says, not necessarily in that order.

But when push came to shove, when he was called upon to be great, to love a woman, to risk his life for a friend, to trust a brother (perhaps hardest of all for an Amberite), Random rose to the occasion.

So, here is Random. Not the best in anything in particular; no rival for his brothers in guile, or power, or the intricacies of politics, Random became king because... Because he is a better man.

Perhaps that's why we love Amber. Because the best man became king. Not the strongest, not the most powerful, not one with the strongest armor. The best man was Random because he was the Amberite who loved best.

RANDOM - CORWIN'S FRIENDLY YOUNGER BROTHER (200 Point Version)

I know damn well that Gérard would have chosen that moment to attack. The big bastard would have strode forward with that monster blade of his and cut the thing in half. Then it probably would have fallen on him and writhed all over him, and he'd have come away with a few bruises. Maybe a bloody nose. Benedict would not have missed the eye. He would have had one in each pocket by then and be playing football with the head while composing a footnote to Clausewitz. But they are genuine hero types. Me, I just stood there holding the blade point upward, both hands on the hilt, my elbows on my hips, my head as far back out of the way as possible. I would much rather have run and called it a day...

Random's tale,
Sign of the Unicorn

Corwin sees Random as an older brother looking at the baby of the family. A nice kid, sometimes misguided, always a little behind, but with all the best and little of the worst family qualities.

Current Objectives. Right now Random wants nothing more than to consolidate his rule over Amber, and to avoid trouble. He'd like to spend more time with Vialle, and he's not exactly happy with all the duties that kingship has imposed upon his life, but he's pretty content. He'll try to arrange to have at least two of his brothers in Amber at all times, just to keep the realm safe and secure.

ATTRIBUTES
PSYCHE - [25 Points]
STRENGTH - [25 Points]
ENDURANCE - [25 Points]
WARFARE - [50 Points]

POWERS
Pattern Imprint [50 Points]
Initiate of the Jewel of Judgement [10 Points]
Power Words [10 Points] - Magic Negation, Chaos Negation, Resume True Form, Defensive Luck and Process Snuff.

PERSONAL SHADOW

Texorami was a wide open port city, with sultry days and long nights, lots of good music, gambling around the clock, duels every morning and in-between mayhem for those who couldn't wait. And the air currents were fabulous. I had a little red sail plane I used to go sky surfing in, every couple of days. It was the good life. I played drums till all hours in a basement spot up the river where the walls sweated almost as much as the customers and the smoke used to wash around the lights like streams of milk. When I was done playing I'd go find some action, women, or cards, usually. And that was it for the rest of the night...

Random, on his old haunt,
Sign of the Unicorn

Random's Shadow Texorami [2 Points]
Personal Shadow [1 Point]
Control of Contents [1 Point]

GOOD STUFF
[3 Points]

RANDOM - AMBER'S JUVENILE DELINQUENT
(300 Point Version)

'He was a homicidal little fink, who I recalled had always been sort of a rebel.'

Nine Princes in Amber

Maybe he's recently turned over a new leaf, but plenty of people remember Random as he was; spoiled rotten, nasty, ill-tempered, and irresponsible. Random was the darling baby of the family, and he grew up into a snarling, resentful miscreant. He's probably the best of his family, but that's like saying he was the most reformed prisoner in the penitentiary.

Current Objectives. Random could be engaged in an all-out campaign to ferret out opposition and dissent. It's not difficult to imagine Amber with both internal secret police and a wide- ranging network of spies. Random's control over these covert activities would be absolute, and he would not hesitate to use them to snoop on the younger Amberites.

ATTRIBUTES
PSYCHE - [50 Points]
STRENGTH - [35 Points]
ENDURANCE - [35 Points]
WARFARE - [75 Points]

POWERS
Pattern Imprint [50 Points]
Initiate of the Jewel of Judgement [10 Points]
Power Words [10 Points] - Magic Negation, Chaos Negation, Resume True Form, Defensive Luck and Process Snuff.
Sorcery [15 Points] - Aside from a full range of spells, Random will have prepared some personal object with the potential to contain spells.
Conjuration [20 Points]

RANDOM - LATENT POWERHOUSE
(400 Point Version)

He was a little guy, maybe five-six in height, weighing perhaps one thirty-five. But he sounded as if he were talking dead serious talk. I felt reasonably sure that he meant it when he said he'd take on two or three bruisers, singlehanded...

Nine Princes in Amber

Since Random consistently displays surprising potential, all the way to the penultimate moment of the *Chronicles of Amber*, it could be that he was more prepared for the task of kingship than he let on.

Current Objectives. Random's degree of control over the throne will mean that he takes much more than a casual interest in the affairs of the younger Amberites. He'll be carefully watching each of the next generation, evaluating and judging, and, most importantly, making sure there is no young Brand in the making.

ATTRIBUTES
PSYCHE - [65 Points]
STRENGTH - [50 Points]
ENDURANCE - [55 Points]
WARFARE - [100 Points]

POWERS
Advanced Pattern Imprint [75 Points]
Master of the Jewel of Judgement [25 Points] - Random's control over the Jewel of Judgement is absolute. He is capable of controlling it from a distance, and of locating it throughout Shadow. At a minimum he can use the Jewel of Judgement to manipulate the weather of Amber, to observe surrounding Shadow, and to affect Shadow Storms.
"Pattern" Sorcery [15 Points] - Aside from a full range of spells, Random will be able to cast spells based on Pattern, and hang spells in the Jewel of Judgement.

GOOD STUFF
[5 Points]

GM TIPS FOR PLAYING RANDOM

He staggered in and immediately pushed the door shut behind himself and shot the bolt. There were lines under those light eyes and he wasn't wearing a bright doublet and long hose. He needed a shave and he had on a brown wool suit. He carried a gabardine overcoat over one arm and wore dark suede shoes. But he was Random, all right—the Random I had seen on the card—only the laughing mouth looked tired and there was dirt beneath his fingernails.

Nine Princes in Amber

Playing Random is pretty straight-forward. He is, and seems to be, a basically nice guy. Friendly, open, and, when he gets the chance, fun-loving. He seems honestly concerned about the welfare of Amber and its family.

On the other hand, Random spent most of his life carousing and having a very good time just enjoying himself. The drudgery part of the job of being King of Amber is bound to get to him eventually. Which means that Random will be happy to oblige anyone who comes to him looking for more responsibility. Amber has, at any given time, oodles of major problems and a hundred times as many little ones. Anyone offering to help is not going to be turned away.

RANDOM AS A FATHER

Random sowed a lot of oats in his wilder days, before he became royal liege of Amber. One of the more interesting aspects of any child of Random is that they automatically become the direct heirs to the Throne of Amber.

Fun-Loving Random. It's not difficult to imagine the kind of "hands on" fatherhood that Random would provide. He's capable of getting down on the floor with a toddler, cuddling a young kid, running around in a family football game, and talking honestly with an adolescent. Plus, he'll give a child plenty of room to grow. In short, Random could give what every youngster needs most, time and affection.

Random the Runaway. Random's first known child was abandoned. He met the woman, quarrelled with her, returned her to her home, and never returned to claim his child after she committed suicide. He could have done the same elsewhere, and a player character could be the unfortunate result.

RANDOM AS A PLAYER IN THE GAMES OF THE AMBER COURT

Being King doesn't mean Random can ignore politics. He has favors to trade, but he needs more information, support, and allegiances than anyone else.

Constant Allies & Enemies: Corwin is probably Random's best buddy. Unfortunately, Corwin is not necessarily reliable, he tends to be moody, and takes off a lot. Not someone Random can pick for Minister of Defense.

Fact is, Random has to rely on the same old crowd that supported Oberon, Eric and Corwin. The trio of Gérard, Julian and Caine are the only ones who can be trusted to stay put and do their jobs. Caine in particular must be a thorn in Random's side, since Caine always needles Random.

Benedict and Bleys are better fighters, but they tend to travel, not a trait you'd appreciate in someone responsible for the Castle's defenses.

QUEEN VIALLE

I heard her footsteps and then the door swung in. Vialle is only a little over five feet tall and quite slim. Brunette, fine featured, very soft-spoken. She was wearing red. Her sightless eyes looked through me, reminding me of darkness past, of Pain.

The Hand of Oberon

Random is the only one of his family that we know to be in a happy marriage. He loves Vialle, and she loves him. When he was imprisoned in the dungeons of Amber, she willingly joined him. Random shares everything with her, tells her secrets that only Amberites know, and values her counsel.

Vialle is a fairly important member of the Court of Rebma. At least important enough for her to be of personal interest to Queen Moire.

In place of her sight, Vialle seems to have a keen insight into people and their emotions. Her Psyche is probably above Amber Rank, though its doubtful that she has any other Powers.

DWORKIN

He was a small man. Tiny, might be an even better word. He was around five feet tall and a hunchback. His hair and beard were as heavy as my own. The only distinguishing features in that great mass of fur were his long, hook nose and his almost black eyes, now squinted against the light.

Nine Princes in Amber

Creator of the Trumps. Creator of Pattern. Hunch-backed with an unkept white beard. Nutty as a fruitcake.

To know him...

No. You can't really know Dworkin. You might talk to him, you might even get him to do you a favor. You can't understand this guy any more than you can understand a force of nature, or a planet.

This genial old geezer looks pretty dwarf-like, what with his long beard and hunchback and all. Don't let the looks deceive you, Dworkin's powers are deep and mysterious, and he likely knows more about Pattern, and Logrus, and Trump, and Magic, than anyone.

Unfortunately, conversation with Dworkin is risky, for he is quite mad. And, when the fits overtake him, he is likely to take the forms of demons, or worse...

Which kind of explains his current digs. Dworkin lives in a cave. A cave located next to the Primal Pattern itself. Occasionally Dworkin is locked in. A monster is always posted at the massive metal door of his cave. The monster is not there to protect Dworkin, nor is it there to protect you from Dworkin. No, the monster is there to protect the Primal Pattern itself, just in case Dworkin decides to erase his creation.

Even when Dworkin is sane, lucid, and more-or-less human (though he might be a dwarf, so the rumor goes), he is still dangerous. He may believe a lie, or not, and never let on. He may seem to help by accident, when it may actually serve a grand scheme.

With Dworkin, you'll never know.

DWORKIN - DEALER OF TRUMP (400 Point Version)

I fingered my cards, weighed the deck in my hand. I could try a contest of wills through them, with either Eric or Caine. There was that power present, and perhaps even others of which I know nothing. They had been so designed, at the command of Oberon, by the hand of the mad artist Dworkin Barimen, that wild-eyed hunchback who had been a sorcerer, priest, or psychiatrist—the stories conflicted at this point - from some distant Shadow where Dad had saved him from a disastrous fate he had brought upon himself. The details were unknown, but he had always been a bit off his rocker since that time. Still, he was a great artist, and it was undeniable that he possessed some strange power. He had vanished ages ago, after creating the cards and tracing the Pattern in Amber. We had often speculated about him, but no one seemed to know his whereabouts. Perhaps Dad had done him in, to keep his secrets secret.

Nine Princes in Amber

Perhaps, as has been sometimes speculated, perhaps Trump is the true power of creation in the *Amber* universe. Pattern might be a construct of Trump. Logrus is another result of Trump Artistry. In fact the entire universe might be an artistic yearning, a piece of emotional creation. This version of Dworkin is, in some ways, the most frightening of all, since characters might exist only because of their images rendered in Trump, and may cease to exist if erased.

Current Objectives. It's hard to imagine that this version of Dworkin would want to do much more than continue his artistic efforts. In all probability he would insist on creating new Trump for each new player character, and would distribute sufficient decks to update all the elder Amberites and to allow each young character to have a deck.

ATTRIBUTES
PSYCHE -   [45 Points]
STRENGTH - [10 Points]
ENDURANCE - Chaos Rank
WARFARE - Amber Rank

POWERS
Advanced Pattern Imprint [75 Points]
Attuned to the Jewel of Judgement [10 Points]
Advanced Logrus Mastery [70 Points]
"Exalted" Trump Artistry [100 Points] - What is Pattern but a drawing, an extension of Trump Artistry into the abstract? Could it be that the warped mind of Dworkin could, like some crazed Van Gogh, transcend the limitations of the real and find a chord of truth? Such a Dworkin could do virtually anything with Trump.
Shape Shift [35 Points]
Power Words [30 Points]
Sorcery [15 Points]
Advanced Conjuration [25 Points] - Dworkin's control over Conjuration allows him to create items in combination with his Trump Artistry. For example, he can create multiple decks from a single original.

BAD STUFF
[+5 Points]

DWORKIN - SCION OF CHAOS (500 Point Version)

'...My flight from Chaos to this small sudden island in the sea of night? My meditations upon the abyss? The revelation of the Pattern in a jewel hung around the neck of a unicorn? My transcription of the design by lightning, blood, and lyre while our fathers raged baffled, too late come to call me back while the poem of fire ran that first route in my brain, infecting me with the will to form? Too late! Too late... Possessed of the abominations born of the disease, beyond their aid, their power, I planned and built, captive of my new self...'

Dworkin speaking as if to Oberon,
Hand of Oberon

As far as we know, Dworkin started out as a Lord of Chaos. Even though an exile, and chained to immutability by his connection with the Pattern, he remains a powerful Logrus Master. His relatives in the Courts of Chaos may have rejected him, but that will not keep him from monitoring events "back home."

Current Objectives. Hard to say. Once Dworkin accompanies Oberon's funeral procession into the Courts of Chaos, he may be inclined to stay there for some time. If and when he returns to Amber, it will likely be to resume some important position in the Court, as a chancellor in King Random's circle of advisors.

ATTRIBUTES
PSYCHE -   [35 Points]
STRENGTH - [25 Points]
ENDURANCE - [25 Points]
WARFARE - [60 Points]

POWERS
Advanced Pattern Imprint [75 Points]

Attuned to the Jewel of Judgement [10 Points]
"Exalted" Logrus Mastery [100 Points] - Somehow Dworkin has taken the understanding of Logrus beyond its ever-changing nature, all the way to the underlying rules that govern its movement. Just as the mathematics of Chaos, orderly in their formulations, predict the unpredictable, so Dworkin found a way to define the rules behind Logrus. It was this knowledge that allowed him to find the true purpose of the Jewel of Judgement, and that showed him how to inscribe the symbol that is the opposite of Logrus; the Pattern.
Advanced Trump Artistry [60 Points]
Advanced Shape Shift [65 Points]
Power Words [10 Points]
Sorcery [15 Points]
Conjuration [20 Points]

DWORKIN - THE OLD SHAPE SHIFTER (650 Point Version)

'Funny. Funny, funny, funny,' he said. 'You come as the young Lord Corwin, thinking to sway me with family sentiment. Why did you not choose Brand or Bleys? It was Clarissa's lot served us best.'

I shrugged and stood.

'Yes and no,' I said, determined now to feed him ambiguities for so long as he'd accept them and respond. Something of value might emerge, and it seemed an easy way to keep him in a good humor. 'And yourself?' I continued. 'What face would you put on things?'

'Why, to win your good will I'll match you,' he said, and then he began to laugh.

He threw his head back, and as his laughter rang about me a change came over him. His stature seemed to increase, and his face luffed like a sail cut too close to the wind. The hump on his back was diminished as he straightened and stood taller. His features rearranged themselves and his beard darkened. By then it was obvious that he was somehow redistributing his body mass, for the nightshirt which had reached his ankles was now midway up his shins...

I regarded a slightly slimmer version of myself.

The Hand of Oberon

Dworkin, as Shape Shifter, will never be safe company. His mind and body are constantly in a state of adaptation and agitation, as he struggles *not* to become a part of the Pattern itself. Dworkin, in this state, can never be considered reliable, as his form, and his personality, can change moment by moment.

Current Objectives. By retreating back to his old cave, near the Primal Pattern, Dworkin can resume a measure of control over his ever-shifting body and mind. He will remain in the proximity of the Pattern for so long as he is able. Although he is mightily interested in the affairs of the player characters, it is very difficult for him to see through their eyes. His view of the universe is just too large.

ATTRIBUTES
PSYCHE -　[100 Points]
STRENGTH - [20 Points]
ENDURANCE - [125 Points]
WARFARE - [40 Points]

POWERS
Advanced Pattern Imprint [75 Points]
Attuned to the Jewel of Judgement [10 Points]
Advanced Logrus Mastery [70 Points]
Advanced Trump Artistry [60 Points]
"Exalted" Shape Shift [135 Points] - Dworkin's major power, Shape Shifting, means that the Pattern is not a separate thing, but just one "aspect" of Dworkin himself. In a very real sense Dworkin is the Pattern.

An even weirder thought, perhaps Dworkin is Oberon, the Unicorn, and all of the infinite inhabitants of Amber and Shadow. All are just Shadows of Dworkin, fragments of him. Scary, huh?
Sorcery [15 Points]

DWORKIN - EMBODIMENT OF THE PATTERN (700 Point Version)

'I am the Pattern,' he said, 'in a very real sense. In passing through my mind to achieve the form it now holds, the foundation of Amber, it marked me as surely as I marked it. I realized one day that I am both the Pattern and myself, and it was forced to become Dworkin in the process of becoming itself. There were mutual modifications in the birthing of this place and this time, and therein lay our weakness as well as our strength. For it occurred to me that damage to the Pattern would be damage to myself, and damage to myself would be reflected within the Pattern. Yet I could not be truly harmed because the Pattern protects me, and who but I could harm the Pattern? A beautiful closed system, it seemed, its weakness totally shielded by its strength.'

He fell silent. I listened to the fire. I do not know what he listened to.

Then, 'I was wrong,' he said. 'Such a simple matter, too... My blood, with which I drew it, could deface it. But it took me ages to realize that the blood of my blood could also do this thing. You could use, you could also change it - yea, unto the third generation.'

The Hand of Oberon

If Dworkin is really a part of the Pattern, he will have a perspective on events that is very wide-reaching. Which does not mean that he will understand the threats to Amber, or their solutions. Dworkin's problem will be that he can see everything, but understand very little.

Current Objectives. Dworkin can never really enter into the social life of Amber, since his mind is constantly on other things. As the servant of the Pattern, he will live in his cave by the Primal Pattern, venturing out only occasionally.

ATTRIBUTES

PSYCHE -   [150 Points]
STRENGTH - [100 Points]
ENDURANCE - [100 Points]
WARFARE - Chaos Rank

POWERS

"Exalted" Pattern Imprint [100 Points] - Because Dworkin really is the Pattern, and the Pattern is Dworkin, there is no separation between Dworkin's thoughts and the response of the Pattern. He must be constantly vigilant, for a single stray concern, a moment of inattention, and his link to the Pattern could bring about change through infinite universes.

Master of the Jewel of Judgement [25 Points] - Dworkin's control over the Jewel of Judgement, now that the Primal Pattern has been mended, is nearly absolute. From anywhere to anywhere, Dworkin can contact the Jewel, transport himself to it, or it to him, or simply work it's powers at a distance. Dworkin can also use the Jewel to examine other versions of Amber, other realities, and any other Patterns that may come into being.

Advanced Logrus Mastery [70 Points]
Advanced Trump Artistry [60 Points]
Advanced Shape Shift [65 Points]
Power Words [10 Points]
Sorcery [15 Points]
Conjuration [20 Points]

BAD STUFF
[+5 Points]

GM TIPS FOR PLAYING DWORKIN

Turning, I saw him just beyond the threshold. About five feet in height, hunchbacked. His hair and beard were even longer than I remembered. Dworkin wore a nightshirt which reached to his ankles. He carried an oil lamp, his dark eyes peering across its sooty chimney.

'Oberon,' he said, 'is it finally time?'

'What time is that?' I asked softly.

He chuckled.

'What other? Time to destroy the world, of course!'

The Hand of Oberon

Wacko? Drunk? Wierd? Inconsistent? Erratic?

Or just a kindly old great-grandfather?

All are possible. If you, as Game Master, ever had the urge to develop your play-acting, Dworkin provides a great opportunity.

"I could not tell whether he spoke literally or metaphorically or was simply sharing paranoid delusions," said Corwin. So feel free to have Dworkin make up all kinds of stories, or to darkly hint at awful things lurking in the corners, or whatever comes to mind. Just remember that anything Dworkin says, even the most insane and paranoid ranting, could easily be the literal truth.

Personally, I love playing Dworkin in fits and starts. He'll be lucid and comprehensible, then suddenly stop and focus his attention elsewhere. He'll suddenly forget who he has been talking to, and will demand their names, or confuse them with someone else altogether. Best of all, it's really disconcerting to the players when Dworkin will break off in the middle of some cataclysmic announcement...

Well, an example would really be better...

GM: There's a long, dark hallway, but you can see an open doorway up a little ways, from which a bright bluish light is spilling out into the hall.

Kevin: We'll approach.

Beth: Yes.

GM: A short, stooped-over figure backs out of the open doorway into the hallway, mumbling something, and holding what seems to be a large mug or coffee cup. What are you doing?

Beth: Yvonne will say hello.

Kevin: I'll look into the open doorway.

GM: The man, who has a long, long beard, gives a start at your greeting, and spills some liquid from the mug onto the floor. Kevin, you can see over the man easily enough, into a room lit by a number of lanterns or bulbs, emitting bright blue light. You also see a large table covered with books, dark brown pots, glassware, and something that looks like a bunsen burner.

Beth: I'm sorry! I'll tell the man. We mean you no harm. What else do I notice about him?

GM: He's wearing something like a shapeless gown or dress, covered with dozens of multicolored stains. He's something less than five feet tall. He looks at you with black eyes, as if totally confused by your words. Kevin, the door is swinging shut, what are you doing?

Kevin: I'll hold it open.

Beth: I'll ask the guy if he's okay.

GM: Beth, try that again in Yvonne's voice.

Beth: Old sir? Are you all right? Can you understand me?

GM: He looks from you down to the floor, and back up at you again. What are you doing?

Beth: Is he showing any signs of recognizing me?

Kevin: Hey, what's on the floor?

GM: You see a bit of steam, right where the old man spilled his cup. There seems to be something moving inside of it. He speaks to Yvonne, saying, "Child, why are you out so late? Is it not past your bedtime?"

Beth: Child? Why is he calling me a child?

GM: I don't know, maybe you should ask him. Kevin, there's definitely something moving around in that bit of steam, and it seems to be growing.

Kevin: I don't like the looks of this. Can I get a better look?

GM: Sure, if you want to lean over. Beth?

Beth: Sir, I am not a child. I am Yvonne, and this is Roderick, what is your name?

GM: Hmmm..., he seems a bit confused, then he sort of comes back into focus. I am Dworkin, of course. Who did you take me for? Has that twin of mine been masquerading as me again?

Beth: Dworkin! Oh, boy... Dworkin, honored elder, I'm pleased to finally meet you.

Kevin: Twin? What twin? Are there more Dworkins?

GM: Beats me Kevin. What are you doing?

Kevin: I'll squat down and look at the steaming spot on the floor.

GM: There seems to be a little winged creature, struggling with tiny arms and hands to free itself from something sticky. Beth, Dworkin looks at you and says, "You are not who you say you are. You are not lying, I can tell, but you are either deceived or you have not been informed."

DWORKIN AS FATHER

Dworkin would be a rare choice as father. Generally he's reserved for characters with a great many problems (a lot of Bad Stuff).

Even if Dworkin isn't necessarily suitable as a father, he still makes for a great guardian. Can't you just see him stealing in somewhere, in the dead of night, and snatching away a tiny baby? Then raising it as his own?

As a father, Dworkin would make a much better grandfather. He'll be distracted most of the time. He'll also be a disturbed monster some of the time, so he'll have to either send the kid away a lot, or else arrange for guardians capable of protecting a child from a shape-shifted Dworkin.

One possibility is to have a child, natural or adopted, of Dworkin's raised by a variety of demonic and supernatural creatures. Rather than just arranging for a human nurse, he might summon and conjure whatever creature seems most suitable for the child's current needs.

DWORKIN AS A PLAYER IN THE GAMES OF THE AMBER COURT

Dworkin has always been rather avuncular in Amber. He helps when he is asked, provides Trump to all who request them, and generally assists the King whenever he is called. Otherwise, he seems rather uninvolved.

If Dworkin gets involved in Amber politics at all, it will either be in the form of demands, ordering the court to do his bidding, or in the form of such subtle maneuvering that his hand will never be seen.

Constant Allies & Enemies: He has spent many hours tutoring Bleys, Brand and Fiona, so he may have some fondness for them. As for the others, he treats them as children. Except, perhaps, Benedict, who he has described as "having the mark of doom."

BENEDICT'S OLDER BROTHERS

'...Benedict is the eldest. His mother was Cymnea. She bore Dad two other sons, also—Osric and Finndo. Then—how does one put these things?—Faiella bore Eric. After that, Dad found some defect in his marriage with Cymnea and had it dissolved—ab initio, as they would say in my old shadow—from the beginning. Neat trick, that. But he was the king.'

'Didn't that make all of them illegitimate?'

'Well, it left their status less certain. Osric and Finndo were more than a little irritated, as I understand it, but they died shortly thereafter...'

Sign of the Unicorn

And that's the only time that Osric, or his brother Finndo, are mentioned in Zelazny's *Chronicles of Amber*.

Of course, we don't know that Osric and Finndo are dead. We only know that Corwin said they were dead. He certainly didn't go into detail. Corwin also said, "There had been fifteen brothers and six were dead. There had been eight sisters, and two were dead, possibly four."

All of which sounds a little shady. Especially considering that in Zelazny's later books some of these "dead" siblings show up looking very lively indeed.

So, perhaps Corwin is lying. Perhaps he's just misinformed.

In any case, Osric and Finndo are perfect foils for any *Amber* campaign. Feel free to use the details that follow. Or, if you prefer, make up your own versions. One possible alteration would be to have Finndo and Osric already allied at the start of the campaign and sharing their considerable resources.

OSRIC

If a player character comes across Dworkin's rendering of Osric, this is what they might see:

'Painted on a plain white background is the portrait of a tall and well-built man, with thinning blond hair and a trim mustache. His dark green eyes are penetrating, looking out as if he were judging one accused of a serious crime. He weighs his spiked mace in one hand, uncertain yet in a state of readiness. Over a quilted vest of dark blue with white markings, he wears a loose, wide-sleeved garment of some softer, lighter material.'

OSRIC - HIDDEN WIZARD OF AMBER (300 Point Version)

The court of Amber assumes that Osric is long-dead, the victim of one Oberon's rages. That's just the way Osric likes it. He lurks outside of Amber, manipulating a network of agents and spies.

ATTRIBUTES
PSYCHE - [55 Points]
STRENGTH - [30 Points]
ENDURANCE - [30 Points]
WARFARE - [35 Points]

POWERS
Pattern Imprint [50 Points]
Advanced Shape Shift [65 Points]
Sorcery [15 Points]

CREATURES AND ARTIFACTS OF POWER
Osric's Mace, True-Strike [16 Points]
- Impervious to Damage [8 Points]
- Destructive Force Damage [8 Points]

Osric's Hidden Agents [4 Points] - Creatures out of Shadow, man-like, but bred for darkness and cold, Osric's creatures are capable of standing motionless for weeks, or even months, in the guise of statues. In this state they overhear everything around them, and they have excellent memories.
- Alternate Named & Numbered Forms [2 Points] - Can shape shift to invisibility, a stone "statue" form, and into a good imitation of some faceless castle servant.
- Named & Numbered [*2 Points]

GM TIPS FOR PLAYING OSRIC AS THE HIDDEN WIZARD

It's possible that Osric could be anyone in the campaign, from a player character's servant, to an elder Amberite, by virtue of his Shape Shifting power.

When revealed, he will be bitter, vengeful, and angry. Not crazy enough to endanger his own life, but not apologetic either.

Constant Allies & Enemies: Since Benedict and Finndo are Osric's full brothers, it is likely that they will be his strongest allies, or his most hated enemies. Osric holds a simmering anger against his father, Oberon, and it may spread to any of Oberon's favorites, especially Corwin.

OSRIC - SULLEN WANDERER (400 Point Version)

He left Amber after a monumental falling-out with his father, Oberon. Since then he has roamed through Shadow, feeling like an outcast, subject to moods ranging from depression to melancholy. He is an unreliable friend and inconstant enemy. About the only thing you can count on is that he'll never return to Amber, never forgive, and never forget.

ATTRIBUTES
PSYCHE - [35 Points]
STRENGTH - [65 Points]
ENDURANCE - [80 Points]
WARFARE - [135 Points]

POWERS
Pattern Imprint [50 Points]
Shape Shift [35 Points]

CREATURES AND ARTIFACTS OF POWER
Osric's Mace, True-Strike [5 Points]
- Extra Hard [1 Point]
- Double Damage [2 Points]
- Sensitivity to Danger [2 Points]

BAD STUFF
[+5 Points]

GM TIPS FOR PLAYING OSRIC AS SULLEN WANDERER

Osric the wanderer is an inconstant character. Though his mood is always bleak, the wrong word from a player character can drive him right over the edge. He may attack, pausing just short of killing. Or he may simply take off into Shadow, seeking those he can kill with a clean conscience.

Constant Allies & Enemies: He has no friends as such. It is unlikely that any still alive even remember Osric, much less hate him.

OSRIC - DUKE OF CHAOS (450 Point Version)

Osric is not just a denizen of the Courts, but one of it's most important insiders. He is married into one of the most powerful families of Chaos, and has put his wife up as a powerful contender for the Throne of the Courts of Chaos.

Of necessity, this version of Osric must be circumspect

about his relationship with Amber and that side of his family. While he may be willing to help his young nieces and nephews, he may be very reluctant to display this compassion to others in Chaos.

ATTRIBUTES
PSYCHE -   [75 Points]
STRENGTH - [29 Points]
ENDURANCE - [35 Points]
WARFARE - [35 Points]

POWERS
Pattern Imprint [50 Points]
Advanced Logrus Mastery [70 Points]
Advanced Trump Artistry [60 Points]
Shape Shift [35 Points]
Power Words [10 Points]
Sorcery [15 Points]
Conjuration [20 Points]

CREATURES AND ARTIFACTS OF POWER
Osric's Mace, True-Strike [16 Points] - The one reminder of his Amber heritage is his Mace, a weapon that causes fear among all the Courts of Chaos. He'll be reluctant to use it unless his life is truly in danger.
• Primal Damage to Creatures of Chaos [16 Points]

GM TIPS FOR PLAYING OSRIC AS DUKE OF CHAOS

Osric is just as manipulative, scheming and two-faced as any of his Amber brothers. However, he saves that side of his nature for dealing with his fellow players in the games for control of the Courts of Chaos. With Amberites he'll simply be sly, and unwilling to commit to any action.

Constant Allies & Enemies: His place in Amber is unimportant to Osric. Only Benedict and Osric, his full brothers, can arouse much of a passion.

OSRIC - PLAYER CHARACTER WITH AMNESIA (125 Point Version)

Since Osric is such an unknown, it's possible to set him up as a character for one of the players in the campaign. Tricky as hell, and recommended only if the player has opted for a fairly large dose of Attribute bids, and a hefty amount of Bad Stuff. The player will set up a character in the normal way, and be completely unaware of the shape shifting that has been done and the memories that have been removed.

ATTRIBUTES & POWERS

Attributes and Powers will reflect the player's bidding and character construction. However, since Osric has an extra twenty-five points, the player will occasionally discover inexplicable abilities.

Walking the Pattern. Whatever has afflicted Osric, so that he is unaware of his true identity, will not be cured by a simple walk on the Pattern.

ALLIES & ENEMIES

This is critical, since some non-player characters, like Dworkin, Benedict, Fiona and/or Brand, may know of Osric's true identity. How will they react to this new version of Osric? And how will the rest of the family deal with the character if and when the connection with Osric is revealed?

Before implementing a player character as Osric, the Game Master should carefully work out the background of the situation. Critical to this is the cause of the amnesia, its cure, and the agent responsible for Osric's plight.

FINNDO

All we know of Finndo is a brief mention of his name. He was said to be elder brother to Benedict, long lost, and certainly dead. Here is how Dworkin's Trump might picture him:

'The card shows a serious man, with a stern look about him, almost puritanical. His eyes are stormy grey, shot through with tiny shards of silver. His hair and thick beard are a mix of hot red and cold iron. He is thick of neck and chest, and looks muscular. The steel ring on his left hand has a large, square stone of indeterminate dull color. He holds his heavy, two-hand sword more like an instrument than a bludgeon. He wears an ornate piece of red armor on his right forearm and has a pair of grey gloves on his belt. His dress is formal, in colors of dark red and grey.'

FINNDO - ANCIENT MASTER OF ARMS
(325 Point Version)

What if Finndo was Benedict's teacher in Warfare? This version of Finndo has him as an even more devastating fighter than Benedict. The obsession with combat has burned into his brain to the point where there is room for little else.

Players may find Benedict acting strangely toward this version of Finndo. Benedict may treat Finndo with respect and a bit of childlike adoration, seeing Finndo still as the older brother Benedict always looked up to. Or with fear and loathing if Finndo was a task-master who always found Benedict falling short of his expectations.

ATTRIBUTES
PSYCHE -   Amber Rank
STRENGTH - [65 Points]
ENDURANCE - [50 Points]
WARFARE - [155 Points]

POWERS
Pattern Imprint [50 Points]

CREATURES AND ARTIFACTS OF POWER
Finndo's Red Forearm Guard [4 Points] - Used expertly by Finndo either to ward off blows, to deflect missile fire, or, when the opportunity presents itself, to strike or scrape.
  • Resistant to Firearms [2 Points]
  • Double Damage [2 Points]
Finndo's Two-Handed Sword [1 Point] - A big sword, it's main quality is that it will not be broken from Finndo's overpowering blows.
  • Extra Hard [1 Point]

GM TIPS FOR PLAYING FINNDO AS ANCIENT MASTER OF ARMS

Finndo has had an epic falling-out with his father Oberon. Hearing of the old King's death, he may very well return to Amber. He is no more likely to claim the throne than Benedict, but he will no doubt give Random something of a hard time, and will expect to be treated with proper respect.

Constant Allies & Enemies: Finndo treats Osric as his strongest ally. Benedict he will greet with affection, but he will expect Benedict to obey Finndo, and to follow his lead. As to his younger siblings, he thinks of them as children who are not quite ready to take on threats in the real world.

The Critical Ally: If this Finndo exists in the campaign, he can be a decisive participant in any conflict. From the players' point of view, swinging Finndo's help can be the equivalent of bringing in the Cavalry to rescue Amber from any big threat.

FINNDO - ENEMY OF AMBER
(450 Point Version)

Burning with hatred for the homeland that betrayed him, Finndo has been waiting, and preparing, for his revenge. He has achieved a warped new view of Pattern and built a circle of Shadow primed for cutting off and attacking Amber.

ATTRIBUTES
PSYCHE -   [40 Points]
STRENGTH - [35 Points]
ENDURANCE - [30 Points]
WARFARE - [172 Points]

POWERS
"Abnormal" Advanced Pattern Imprint [90 Points] - Finndo has concentrated on changing the Pattern in his Personal Shadows, warping them so that they no longer exactly fit the curves and shapes of Amber's Pattern. This means that only initiates of his warped version can walk or Hellride to his Shadows. Those with Advanced Pattern can, if they study things properly, gain some access. Any natural child of Finndo will have an affinity for this warped Pattern, crossing its boundaries, and commanding its creations without special effort.

CREATURES AND ARTIFACTS OF POWER
Finndo's Pattern Warping Ring - [16 Points] - Within the realms of Finndo's Warped Pattern, and to the creatures molded by it, the ring serves as a counterpart to the Jewel of Judgement.
  • Impervious to Damage [8 Points] - No natural force or material can damage the ring, nor can it be affected by Magic or any lesser Power.
  • Able to "Mold" Pattern Reality [8 Points]
Finndo's Dogs of War [34 Points] - Though called dogs, these are actually dog-like men, averaging seven feet in height with canine faces and lean, muscular bodies. There are few of them, but each is capable of leading armies through the warping of Finndo's Pattern.
  • Immense Vitality [4 Points]
  • Combat Mastery [4 Points]
  • Resistant to Normal Weapons [1 Point]
  • Warped Pattern Walk [8 Points]
  • Named & Numbered [*2 Points]
Finndo's Soldiers [12 Points] - Throughout Finndo's Shadows his people wait, and train for battle, in the hopes that he will one day call upon them.
  • Double Normal Stamina [2 Points]
  • Combat Training [1 Point]
  • Shadow Wide [*4 Points]

PERSONAL SHADOWS
De'alund - FINNDO'S CIRCLE OF SHADOW [21 Points] - Like a great ring of islands, surrounding a mighty continent, Finndo's Shadows lay at the outskirts of Amber, each connected to Amber's seas, but each accessible only to those with the right key, the vision of Finndo's Warped

Pattern.
- Primal Shadow [4 Points]
- Restricted Access [2 Points]
- Control of Contents [1 Point]
- Horde [*3]

GM TIPS FOR PLAYING FINNDO AS ENEMY OF AMBER

Licking his wounds, and cold with anger, Finndo has vowed that his father will one day fall.

Does Finndo know of Oberon's demise? The recounting of Oberon's self-sacrificial repair of the Pattern, at the conclusion of the Patternfall War, could soften his view toward his father and the rest of his relatives. This makes it possible, though still difficult, for players to make some kind of reconciliation.

As for the other children of Oberon, he has little respect for them, feeling them to be nothing more than Oberon's afterthoughts. Should he not be offered the Throne, he will demand it, his right as firstborn son.

Constant Allies & Enemies: Most hated is his brother Benedict, whom he accuses of betrayal. Although alienated, he will seek reconciliation with his long-lost brother Osric.

FINNDO - LORD OF CHAOS (450 Point Version)

The eldest son of Oberon, Finndo was born and raised in the Courts of Chaos. Irritated by his father's game-playing and manipulation of his little brothers and sisters, Finndo left. Finding things more comfortable near the Courts, but not quite in them, he has a realm of his own.

Finndo has a fondness for his younger brothers, and feels kindly toward their children. He will aid them when he can, and will welcome them to his realms and his table. His standing and reputation in the Courts of Chaos is such that he can intercede on their behalf.

The ring he wears upon his left hand is the seal of the High Lord of Appeals. In Chaos he presides over a special judicial court, or docket, where those accused of certain crimes against the Courts of Chaos may enter their last appeal. By appointing an outsider, one who carries the Blood of Amber, and who can never be king in Chaos, to this position, the royal line of Chaos assures that this court of final judgement will be impartial and exempt from the politics of those jockeying for the throne.

ATTRIBUTES
PSYCHE -   [25 Points]
STRENGTH - [50 Points]
ENDURANCE - [25 Points]
WARFARE - [155 Points]

POWERS
Pattern Imprint [50 Points]
Logrus Mastery [45 Points]
Shape Shift [35 Points]
Sorcery [15 Points]
Conjuration [20 Points]

CREATURES AND ARTIFACTS OF POWER
Finndo's Red Forearm Guard [18 Points] - In addition to its function as a piece of armor, the guard is a control and summoning device for his trained creatures of Chaos. When he holds out his right arm, the creatures will return to him, in winged form, landing upon the guard.
- "Augmented" Psychic Force [8 Points]
- Able to "Mold" Shadow Creatures [2 Points]
- Inflicts Shape Shifting [8 Points]

Finndo's Two-Handed Sword [1 Point] - as above.
- Extra Hard [1 Point]

The Kinsman Ring [12 Points] - One of a number of rings, each worn by a particular contingent of the Courts of Chaos. They use the rings for communication, and each has an extra which can be used for carrying messages (in magical form) and to identify their messengers.
- Extraordinary Psyche [4 Points]
- Racks Named & Numbered Spells [2 Points]
- Named & Numbered [*2 Points]

BAD STUFF
[+1 Point]

GM TIPS FOR PLAYING FINNDO AS LORD OF CHAOS

A gentleman of Chaos, Finndo will be formal but friendly to any visiting Amberites. He'll not welcome any discussion of Oberon, still feeling some pain from their parting so long ago.

Constant Allies & Enemies: Finndo will be open and welcoming with his full brothers Benedict and Osric. Powerful members of the Courts of Chaos stand with Finndo, reporting to him of all activities and movement. By the same token, Finndo has bitter enemies among the Logrus users, those who hate all things related to Pattern, or who are resentful of his political power in royal circles.

DARA

She stood about a dozen paces from me, a tall, slender girl with dark eyes and close-cropped brown hair. She wore a fencing jacket and held a rapier in her right hand, a mask in her left. She was looking at me and laughing. Her teeth were white, even, and a trifle long; a band of freckles crossed her small nose and the upper portions of her well-tanned cheeks.

The Guns of Avalon

She looks like a feisty teenager. A tall, slender girl with dark eyes and short brown hair, with a cute splash of freckles across her nose. Really cute.

Except that a minute later she can change from co-ed cute into one of the keepers of hell, complete with horns, hooves, tail, and that freckled snub nose transformed into a sulphur-snorting snout.

She's one of the rare individual's who has both the heritage of Amber, and the blood of the royal family of the Courts of Chaos. On her mother's side she is Benedict's granddaughter, or maybe great-granddaughter, or something like that. On her father's side, she is one of the dozen or so heirs to the throne of Chaos.

Dara is also Corwin's one true love, or maybe just Corwin's one great enemy. In either case, she is the mother of Corwin's first known son, Merlin.

That role of motherhood was the result of a lifetime of preparation, and an innate sense of guile, that allowed her to deceive Corwin as to her true origins. And the same twisted tactics that lead her to penetrate Amber, to slay its guards, walk the Pattern, and become, potentially, Amber's greatest enemy.

But Dara is not someone who can be ordered about, by Chaos, or by Amber. She has her own will, her own alliances, and is perfectly capable of creating, if she sees the need, a third base of power.

Politics aside, Dara is also brilliant when it comes to the use of the sword, to be ranked with the best of Amber.

DARA - DESCENDED OF BENEDICT (150 Point Version)

> *I continued to study her face.*
> *Yes, it was possible...*
> *'I am hurt,' I said, finally.*
> *'Why?' she asked.*
> *'Benedict didn't give me a cigar.'*
> *'Cigar?'*
> *'You are his daughter aren't you?'*
> *She redened, but she shook her head.*
> *'No,' she said. 'But you are getting close.'*
> *'Granddaughter?' I said.*
> *'Well... sort of.'*
> *'I am afraid that I do not understand.'*
> *'Grandfather is what he likes me to call him. Actually, though, he was my grandmother's father.'*
>
> *The Guns of Avalon*

Dara, having joined the winning side in the Patternfall War, now stands to make her inheritance good in Amber. She will join up with Benedict, become a member of his family, and his apprentice. In matters of family politics she will side with Benedict, and will turn away from the affairs of the Courts of Chaos.

ATTRIBUTES
PSYCHE -   Chaos Rank [+10 Points]
STRENGTH - Chaos Rank [+10 Points]
ENDURANCE - Amber Rank
WARFARE - [70 Points]

POWERS
Pattern Imprint [50 Points]
Shape Shift [35 Points]
Sorcery [15 Points]

GM TIPS FOR PLAYING DARA AS DESCENDED OF BENEDICT

A good liar, Dara. She is also a good swordswoman, and she longs for action. She will not spend a lot of time hanging around if the possibility of action is at hand. With the younger generation (player characters) she will be friendly, trying to fit in as one of their group, since she doesn't really feel like she belongs with the elders.

Constant Allies & Enemies: She will do everything possible to cultivate her relationship with her grandsire Benedict. She has been bruised by Corwin. It was her

own doing, of course. Perhaps, in time, she will see that. For the time being she has no other serious enemies outside of the Courts of Chaos.

DARA - PRINCESS OF CHAOS (175 Point Version)

> *'She told me that Amber, in the fullness of its corruption and presumption, had upset a kind of metaphysical balance between itself and the Courts of Chaos. Her people now had the job of redressing the matter by laying waste to Amber. Their own place is not a shadow of Amber, but a solid entity in its own right. In the meantime, all of the intervening shadows are suffering because of the black road. My knowledge of Amber being what it was, I could only listen. At first, I accepted everything that she said. Brand, to me, certainly fit her description of evil in Amber. But when I mentioned him, she said no. He was some sort of hero back where she hied from. She was uncertain as to the particulars, but it did not trouble her all that much. It was then that I realized how oversure she seemed about everything—there was a ring of the fanatic when she talked...'*
>
> Martin, speaking of Dara,
> *The Hand of Oberon*

Dara, having tried to ally herself with Amber and Corwin, has now turned her back on them, and even on the human form. She will return now to Amber, with the added benefit of her experiences, and her knowledge of Pattern. There she will go back to the games of power, and resume her place in line for the throne.

ATTRIBUTES
PSYCHE -   [5 Points]
STRENGTH - Chaos Rank [+10 Points]
ENDURANCE - Amber Rank
WARFARE - [50 Points]

POWERS
Pattern Imprint [50 Points]
Logrus Mastery [45 Points]
Shape Shift [35 Points]

GM TIPS FOR PLAYING DARA AS PRINCESS OF CHAOS

Having been burned by Corwin of Amber, the murderer of

her Uncle Borel, she is not given to treating any Amberites with kindness. She is still young, still rash, and still filled with pride.

Constant Allies & Enemies: Her great enemy, in her own eyes, is Corwin. As an aspirant to the throne of the Courts of Chaos, she must now also regard all of Amber with suspicion, at least outwardly. Yes, there is a peace between Amber and Chaos, but the Courts will never have as Queen anyone who shows divided loyalty.

DARA - QUEEN OF SHAPE SHIFTERS (250 Point Version)

'I am Dara. Dara of Amber, Queen Dara. I hold this throne by right of blood and conquest...'

Sign of the Unicorn

Dara has never, to the very end, denied her ambitions. She went to the extent of seducing Corwin, mothering his child, murdering the guards of Castle Amber, and conspiring with Brand to destroy the ruling family of Amber. She is, in short, the most serious and threatening external enemy that Amber has ever known.

Now, after the Patternfall War, she sees herself as the heir to the throne of Amber. She is, after all, the only direct descendent of Benedict, the eldest surviving son of Oberon. She also held Oberon's signet ring, a sign of the old King's approval. She will not contest Random, but, should he disappear, she will claim his place instantly.

All this makes Dara doubly dangerous now. She will attempt to work her way back into the family, perhaps by associating with the younger generation. Learning much about Amber, she seeks to learn even more. She may, if she finds it possible, get close to Dworkin, Amber's greatest secret holder.

ATTRIBUTES
PSYCHE -   [15 Points]
STRENGTH - Amber Rank
ENDURANCE - [35 Points]
WARFARE - [50 Points]

POWERS
Pattern Imprint [50 Points]
Advanced Shape Shift [65 Points]
Sorcery [15 Points]
Conjuration [20 Points]

GM TIPS FOR PLAYING DARA AS QUEEN OF SHAPE SHIFTERS

Sly, and given to making up very plausible stories. She will use whoever she comes across, being pleasant and talkative. At the back of her mind, though, she will be looking for any weakness that she can use.

Constant Allies & Enemies: There is no one in Amber who she hates, save perhaps Corwin. She needs allies, and she knows it well.

DARA, IN HER OWN VOICE

'All right.' She sighed. 'All right. You are by now aware of Brand's plans—the destruction and rebuilding of Amber...?'

'Yes.'

'This involved your consent and co-operation.'

'No,' she said. 'We did not know who he intended to use as the—agent.'

'Would it have stopped you had you known?'

'You are asking a hypothetical question,' she said. 'Answer it yourself. I am glad that Martin is still alive. That is all that I can say about it.'

'All right,' Random said. 'What about Brand?'

'He was able to contact our leaders by methods he had learned from Dworkin. He had ambitions. He needed knowledge, power. He offered a deal.'

'What sort of knowledge?'

'For one thing, he did not know how to destroy the Pattern—'

'Then you were responsible for what he did,' Random said.

'If you choose to look at it that way.'

'I do.'

She shrugged, looked at me.

'Do you want to hear this story?'

'Go ahead.' I glanced at Random and he nodded.

'Brand was given what he wanted,' she said, 'but he was not trusted. It was feared that once he possessed the power to shape the world as he would, he would not stop with ruling over a revised Amber. He would attempt to extend his dominion over Chaos as well. A weakened Amber was what was desired, so that Chaos would be stronger than it now is—the striking of a new balance, giving to us more of the shadowlands that lie between our realms. It was realized long ago that the two kingdoms can never be merged, or one destroyed, without also disrupting all the processes that lie in flux between us. Total stasis or complete chaos would be the result. Yet, though it was seen what Brand had in mind, our leaders came to terms with him. It was the best opportunity to present itself in ages. It had to be seized. It was felt

that Brand could be dealt with, and finally replaced, when the time came.'

'So you were also planning a double-cross,' Random said.

'Not if he kept his word. But then, we knew that he would not. So we provided for the move against him.'

'How?'

'He would be allowed to accomplish his end and then be destroyed. He would be succeeded by a member of the royal family of Amber who was also of the first family of the Courts, one who had been raised among us and trained for the position. Merlin even traces his connection with Amber on both sides, through my forebear Benedict and directly from yourself—the two most favored claimants to your throne.'

'You are of the royal house of Chaos?'

She smiled.

I rose. Strode away. Stared at the ashes on the grate.

'I find it somewhat distressing to have been involved in a calculated breeding project,' I said, at length. 'But be that as it may, and accepting everything you have said as true—for the moment—why are you telling us all of these things now?'

'Because,' she said, 'I fear that the lords of my realm would go as far for their vision as Brand would for his. Farther, perhaps. That balance I spoke of. Few seem to appreciate what a delicate thing it is. I have traveled in the shadowlands near to Amber, and I have walked in Amber herself. I also have known the shadows that lie by Chaos's side. I have met many people and seen many things. Then, when I encountered Martin and spoke with him, I began to feel that the changes I had been told would be for the better would not simply result

in a revision of Amber more along the lines of my elders' liking. They would, instead, turn Amber into a mere extension of the Courts, most of the shadows would boil away to join with Chaos. Amber would become an island. Some of my seniors who still smart at Dworkin's having created Amber in the first place are really seeking a return to the days before this happened. Total Chaos, from which all things arose. I see the present condition as superior and I wish to preserve it. My desire is that neither side emerge victorious in any conflict.'

The Courts of Chaos

MARTIN

He was several inches taller than Random, but of the same light build. His chin and cheekbones had the same general cut to them, his hair was of a similar texture.

The Hand of Oberon

A bit bigger, broader in the shoulders, maybe with a bit less of the low-down scoundrel look. Martin still looks like his father, is still recognizable as the son of Random.

He is somewhat leery of Amber. His first experience with his relatives, other than some training from his Uncle Benedict, very nearly killed him. So he's been made, shall we say, a little sword-shy.

He's also developed a certain rapport with a select few from the Courts of Chaos, among them his cousin Merlin, and Merlin's Mom, Dara, Princess of Chaos. Otherwise, Martin has never seemed particularly special.

MARTIN - HEIR TO AMBER (120 Point Version)

The man I regarded upon was half familiar - meaning of course that he was also half strange. Light, straight hair, a trifle sharp-featured, a small smile, somewhat slight of build.

The Hand of Oberon

Now that he has been united with his father, Random, and brought back to the Court of Amber, he will begin to learn his new job. That means learning to be Heir to the Throne, and successor to King Random.

ATTRIBUTES
PSYCHE -   [15 Points]
STRENGTH - Amber
ENDURANCE - [15 Points]
WARFARE - [40 Points]

POWERS
Pattern Imprint [50 Points]

GM TIPS FOR PLAYING MARTIN AS HEIR TO AMBER

The big question about Martin is whether or not he is heir to the throne of Amber. His father is certainly Random, the current King. That his mother, *Morganthë*, daughter of Queen Moire, means that he is also a direct heir to the throne of Rebma.

Unfortunately, Martin's history up to this point makes him more of a pawn than a king. Martin's elders distrust him with some good reason. At some point, most likely after Brand's attack, but perhaps earlier, Martin allied himself with Dara and Merlin, both ace agents of the Courts of Chaos. There are extenuating circumstances, but elder Amberites, suspicious as they are, will never forget.

Constant Allies & Enemies: Having been stabbed by Brand, Martin will always have a lot of sympathy from his elders. He'll never entirely trust any of them, having once been treated as a handy bottle of Pattern eraser fluid.

MARTIN - HELLRIDER & VAGABOND (150 Point Version)

'After I left Benedict's, I traveled for years in Shadow,' he said. 'Those were the happiest times I have known. Adventure, excitement, new things to see, to do... In the Back of my mind, I always had it that one day when I was smarter and tougher—more experienced—I would journey to Amber and meet my other relatives...'

The Hand of Oberon

Having spent a lot of time wandering through Shadow, Martin has developed quite a taste for it. His every dream and fantasy can be fulfilled, and there are an infinite variety of pleasures and adventures yet to be experienced.

Martin may come to Amber only rarely, and then only if someone insists upon it. He's not too keen on Trump calls anyway, so it takes a lot of effort even to get in touch with Martin.

ATTRIBUTES
PSYCHE -   [15 Points]
STRENGTH - [35 Points]
ENDURANCE - [5 Points]
WARFARE - [40 Points]

POWERS
Pattern Imprint [50 Points]

PERSONAL SHADOWS
Martin's Shadow Sanctuary [2 Points] - Having had one bad experience with Trump, Martin had found a place where he can retreat and where no Trump will bother him.
- Personal Shadow [1 Point]
- Communication Barrier [1 Point]

GOOD STUFF
[3 Points]

GM TIPS FOR PLAYING MARTIN AS HELLRIDER & VAGABOND

By no means the liar that his family expects, Martin seems to be a very straightforward young man. He is certainly capable of involving himself in the various family plots, he's just not devious enough to keep things completely covered up. Therefore, he'll just stay away. As he says, he needs experience.

On the other hand, Martin will welcome contacts from the player characters, members of his own generation. He's a friendly guy, Martin, and he'd like to be part of the Amber brat pack, when one gets together.

Constant Allies & Enemies: Random, formally an absent father, has very much attempted to make amends. Still, Martin may have some lingering doubts about the man who may have caused his mother's death.

MARTIN - STUDENT OF CHAOS (150 Point Version)

...nothing that I was telling Random could really be used against us, and I strongly doubted that Martin could do us much damage if that was his intention. No, more likely he was being as cagey as the rest of us, and for pretty much the same reason: fear and self-preservation. On a sudden inspiration, I asked him, 'Did you ever run into Dara again after that?'

He flushed.

'No,' he said, too quickly. 'Just that time. That's all.'

'I see,' I said, and Random was too good a poker player not to have noticed; so I had just bought us a piece of instant insurance at the small price of putting a father on guard against his long-lost son.

The Hand of Oberon

We know that Martin was helped by both Dara and Merlin. Is it far-fetched that he may have become something of a pupil in the Courts of Chaos? By this point he has come to know the basics of Shape Shifting, but he is still working his way toward an attempt at the Logrus. He may also be casually studying Trump.

ATTRIBUTES
PSYCHE - [35 Points]
STRENGTH - Amber Rank
ENDURANCE - Amber Rank
WARFARE - [15 Points]

POWERS
Pattern Imprint [50 Points]
"Partial" Trump Artistry [10 Points] - Martin has picked up enough knowledge and sensitivity to Trump to be able to check on incoming calls. At the first twinge of contact he'll fan through his deck, attempting to identify the caller.
Shape Shift [35 Points]
Power Words [10 Points] - Magic Negation, Psychic Disrupt, Pattern Negation, Trump Disrupt and Burst of Psyche.

BAD STUFF
[+5 Points]

GM TIPS FOR PLAYING MARTIN AS STUDENT OF CHAOS

Having a fair degree of ambition, and the birthright to a couple of thrones, Martin is now working on increasing his Powers. He may be a little clumsy in the execution, he may make a fair number of mistakes, but he'll be constantly pushing to see just how much he can do.

MARTIN AS A FATHER

Martin seems a bit young, by Amberite standards, to have fathered any children as yet. Still, in his forty or fifty years it may have happened. There are three possibilities. First, in his early years in Rebma, where he was a member of the royal court, he could easily have left a child with some serving wench. Second, he could have become a father in his days of Hellriding, when he may have left a child virtually anywhere out in Shadow. Third, he might have been inveigled by the same kind of "breeding program" that involved Corwin, in a match up with a Lady of Chaos.

A final note on Martin. In Zelazny's later series Martin has appeared in rather different form. His new look is decidedly punkish, complete with Mohawk hairstyle and leather jacket. Even more strange is the electrical socket built into his neck. Phage Press will hold off on an exact description until the book, *Shadow Knight*, but each Game Master has the option of coming up with their own high-tech version of Martin.

MERLIN

...The second rider was rapidly approaching. He was not pale like the first. His hair was dark and there was a color in his face. His mount was a properly maned sorrel. He bore a cocked and bolted crossbow...

...He was beardless, slim. Possibly light-eyed within the squint of his aim. He managed his mount well, with just the pressure of his legs. His hands were big, steady. Capable. A peculiar feeling passed over me as I beheld him.

The Hand of Oberon

Is there such a thing as a nice guy in all of Amber? Maybe. If there is, his name is Merlin.

Son of Corwin of Amber, and Dara of Chaos, Merlin was bred to be a king, and raised to be a pawn. Tall and slim, with Corwin's looks, he was given a gentleman's upbringing in the courtly graces of magic, weapons, poisons, riding, and dancing. He also is heir to some pretty major titles. And he doesn't give a damn about any of it. Merlin simply has no ambitions, for courtly life, rulership, or much of anything other than satisfying his own curiosity.

In Amber, Merlin is the second of the next generation, and the first son with a Chaos heritage. Even during his upbringing in the Courts, he was always very attracted to his Amber side, to the point of risking his life to take the Pattern, when he was already an initiate of the Logrus. And, to make his mastery of the powers complete, Merlin is also a Trump Artist, and probably a Sorcerer.

By the way, the Trump of Merlin shown here is a self-portrait. Merlin's style can also be seen in the Trumps of Dara and Martin.

MERLIN - CORWIN'S SON (100 Point Version)

'I have no idea what it was like,' I said, 'growing up in the Courts.' He smiled for the first time.

'And I have no idea what it would have been like anywhere else,' he responded. 'I was different enough to be left to myself a lot. I was taught the usual things a gentleman should know - magic, weapons, poisons, riding, dancing. I was told that I would one day rule in Amber...'

The Courts of Chaos

Merlin has already expressed a great desire to know Amber and to walk Shadow. He'll likely spend time at Castle Amber, Shadow Earth, and anywhere else his spirit takes him. He'll also want to spend time with his father, to make a relationship that he has always missed.

ATTRIBUTES

PSYCHE - [4 Points]
STRENGTH - Chaos Rank [+10 Points]

ENDURANCE - Chaos Rank [+10 Points]
WARFARE - Chaos Rank [+10 Points]

POWERS
Trump Artistry [40 Points]
Logrus Mastery [45 Points]
Shape Shifting [35 Points]

ALLIES
Merlin's Amber Court Devotee [6 Points] - This, of course, turned out to be Corwin, Merlin's father. And that parentage later allowed him to take on Pattern Imprint along with his other powers.

GM TIPS FOR PLAYING MERLIN AS CORWIN'S SON

Merlin, as the son of Corwin, ex-regent of Amber, and Dara, one of the heirs to the Throne of the Courts of Chaos, is going to be pretty hot property in either Court.

Constant Allies & Enemies: From the point of view of certain manipulative souls, Merlin is just the perfect pawn. He's got royal blood on both sides, a strong claim to both thrones, plus he's young, inexperienced, and pretty naive. What a combination! The more power-hungry Amberites must be salivating every time Merlin walks by.

MERLIN - THE ULTIMATE NEUTRAL (150 Point Version)

'...I was told that I would one day rule in Amber. This is not true anymore, is it?'

'It does not seem too likely in the foreseeable future,' I said.

'Good,' he replied. *'This is the one thing I did not want to do.'*

Merlin & Corwin,
The Courts of Chaos

Whatever Merlin's upbringing in the Courts of Chaos, it has left him with a strong distaste for the whole political scene. Now that the war is over he intends to use his dual heritage as a way of avoiding responsibility in either court.

ATTRIBUTES
PSYCHE - [4 Points]
STRENGTH - Chaos Rank [+10 Points]
ENDURANCE - Chaos Rank [+10 Points]
WARFARE - Chaos Rank [+10 Points]

POWERS
Pattern Imprint [50 Points]

Trump Artistry [40 Points]
Logrus Mastery [45 Points]
Shape Shifting [35 Points]

ALLIES
Merlin's Amber Court Devotee [6 Points] - **as above.**

GM TIPS FOR PLAYING MERLIN AS THE ULTIMATE NEUTRAL

Why should Merlin get involved in any family squabble? Or, for that matter, why should he stand **anywhere** in the middle when Amber and the Courts of Chaos start pushing at each other?

Merlin will be interested in meeting and making friends with the player characters in the younger generation, just so long as they are intent on having fun and doing neat stuff. As soon as they turn toward the boring topic (from Merlin's point of view) of politics, Merlin will be off to do something else.

Constant Allies & Enemies: Unlike Martin, who will probably always be viewed with some suspicion, Merlin was not stained by any of the political intrigues of the Patternfall War. No matter what stance he took, he could always claim it was out of loyalty (loyalty to who, is a question that is not asked of royals).

MERLIN - ASPIRING LORD OF CHAOS (225 Point Version)

'What of Merlin? Where is he now?'
'They have him,' she said. *'I fear that he may be their man now. He knows his parentage, but they have had charge of his education for a long while...'*

The Courts of Chaos

Merlin affected an innocent, boyish tone when he met with Corwin at the edge of the Abyss. It certainly paid off in that Corwin spilled his guts, telling Merlin the entire story of the Patternfall War from the Amber point of view.

If it was an act, it worked to get Merlin what he wanted. If it was an act, then Merlin is not the wide-eyed boyish figure that he claims to be. Instead, he may not have resigned from the role designed for him; King of the unified Amber and Chaos.

ATTRIBUTES
PSYCHE - [35 Points]
STRENGTH - Chaos Rank [+10 Points]
ENDURANCE - Chaos Rank [+10 Points]
WARFARE - Amber Rank

POWERS
Pattern Imprint [50 Points]

Logrus Mastery [45 Points]
Trump Artistry [40 Points]
Shape Shifting [35 Points]
Sorcery [15 Points]
Conjuration [20 Points]

ALLIES
Merlin's Amber Court Devotee [6 Points] - as above.

BAD STUFF
[+1 Points]

GM TIPS FOR PLAYING MERLIN AS ASPIRING LORD OF CHAOS

Merlin is really a rotter, and he's just been biding his time. He does have that old claim on the twin thrones, and might be thinking, long-range, that it's only a matter of time.

Powers in *Amber* are dangerous. Not only can they kill you, they can also give you an addiction. Like anything that feels good, you can get hooked on Power. Most folks

TRUMPS BY MERLIN

Most of the characters are based on the Trump executed by Dworkin. Three cards; Dara, Martin and Merlin, are all Trumps painted by Corwin's son, Merlin.

are content to stick with one Power, like most alcoholics prefer one kind of drink. Merlin seems to be one of those guys who has to sample a bit of everything. Not content with Logrus, he wants Pattern as well. He's got Trump, Shape Shifting, and Magic. Plus, in the recent books, Merlin seems fascinated by any new source of Power that he happens to come across.

A final note on Merlin. Zelazny has a second *Amber* series with Merlin as the main character. In those books Merlin is a spell-caster, Logrus Master (perhaps even at the advanced level), Initiate of the Pattern, and Trump Artist. He also carries a living strangle cord, named *Frakir*, a Pattern Creature/Artifact named *Ghostwheel*, and some very interesting relatives. To find out more, read *Trumps of Doom, Blood of Amber*, *Sign of Chaos, Knight of Shadows*, and *Prince of Chaos*. Or pick up a copy of the Phage Press resource book, *Shadow Knight*.

THE ELDER AMBERITES

Having read about the twenty Amberites described in this book, the question for the Game Master is, what now?

Mix 'n Match. The main reason there are three versions of each of the characters is so each campaign will have a unique version of each elder Amberite. Feel free to select from the versions given here, or to customize any of the characters. For example, none of the versions of Fiona list any artifact or device for hanging spells. Fiona, being probably the foremost spell caster in Amber, it's likely to have many such devices. Their exact nature varies from campaign to campaign.

Build Your Own. Whether you decide to fill in a few of the missing relatives mentioned in the *Chronicles*, or to create an entirely new cast of characterc, you can use these descriptions as a model.

The next step is figuring out just how to role-play this den of cut-throats. Here are a few tips:

A Murderous Bunch.

Early on in the very first book of the *Amber* series, we get a peek into the attitude of the elder Amberites. Corwin has been driving a car through Shadow, and had unwittingly ended up on the wrong side of the road, narrowly avoiding an accident with a gasoline tanker. The truck driver is, shall we say, a bit upset, calling Corwin, among other things, "a friggin' menace!" Then Random steps up, gun in hand.

The guy turned about and started to run, a look of fear widening his eyes and loosening his jaw.

Random raised his pistol and took careful aim at the man's back, and I managed to knock his arm to the side just as he pulled the trigger.

It scored the pavement and ricocheted away.

Random turned toward me and his face was almost white.

'You bloody fool!' he said. 'That shot could have hit the tank!'

'It could also have hit the guy you were aiming at.'

'So who the hell cares? We'll never pass this way again, in this generation. That bastard dared to insult a Prince of Amber! It was your honor I was thinking about.'

'I can take care of my own honor,' I told him, and something cold and powerful suddenly gripped me and answered, 'for he was mine to kill, not yours, had I chosen,' and a sense of outrage filled me.

He bowed his head then, as the cab door slammed and the truck tore off down the road.

'I'm sorry, brother,' he said. 'I did not mean to presume. But it offended me to hear one of them speak

to you in such a manner. I know I should have waited to let you dispose of him as you saw fit, or at least have consulted with you.'

Nine Princes in Amber

Notice that Corwin doesn't say anything to Random about mercy or about human kindness or about how killing a guy for being rude is a little psycho. No, Corwin acts like he owns the truck driver, and that Random is out of line for shooting at his property.

There is also a punch line to this little story. Random is the nicest guy in the family. Everybody else we meet is worse. Corwin sums it up nicely, saying of Amber, "All my relatives here had too many centuries of cynicism or wisdom..."

A Tradition of Machiavellian Planning.

There was a strange tone of admiration in the voice of this woman who was admitting she'd just tried to sell me out to my enemy, and still would—given half a chance—as she talked about something she thought I'd done which had thrown a monkey wrench into her plans. How could anyone be so admittedly machiavellian in the presence of a proposed victim? The answer rang back immediately from the depths of my mind: it is the way of our kind...

Nine Princes in Amber

Plots and schemes and cabals. All the elder Amberites are constantly scheming and plotting. They can't help it.

Let's look at a "real" example of what it means to be "Machiavellian." This is one of my favorite tales from Niccoli Machiaveli's book, *The Prince*.

Assume that you, the Prince, have just conquered your enemy, and, in so doing, you have come to control a foreign province. The people of the province hate you and everything you stand for. They fought bitterly against your forces, and now, as the conquered they are snarling and already raising the call for freedom and rebellion.

How do you best deal with the situation?

Start by finding the most ambitious, brutal, nasty, cruel and heartless, yet still efficient, officer in your command. We'll call him the *Commander*. The Commander should be someone who is completely unfit for ruling a civilian province, someone who knows only military discipline, and who will ruthlessly quell any resistance against the new order.

Then assign the monster to the newly conquered province, with a mandate to maintain order, and full authority to act as sheriff, judge, jury and executioner.

Wait a few months.

Now what do you have?

An outraged province, but one where the most vocal leaders have been executed and where rebellion has been suppressed by military force.

Now you, the Prince, come to the province.

Someone will inevitably blab, coming forth to complain of their treatment at the hands of the Commander. You react with astonishment. "It could not be so," you say, "surely no one else could have witnessed such atrocities!"

Seeing that you have an open ear, the people of the province will run to you with their stories of the Commander.

Shocked, you have him arrested. As the tales mount up, higher and higher with their descriptions of his nasty behavior, you organize a public trial.

Eventually, moved to tears by the Commander's actions, you order his public execution.

Of course, your new governor is a saint in comparison. Skilled, smooth, and trained in the post.

The end result is that you have done away with the Commander, a potential threat to your own rulership. You have brutally pacified a province, killing off a great many potential rebels. You have turned an unruly possession into one that will peacefully submit to reasonable rulership. Last, but not least, you have made yourself a hero in the eyes of the people, and have become their champion by striking down the monster they most feared.

That's what it means to be Machiavellian.

Elder Amberites put Machiavelli to shame. They are conniving, low-down, immoral and sublimely sneaky. In fact, they are rather proud of it.

Apologies.

This is part of another little game the elder Amberites play. They speak to each other as if they received a point every time they could force one of their relatives into apologizing. Since they play the game for thousands of years, and treasure their "points" forever, they are very reluctant to give away any apologies.

All of which is just another way of saying that elder Amberites hate to apologize. Especially to their siblings, but they won't be any too happy about apologizing to anyone younger.

So forget having one of them say, "I'm sorry," or "I apologize."

Instead work on substituting some kind of dodge for any apology that an elder Amberite might be cornered into saying. The words, *probably, perhaps, perchance, possibly,* and *presumably* are all good qualifiers for any genuine expression of regret. Rather than saying "I'm sorry," consider responding with one of the following:

• "I might owe you an apology."

• "I should probably be remorseful."

• "Does my action warrant repentance?"

• "If that were true, then I would be sorry."

• "Maybe I should consider offering you my regrets."

• "It is unlikely that anything I do would be a sufficient act of contrition"

If a player character fails to play the game properly by

handing out apologies too freely, they're sure to get some sour looks. For one of the Blood of Amber to offer a free apology, is sort of like cracking jokes in the middle of a serious, high stakes poker game.

The Distrust of Innocence.

'I am going to tell you something Benedict should have told you long ago,' I said. 'Never trust a relative. It is far worse than trusting strangers. With a stranger there is a possibility that you might be safe.'

'You really mean that, don't you?'

'Yes.'

'Yourself included?'

I smiled.

'Of course it does not apply to me. I am the soul of honor, kindness, mercy, and goodness. Trust me in all things.'

Corwin & Dara,
The Guns of Avalon

Most Amberites are suspicious of any kind of goodness. A kind, honest character is more likely to invoke distrust than any shady manipulator.

Why?

Well, the shady manipulator they know. They understand that. It makes sense.

They see honesty as some kind of pose, a cover for devious behavior. Elder Amberites are likely to distrust anyone who seems straightforward.

They won't be happy until they are dealing with cynicism, or until the character has been to the absolute brink of destruction, and, with nothing to gain, and everything to lose, has stuck to the honesty gig.

Even then, most elder Amberites will still have their doubts.

Being Smarter Than You Are.

Here's a shocker.

You, the Amber Game Master, are not as smart as any of the elder Amberites. Yet it's your main job to role-play them accurately. You have to hatch the kind of plots that this devious lot might come up with, and plan grand strategy, and appear to be always one step ahead of the player characters.

Amberites routinely engage in manipulations so subtle that their actions are completely invisible. For example, having put some scheme into motion, an Amberite might communicate with his confederates with a signal as slight as the arching of an eyebrow, or the tiny adjustment of a bouquet of flowers.

So how the Hell are you, a mere Shadow dweller, supposed to emulate all that?

Trickery. Pure Game Master trickery.

Backwards Planning. When things are winding down to a close, and it's obvious how a phase of the campaign will end, have elder Amberites show up at the last minute.

Having been unavailable throughout the entire story, they'll appear at this point of closure, having arranged for everything necessary to defeat the threat, their explanation being that they had planned for that contingency all along.

Assuming the Best. When in doubt assume that the elder Amberites know exactly how to deal with every problem.

Free Peeks at the Universe. It's impossible to categorize all of Dworkin's powers, or Oberon's, or Fiona's. Still, it's pretty safe to assume that they can see a lot farther than anyone else. From a Game Master's view, it's as if some of the elder Amberites get free peeks at what is happening in the universe.

Implementing Elder Amberite Fore-Knowledge. Here's an example of how a Game Master can "retrofit" things to show off an elder Amberite's brilliance.

GM: Juderal is still right behind you!

Cindy: Blast! Farley, what should we do?

Mick: Let's head for Farley's apartment on Shadow Earth. Maybe his power will be weaker there.

GM: What are you doing?

Mick: I'll Shift Shadow, cutting through to Shadow Earth, and taking whatever shortcuts are available to get to my apartment fast.

GM: Cindy? What is Dorell doing?

Cindy: Dorell will follow close on Farley's footsteps, but keeping her sword drawn and with an eye over her shoulder at our pursuer.

GM: Okay, just a few minutes later you are both at the end of the hallway, just outside of the door of Farley's Boston apartment. What now?

Cindy: I'm watching back the way we came.

Mick: I'll get my key and let myself in.

GM: You get the door open. Dorell, you see the pink-armored figure materialize just down the hall.

Cindy: I'll poke Farley and point my sword at Juderal.

GM: Mick?

Mick: I'll pull my machine pistol, and get it ready to fire.

GM: Just as you both prepare for battle, you see Juderal look past you, his eyes widening in shock or fear. He takes a step back and fades away. What are you doing?

Mick: I turn around. What was he looking at?

Cindy: What's here?

GM: You both turn around and come face to face with your Uncle Caine. He's smiling at you. What are you doing?

Cindy: Caine! That stinking, putrid son of...

GM: Are you saying that out loud?

Cindy: No, but Dorell is sure thinking it!

GM: He tuts at you, as if he can hear your thoughts, and then turns to Farley. "I'm glad you managed to get away."

Cindy: Wait a minute. Did Caine just get here?

GM: No, you see the remnants of a meal, and a half-open book on the table. Mick, you see two bottles of your best wine have been opened and emptied. The smell of his tobacco is pretty strong. You'd say Caine must have been here for a couple of hours, perhaps even longer.

Cindy: But we didn't even know we'd be coming here until a few minutes ago!

Mick: Farley is going to come right out and ask. Caine, how

could you possibly know we were going to be coming here?

GM: Caine shrugs and says, "just a hunch. Why don't you tell me about your experiences?"

Elder Amberites & Folk of Shadow.

Imagine that you are out camping in the woods. There is a stream of what looks like fresh water running nearby. You have brought along a bottle of iodine solution used for water purification. Taking a bucket of water from the stream, before you start cooking, would you have any problem putting a drop of the water purifier into the bucket?

No?

Even if you did have qualms for some reason or other, it is doubtful that you were reluctant to use the solution for fear of killing the bacteria and viruses floating around in the water.

People can kill zillions of microscopic life forms without blinking, without considering the moral implications, without even realizing that they are doing it.

That's the way elder Amberites feel about Shadow dwellers.

Sure there are millions, or billions of them swarming around their Shadows. Yes, they probably feel pain and all that. Still, if there's a reason to wipe out a Shadow or two, just to make things more convenient, then consider it gone.

An elder Amberite has no more problem destroying countless humans, than humans have qualms about destroying countless germs.

Elder Amberites & The Younger Generation.

I looked at him. The muscles in his jaws had tightened and his eyes narrowed. For a moment, all of our faces fled across his, like a riffling of the family cards. All of our egoism, hatred, envy, pride, and abuse seemed to flow by in that instant—and he had not even set foot in Amber yet. Something snapped inside me and I reached out and seized him by the shoulders.

'You have good reason to hate him,' I said, 'and the answer to your question is 'yes.' The hunting season is open. I see no way to deal with him other than to destroy him. I hated him myself for so long as he remained an abstraction. But—now—it is different. Yes, he must be killed. But do not let that hatred be your baptism into our company. There has been too much of it among us. I look at your face—I don't know... I am sorry, Martin. Too much is going on right now. You are young. I have seen more things. Some of them bother me—differently. That's all.'

The Hand of Oberon

Unrealistic Hopes & Ambitions. In the quote above, Corwin is looking into Martin's eyes, and he doesn't like what he sees. Sure, it's just the kind of expression he'd expect to see in Random's face. It's just that Corwin, like many of his siblings, hopes for something better from the next generation. It doesn't make any sense, but parents always hope for the best from their children.

Denial. It doesn't matter how much an Amberite might love one of their younger relatives, it's rare for them to display such emotion openly. All part of the game of distrust. Basically the elder Amberites regard all the younger generation, their children, as potential hostages and pawns. So, often as not, the Amberites will deal with all the player characters in a cool manner.

Experience is the Best Teacher. We all wish we could tell a child that fire hurts, and we wish that they would just believe it. Fact is, it doesn't work. Sometime in your young life you had to get burned. It had to hurt. Then, with a basis of comparison, your parents could teach you that other things hurt. Elder Amberites have learned that lesson well. They don't tell the young Amberites, they show them what will hurt.

Family Oriented. For whatever reason, Amberites are very family oriented. They are, for example, incurable gossips. Even the worst of enemies will put their differences aside to chat about the latest dirt on their siblings.

Uncles & Aunts, Mothers & Fathers. One of the strongest forms of stress is parental. Do you know anyone who is completely comfortable with their parents? Or, for that matter, with the entire generation of elders who stood around and watched diapers being changed, oatmeal being worn, and *ouched* over bumps and bruises. In every family, the younger generation regards the elder with a kind of mythic respect.

Amberite parents are a million times worse. Because they are for all intents and purposes, gods. The Game Master's job is to work with the player, giving some insight into how their parent treated them *from the player character's immature point of view*. Once that's established, it's possible for the player characters to build complex relationships with the elder Amberites.

At the start of a campaign, the elder Amberites should be bad fathers, neglectful mothers, and evil step-parents. In other words, the player characters should see the elder Amberites pretty much the way teenagers view their parents, not as fully formed people, but as pieces of family furniture that they have to deal with.

TECHNIQUES OF ROLE-PLAYING

Amber, in the years of play-testing, has been as much a matter of style as any set of specific rules.

A GAME MASTER'S GUIDE TO ATTRIBUTE AUCTIONS

So, what should your objective be, running an Attribute Auction?

Get the players to spend their points!

Let's look at Zelazny's first Amber book, *Nine Princes in Amber*. What were the most dramatic battles? Why the two between Corwin and his arch-rival Eric. The first with swords (Warfare), and the second, mind to mind over a Trump (Psyche). Why were they dramatic?

First, they were close calls, where Eric started off believing himself to be superior, ended up surprised by his brother's improvements, and then withdrew.

Second, they were bitter. These were old enemies, nurturing the grudges of many a battle. We see a battle between two brothers, each constantly looking for an advantage over the other.

Now consider it from a game point of view. Why did Eric and Corwin waste so many points over things like Warfare and Psyche? Because they are brothers. Because they have a rivalry. Because they care about it.

Your job, as auctioneer, is to get those same kind of bitter rivalries going. You're looking for your game to feature the same kind of interesting, intense, and long-term rivalries between the player characters.

OPTIONAL: Limiting Bad Stuff.

Sometimes characters end up with unreasonable amounts of Bad Stuff, usually by over-bidding. While some Game Masters enjoy this, other prefer to impose controls and limits. If you want to limit the amount of Bad Stuff then announce your limits *before the Bidding War begins*.

Auction Time Limits

There are no absolute time limits. Take your time. Lots and lots of time. Ideally, you should schedule an entire game session for the auction. Don't rush through. Stall by entering all the figures carefully. Be very careful to check with each player before you close off one bidding session and move on to the next. That means pointing at each person in turn, and reviewing their situation, just to make sure they don't have something extra to add.

The Most Important Attribute Auction

Every Attribute Auction is important. Every Attribute Auction is the most important one.

Tips on Auctioning Psyche

Here's a few lines to sprinkle in during a Psyche Auction:

- "Remember, mind to mind battles are absolute. Once you've entered into a Trump Contact, it's pure Psyche that determines the winner."
- "Psyche is the most important Attribute for manipulating the powers of Pattern, Logrus, Trump and Magic. "
- "Don't forget Trump contacts. You'll get them all the time! Every time you get a Trump contact you run the risk of a Psyche battle. In fact, if you've got a lousy Psyche you may not even be able to resist taking a Trump call."

Tips on Auctioning Strength

The main thing about Strength is that it's a sure thing.

- "Don't forget, Strength is an important complement to Warfare! After all, you can have a wonderful Warfare rank and have your weapons shattered, along with your bones, if you can't match your opponent's Strength."
- "Strength is what makes all the difference in any hand to hand combat. You don't just buy muscles, you also get the wrestling and martial art skills."
- "Strength is also your resistance. A character with high Strength is covered in a layer of muscle, and able to withstand the impact of falling, crashing, or bashing, aside from being able to break things."

Tips on Auctioning Endurance

The most important thing to stress is that Endurance is the 'battery' that drives all the other Attributes.

- "Endurance has two components. Stamina, which lets you keep going against exhaustion, fatigue, and anything else. Then there's healing and regeneration. The better your Endurance, the faster you'll recover from any wound."
- "Endurance is critical for anyone wishing to engage in Shape Shifting. The stress of changing your body can only be matched by a good, high, ranking in Endurance."
- "If you're planning to use the great Powers, especially in their Advanced forms, then you better consider getting Endurance. After all, remember how drained Corwin got using the Jewel of Judgement. And he had great Endurance!"
- "Bottom line, in any battle where things are close, Endurance is the tie-breaker. Whether you're locked sword-to-sword, hand-to-hand, or mind-to-mind, the character with the best Endurance will win in the end."

Tips on Auctioning Warfare

- "Warfare covers everything military. From swords to guns to tanks to strategic nuclear bombers. A wizard of Warfare knows every weapon, and every way it can be used."
- "It's not just the sword fighting! Warfare also gives the character the tactical and strategic skills that are useful in any situation. After all, Warfare also determines how well you plot and plan, and how well you play chess."

Bidding War Alternatives.

Playing Catch-Up. Once the Attribute Auction has been held, you'll sometimes have a straggler wanting to get into the game. If you decide to let the new player in, you let them create a character as if they had been present, but hadn't bid on anything. Then they have to match the player bids.

Blind Bidding. The idea here is to run an Attribute Auction over a week or two with players submitting bids separately, by phone, or mail, over computer modem, or in person. It's usually best in these cases to run all four Attribute Auctions simultaneously.

Alternate Rules. Some Game Masters may want to allow for more or less points. For example, a campaign based on the Courts of Chaos may call for all players to start with their Attributes at Chaos level (+10 points), with 140 points for each player. Then there could be ranks between Amber and Chaos.

Playing Elder Amberites. Though it's not recommended for a campaign, there's no reason why a Game Master can't run a scenario with elder Amberites as players.

Character Quizzes.

A way of getting players in touch with their characters is by means of Character Quizzes. In addition to being a handy way of getting started, character quizzes can also be used later on, as player characters change and grow, to help both the Game Master and the players get a better picture of the character's personality.

Don't expect every player to respond to every question. To any given question some players will find a lengthy interesting response for their characters. Others will feel that the question doesn't apply to their character, or that the answer is simple.

Pick a handful of questions below. Then either read them to the players, or type them up and distribute them. Feel free to add in other questions, especially about things that might help to develop aspects of your campaign.

- Your character is plagued by a recurring nightmare/dream. Describe the dream.

- Your character has need of a horse for a long journey. Describe the steed's size, coloring, build, training, and personality.

- From your character's point of view, in relations with family, is it better to be loved than feared? Or better feared than loved? Would the answer be any different with any other group of people? If so, who?

- What is your character's favorite food? Drink? What food does the character hate?

- Take a Devil's Advocate stand. Describe what you (the player) hate about the character. What are the good reasons for other characters to dislike/hate the character? What little, minor bad habit does your character have that would annoy anyone after awhile?

- The character's first, or most memorable, love affair. Can be anything from a pre-teen crush, to a long-term romance. Be sure to describe the object of your character's affection, along with the changes that your character went through in terms of feeling and thoughts about that person. Did it start out as a maddening hatred? Instant love? Who fell for who first? Was love a surprise? How did it all end? How does everyone, including you, your lover, rivals, parents, friends, observers, feel about it now?

- You return to a comfortable, way-shadow inn that has housed you in comfort numerous times over the years. You find that the kindly family has just been threatened, beaten up, and robbed. You track down the guilty bandit leader and capture him. What will you do now?

- Your character gets into a discussion about death and dying. How would your character describe the perfect death?

- Going back to the guilty bandit, assume that you decide that banishment in a far-off shadow is an appropriate punishment. Describe the shadow you would seek for this purpose.

- What was your character like at the age of ten (fifth grade)? Was your character a wimp, a bully, a nerd, a snitch, or a klutz? Popular or not? Write up a little description, and, if you like, a little day-in-the-life story about your character at that age.

- Did your character ever have a pet? What kind of pet? Where is it now, or what happened to it?

- Assume that your character needed/wanted to spend some time, incognito, in a place like the Mythical Old West of cowboys, indians, gunslingers, pioneers and prospectors. Assume it's around 1875, after the Civil War. Your stay will be for a matter of years. Describe what role and appearance your character would adopt.

- Which would your character describe as the greater evil, Murder of a Shadow person, or the maiming of an Amberite?

- What would provoke you to murder a Shadow dweller? A lie? An insult? An attack? A crime against one of your friends? Nothing?

- How do you think your character's father would describe your character? And how would that description differ from the description given by your character's mother?

- Describe your character's perfect "date." What events would it involve? With a group, or just one-on-one?

- Describe your character's personal bedroom or suite. Is it located in Castle Amber? Describe the contents and the various rooms. What is the style of the furniture? Describe your favorite reading spot, and your favorite chair.

- Assume that your character has just finished a month-long struggle against hard odds, working toward something of great value to the character. Suddenly, the final stage of the opposition turns out to be stronger than you expected and you are attacked where you are weakest. It may not be a battle to the death, but it will definitely determine whether or not your quest is a success.

    Unexpectedly, you receive a trump call from a senior relative offering assistance. This relative is a champion in the field of battle where you are currently struggling (in other words, if the contest is strength, the rescuer is Gérard, if magic, then Fiona, etc.) and, if asked, will either move through to help or leave you to your challenge.

Do you accept the help? Or do you reject it?

What if the battle involves your greatest attribute (Warfare, Pattern, Psyche, etc.) or ability (Pattern, Trump, etc). Would your decision be the same?

Finally, what if the battle was a matter of life and death. And failure would equal your own death. Would your answers still be the same?

- Which self-improvement course does your character need more; "Assertiveness Training," "Ten Steps to a New Love Life," "How to Win Friends and Influence People," or "Organizing Your Life." Why? Which course would your character be qualified to teach?

- How would your character describe his or her perfect mate/spouse? Do you, the player, agree? How is your character mistaken? Overly idealistic? Or is your character looking for all the wrong things?

- If your character uses servants, which servant is most trusted? Most liked? Most indispensable? Describe each of them.

- Where does your character get the laundry done?

- If your character were suddenly plunged into normal, non-Amberish life, right here in the prosperous west of the late 20th Century, how would the character most likely live? What profession would s/he pursue? Describe the probable life style (vagabond, married with children, socialite, etc.) of the character.

- Everybody has some little item that they regard as somehow sacred. A piece of clothing, a memento of some event, whatever. What is your character's sacred item? Tell the story of how your character found it, or why it came to be important.

- A relative, one you respect but do not fear, has committed the same breach of etiquette three times. Each time, in front of others, this person has treated you as an inferior, giving you orders instead of making requests. And your performance has been criticized in front of others. You can arrange things so that you'll not be bothered again. Or you could confront the person. Or have a third party intervene. What would you do? If you would have to explain your actions, what would you say?

- Midway through a pleasant meal, with a very enjoyable companion, you receive a contact from the aforementioned enemy. Frankly, you are sick of the whole business, and you don't relish becoming a pawn in someone else's game. How would you deal with the situation?

- Your character is the butt of a practical joke. Could your character see the humor in it? Would your character get even? Does getting even mean staging another practical joke?

- Your character, stuck for a time on Shadow Earth, is ready for companionship. The problem is that most potential dates are too easily molded or formed from your unconscious control over Pattern. You would prefer someone new, someone different, yet someone who might find you attractive. Then a rather mundane solution strikes you! How about an advertisement in the "Lonely Hearts - Romance" section of the classified section? Write up your character's ad.

- At one point Corwin said; "...It is an academic, though valid philosophical question, as to whether one with power over Shadow could create his own universe. Whatever the ultimate answer, from a practical point we could." How would your character respond?

- Does your character sleep well? If the character has dreams, what kind of dreams are they?

- To your character, does revenge mean (1) "an eye for an eye," (2) "repayment with interest," (3) "the only good enemy is a dead enemy," or (4) something else? In exacting this revenge, would your character serve it hot, as soon as possible after the offense, or cold, awaiting a perfect place and time?

- What emotions can your character express in public? Sorrow? Anger? Sadness? Humor? Disappointment? Joy?

- You've been beaten, bruised, and battered in a recent adventure, not to mention frustrated by a rather vile ending to the affair (a scorched earth situation where you won but your objectives were destroyed). You now find that you have several weeks to recover your health and composure. Since you only wish to Shadow Shift once, what Shadow would you pick for your retreat, and what activities would you pursue?

- Strange as it seems, you find yourself in need of the service of a hunter-killer beast from the realm of Chaos. Since it can take any form you request, and, in this case, it will be something of a calling card, how would you like it to look? Are there any particular powers, attributes, or qualities that such a thing should have?

- What if your character had the chance to put a single question to DWORKIN? What would be your question? Is there another Amberite, player character or non-player character, living or dead, you'd rather question? Who do you want to ask, and what is your question?

- The character's first battle, first hunt, or first killing of another person. Try to make it as vivid as possible, including the character's age, training and inexperience, plus details on the surroundings (weather, terrain, uniforms or clothing of self and others).

- Describe your character's voice. Would you say your character speaks with formality? Casually? Does the character have any favorite expressions/curses?

CHARACTER BACKGROUND

After the points are worked out, and the players have filled in secondary details like their character's physical descriptions, the job of developing character personalities gets started. One of the best ways to get the player focusing on the character's state of mind is a bit of role-playing.

Encounter on the Streets of Amber.

One favorite way of getting started with a new character is to run the player through a simple confrontation on a street in the City of Amber.

Anytime the player asks for details about the setting or the scenario, the answers should be based on the character's Good Stuff and Bad Stuff. For example, if asked about the weather, a character with Good Stuff will encounter a sunny day with a fresh breeze lightening the smells of the city, Zero Stuff might experience a partly cloudy sky with a light wind, and for Bad Stuff you might inflict a constant rain, leaving the character with sopping hair and soggy boots.

You start by telling the player something like the following:

"You're wandering down a street down near the dock in Amber when someone bashes right into your side. What are you doing?"

Whether the player character is knocked down, has to stagger back, depends on the size of the player character. The drunk is about 200 pounds, so small characters of under 140 pounds, will be knocked down, bigger characters will stumble, and anyone of 250 pounds or over will cause the drunk to bounce.

Most likely the player will want more information.

"Looks like the guy who bashed into you is a drunken sailor. He's big, smelling of rum, none too clean, and wearing stained leather clothing and a cape of sheepskin. As he's steadying himself, he yells out drunkenly, `Look where you're going.' What are you doing?"

There is really no "right" solution to this little problem. The man is a drunk, both rude and abusive, and he's not shy about picking a fight or bullying anyone who tries to avoid one. Any attempt by the player character to smooth things over will be met with ridicule and a curse.

"Stinking Amberites, he mutters, I knew you were all a bunch of stinking cowards."

Walking away, or avoiding the encounter is easy enough. If the character does nothing, or leaves, read the following:

"He stumbles back into the bar, yelling something you can't hear very well, except for the word Amberite spoken as a curse. There is a roar of laughter in response to his words. What are you doing?"

If the player character stays with the drunk, either to clear the air, talk, or fight, then things should get ugly. Here's a typical insult the drunk might be provoked into delivering:

"The whole Castle full of you Amber scum aren't worth a bucket of warm spit, he yells."

Be creative in coming up with insults, in particular coming up with slights that prick the character's weaknesses. Feel free to mess around, having the drunk shift from sarcastic apology to abusive insults, to craven fear and retreat.

If the player character reacts with combat, on the spot, or later, the drunk will attempt to pull out a nasty-looking dagger from under his cloak. He fights at a little better than Chaos Rank (Warfare), and he'll sober up quickly if faced with a little action. The rest of his Attributes are merely Human, so he should be no particular threat to any player character.

Leaving the drunk alone, unharmed, is also an opportunity for conflict. Skip ahead in time, and read the following:

"A few days later you're passing by the street where you ran into the drunk. You notice people seem to be whispering behind your back. They're sharing some joke and laughing at you! What are you doing?"

Investigation will show that the drunk is telling the tale of the encounter to everyone who will listen. Describing the character as a coward and a chump. Tracking down the drunk is easy enough, since he's loudly spreading his lies in every tavern in the area.

More details, such as the drunk's name, occupation, origins, or whatever, are up to the Game Master. In *Amber*, where players can drag the whole campaign to an entire new world in an instant, you're going to have to learn how to make up information as you go along.

In this, and every *Amber* encounter, the idea is to make things sound good. Use whatever trick you need to make things seem like you worked everything out in advance. If you need it, carry around a name list. When you think of an interesting name, add it to the list. Then, whenever you need a new name, you can just pull one off the list. Put yourself in the drunk's place, imagining what your life might be. Use what you know about the player character, their background and history, to come up with details. Or, to be really perverse, picture how an elder Amberite, someone like Bleys, would "play" the role of the drunk.

The point here is to give the player a chance to make some decisions about their character. Is the character passive or aggressive? Combative or tolerant? Given the option, how will the player character deal with revenge? When things come to the crunch, will the player character commit murder? Maiming and mayhem? Or will the player character be merciful? Or creative and vengeful?

By giving the player a chance to react, you're giving them a chance to explore their character's personality. As with most situations, a good Game Master is seeking to stress the player, pushing them to extreme measures, and helping them to understand their character.

THE RULES OF ENGAGEMENT

A Little History.

Back in World War I, in the war for East Africa, the Germans would send raiders across the desert, to harass and loot the better-equipped, but poorly defended, British outposts.

The survival of these commando missions, and the men who performed them, depended on an exact knowledge of the whereabouts of fresh water holes. In the African heat, that water was the difference between life and death, because a light, fast-moving force couldn't carry enough water. This fact was obvious to both sides.

Eventually the Germans were forced to abandon their raids.

One by one, all the critical water holes were found marked with Skull-and-Crossbones signs reading "POISON!" in several languages, and surrounded with the bloated bodies of dead animals.

After the war, the Germans protested. "The use of poison," they said, "is expressly prohibited by the Geneva Convention. You have broken the rules of conduct and should be charged as war criminals!"

The British protested that they were completely innocent. They never used poison.

As they pointed out, "Where in the rules of the Geneva Convention does it prohibit posting signs and scattering around a few dead animals?"

If you want to keep someone from using a water hole, it doesn't matter if it contains poison. What matters is whether or not you can get them to *believe* that the hole is poison.

Nothing in the rules prevented the British from lying.

The Game Master's Rules of Engagement.

There are "Rules of Engagement" in role-playing as well. They define what a Game Master can and cannot do. Abiding by the Rules of Engagement assures the players that the role-playing will be conducted fairly.

However, fair doesn't necessarily mean honest. The British forces succeeded in making the Germans believe their water was poisoned, *without breaking the rules!*

It's up to the Game Master to feed the players everything they see, hear, smell, taste, feel, sense and remember. According to the Rules of Engagement, these must always be reported honestly.

Game Masters don't cheat, because that would be breaking the rules. There's nothing in the rules prohibiting lies.

1. Senses and Memory Must be Truthful.

All players operate blind, depending on the Game Master for eyes, ears, nose, tongue and skin. In addition the Game Master has to report on what the character remembers, senses with Psyche. All these things must be described to the players factually.

Exceptions. The senses can be wrong. Characters can get ill, see illusions, end up in dark places, and otherwise end up with wrong or missing impressions. Memories could be missing, vague or biased, or even wrong if the character's mind has been tampered with. Still, the Game Master must report the memory exactly as it would come into the character's mind.

2. Players Must Control Their Own Actions.

It is unfair to impose an unwanted action on a player. All the actions of the character are based on the decisions and will of the player.

Exceptions. If a character is cold, then it's okay to impose goosebumps and shivers, a sick character may do all sorts of icky things involuntarily. However, if a character has been presented who is tough as nails then having them flinch at the sight of a spider or a snake might not be fair. Leave it up to the players to role-play their weaknesses whenever possible.

In Combat blades flash, characters move back and forth, and thousands of actions can take place in a single minute. When there's no reasonable way to get feedback or approval from the player for all the minute details it's up to the Game Master to make logical assumptions about the character's movements. Again, the best bet is to get a feel for how the player wants the character to handle things. Does the character usually flinch from too close an edge? Or does the character automatically parry anything that comes straight in? These questions are best answered by the experience of role-playing the details when the character is first created.

3. Death and Final Solutions Must be Fair.

There's nothing so irritating for a player than to have a character be penalized for a single unexpected wrong move. There should be no "button of death" that instantly wipes out a character. Not unless the character has repeatedly been warned of the possibility.

Exceptions. There really are none. Sometimes, for the sake of a story line, a character will start a scenario by being unfairly zapped, kidnapped or taken out as a way of setting up the action. This should be only as a ploy to get things moving, and should not permanently affect the character.

4. All Problems Must Have Solutions.

All the threats, villains, monsters and disasters thrown at the players should be beatable and solvable. A good rule of thumb is that there should be at least three "outs" for the player characters in any dilemma. Two solutions that the Game Master has come up with before hand, and an open mind to allow the players the chance to come up with creative solutions of their own.

Exceptions. Sometimes story elements are designed without direct solutions as a way of forcing the players to find another approach. Johnny Hill, one of my great Game Mastering mentors, once threw a "monster" at my role-playing group that seemed completely unbeatable and indestructible. It seemed to be a problem without a solution, until we figured out that we were holding the fragile little jar that contained the monster's soul.

5. Bias is Prohibited.

Good Game Masters have no friends, no enemies, no prejudices and no preferences. It's best to see through the players, and look only at the characters.

Exceptions. There are no exceptions.

GOOD GAME MASTERS CREATE GOOD ROLE-PLAYERS

It's no surprise that the very best Game Master are blessed with great role-players. Game Masters don't find great role-players, they make them.

Too often Game Masters complain about players who aren't good role-players. That's like a teacher complaining about students that don't learn.

They're both bogus complaints. It's the job of the teacher to teach. Not just the smart kids, or the ones who come fully prepared, but everybody who walks in the door.

It's the job of a Game Master to present role-playing opportunities. If a player doesn't catch on, don't assume it's the player's fault. The player may simply lack experience or skill. Bottom line, it's the Game Master's job to challenge every player with problems and situations that will stretch their limits.

Look at if from the other way around. Role-playing is about challenges. If a player succeeds every time, solves every problem, and is never frustrated, then the Game Master isn't trying hard enough. Players can never feel the exultation of victory if they don't realistically fear defeat.

Teaching Role-Playing. No, you don't need to give lectures to teach role-playing. Role-playing is learned by doing. When role-players lack skills, the best way to give them those skills is by throwing them into repeated situations where they can develop the skills.

Here are a few of the problems players present, along with some suggestions for how to deal with them:

Monster Bashers. Some players think that role-playing is just about hitting things. They want to take out their aggressions on anything that moves. If it doesn't move, they want to see it broken into pieces, or at least smashed flat.

Such players can be weaned of their overly violent tendencies. It takes a little time, and a lot of Game Master patience, but mostly it's a matter of role-playing situations where "shoot first, ask questions later" tactics just don't work.

Here are a few possible ways of dealing with such a player:

GM: Something jumps right at you. What are you doing?
Player: I'll skewer it with my sword.

- "You see a frightened looking serf, dressed in rags, looking down at the place where your sword enters their chest. `I've come to warn you...' she says. And then blood comes from her mouth."
- "It gives a squeal of fear and terror. You see that you've just hit some kind of big puppy."
- "Your weapon is knocked from your hand and something catches you by the throat, raising you up off the ground. `Why you want hurt Monjo?' it asks in a hurt tone of voice."
- "`Whoa, big fella!' says the red-haired stranger, `If it's killing you want, I'm sure my father Bleys can accommodate you...' Then he calls out, `Dad, there's someone here to see you'"

Rule Lawyers. For some players the rules are everything. They lose the fun of role-playing by focusing on the minutia of points and system technicalities. Aside from suggesting that they might be better off joining a war gaming club, it's important to side-step their questions and concerns, putting their focus where it should be, back on the role-playing.

Here's an example of one way to deal with folks who dwell on the rules. Willy is intent on arguing about some minor detail, and the Game Master is sticking to the role-playing.

GM: A fist-sized object comes flying into the room, shattering the window and splashing right into the big soup bowl. What are you doing?
Willy: Greymer is here with me, didn't I have any warning?
GM: Your lynx, Greymer, looks startled and his fur is standing straight up on end. What are you doing?
Willy: Greymer is supposed to be psychically sensitive to danger. Why wasn't I warned?
GM: Greymer takes off for the door, so fast you'd think his tail was on fire. What are you doing?
Willy: I don't get it. Greymer should tell me about this stuff. Why didn't I expect this? Is this fair?
GM: Are you asking me?
Willy: Yeah, I'm asking you.
GM: What's the question?
Willy: Why didn't Greymer warn me of this attack?
GM: Are you asking as Willy, or as Garvin?
Willy: What's the difference? Just answer the question.
GM: If you are asking as Willy, I was going to suggest that we get to it some other time, after the session. As to Garvin, how is Garvin going to find out the answer?
Willy: Okay, okay, Garvin asks the question.
GM: Who is he asking?
Willy: I'll ask Greymer.
GM: Greymer left right after the hand grenade landed in the soup. Seems the lynx realized there was some kind of danger here.
Willy: Hand grenade? What hand grenade?
GM: Are you looking in the soup bowl?
Willy: No! I'm running out of here...

Indifferent Players. Players who don't care about their characters often interfere with the tone of the campaign. This especially irritating for other players who care deeply about their characters and who can get quite annoyed at someone saying, "Oh, it's no big deal if we get killed..."

The best solution is to establish a firm emotional connection between the player and their character. That means role-playing through events that allow the player to make decisions for the character.

Getting Along. Often a player can get a better grip on a character just by role-playing through the day-to-day frustrations that all of us face. Because the Amberites are so powerful, the "rush" that a player gets from being able to cut through ordinary problems helps them get attached to the character.

- "As you walk up to the door you see the "closed" sign being posted. It looks like they've already locked up for the day. What are you doing?"

- "Three guys step in front of you. They demand you give them money. What are you doing?"
- "You're alone, and you're feeling hungry for something special. What are you doing?"

Young Versions of Player Characters. Taking a player back to a character's youth can be a wonderful role-playing experience. It gives them a chance to form the character at an early age and to see the possibilities of role-playing an interesting personality, not just a bundle of Attributes and Powers. A good age for this kind of thing is twelve. The character should be relatively normal for that age, but with their specialties turning into promising strengths. Here are some examples of creating early versions of Amberites:

Peggy's character **Iresa**, with her tremendous Strength, would have a younger version capable of performing outrageous feats of strength. Otherwise she will be above average, but with no particular heroic abilities. However, as with most twelve-year-old girls, she'll be reluctant to do anything that sets her apart from her friends.

Willy's character **Garvin**, as an adult, is 5'10" tall, and about 175 pounds. Garvin's main adult Attributes is Psyche, having put fifty-two points into it. Even at the age of twelve his Psyche will be something special. His weakest Attribute, Strength at Chaos Rank, means he'll be normal for a twelve year old kid. The other two Attributes, Endurance and Warfare, are both Amber, which means he'd be considerably better than normal for his age when it comes to stamina and reflexes. Another aspect of Garvin's character is his Advanced Pattern. While he wouldn't have even a trace of the powers of Pattern, his blood would carry the potential. Likewise, his eventual study would indicate that he'd be a really intelligent, at least somewhat studious young man.

Alex's character **Harick** is a huge adult. He'll be a fairly tall and awkward pre-teen. His first place ranks in the Warfare and Endurance Attributes means that he'll be naturally quite athletic. However, with eight points of Bad Stuff, Alex will also get every curse of adolescence in full measure. Clumsiness, broken voice, pimples, overweight, etc...

GAME MASTERING TECHNIQUES

There are two "arts" to being an *Amber* Game Master. The first is the art of creating a story, covered back in the section on "Campaign Building." Art number two is the art of interacting with player and making things interesting from one minute to the next.

Verisimilitude.

Yeah, I know, I can barely pronounce it myself.

The word "verisimilitude" means, according to the dictionary, "the quality of appearing to be true." That's one of a Game Master's main objectives, trying to get the players to believe in what is happening in the campaign. To get players into accepting the events of the role-playing scene as "truth."

Don't Say It, Show It! The first rule of making things believable is, as they put it in Hollywood, "Show it, don't tell it." What that means to a Game Master is that you should never tell the player anything about the universe. Instead you describe what the character sees, hears, smells, feels, and tastes.

For example, you could tell the player that a certain little girl is really rotten. Or you could read the following:

- "You peek into the room quietly, expecting to see both children asleep. Instead, you see that the sweet little girl is up, leaning over her older brother's bed as he lies sleeping. She's lifting his blanket with one hand, and in the other she's holding a jar that seems to contain some kind of bee or wasp."

If you want to present a character as nice, show them doing something nice.

In the same way, trying to tell a player that their character is mad, or sad, just isn't as "real" as role-playing through whatever circumstances it takes to bring them to that emotion.

Even when you have to tell the player about their character's emotions, it's best to present it from the character's point of view. For example, "so-and-so told you," or "you read somewhere," or "you saw," are always superior to saying "you know." Here are a three variations on communicating a "remembered" emotion with a player:

Sample Memory with Told Emotional Memory. Peggy's character Iresa has a memory associated with a name the group has encountered. In this case the Game Master just gives the player a thumbnail sketch of where and when and what are the feelings recalled.

GM: That name, Josek, seems to mean something to you. Yes, you're pretty sure that you knew somebody of that name when you were a kid.

Peggy: I do? How old was I? What was Josek to me? Tell me about it.

GM: You were ten or twelve, off in that boarding school that you hated so much. Back when your mother had the nervous breakdown, and, as usual, when your father was off on one of his frequent absences. You remember Josek as a kid of your age, who frequently humiliated you. You hate him.

Sample Memory with Shared Emotional Memory. Again we'll take a look at Peggy's character Iresa and her relationship with has a memory associated with a name the group has encountered. In this case the Game Master just gives the player a thumbnail sketch of where and when and what are the feelings recalled.

GM: That name, Josek, seems to mean something to you. Yes, you're pretty sure that you knew somebody of that name when you were a kid.

Peggy: I do? How old was I? What was Josek to me? Tell me about it.

GM: You were ten or twelve, off in that boarding school that you hated so much. Back when your mother had the nervous breakdown, and, as usual, when your father was off on one of his frequent absences. Iresa was a perfect athlete, but had some problems with her academic studies.

In particular you were struggling with your compositions, and with math. Also, since your old school hadn't covered classic languages, you were way behind in those subjects. Does that sound right to you?

Peggy: Yeah, I think so. Still, Iresa would have worked hard, she would have been trying.

GM: Exactly, she was trying very hard to keep up. Josek was a smarty-pants who made your life miserable. He was always way ahead of you in every subject, and he would taunt you with his good tests and grades. He also used to make up clever little Latinate riddles with your name. The whole school thought they were just hilarious, and you even heard teachers telling them in class.

Peggy: Was he just picking on me?

GM: Yes, he never made up riddles about anyone else.

Peggy: What a snot!

Sample Memory with Role-Played Emotional Memory. For the most involved possibility, the Game Master will role-play Peggy through a critical memory of her relationship with Josek. The background starts the same, but in this case the Game Master takes Peggy into the past.

GM: That name, Josek, seems to mean something to you. Yes, you're pretty sure that you knew somebody of that name when you were a kid.

Peggy: I do? How old was I? What was Josek to me? Tell me about it.

GM: You were ten or twelve, off in that boarding school that you hated so much. It's back when your mother had the nervous breakdown, and, as usual, your father was off on one of his frequent absences. Does that sound right to you?

Peggy: Yes, that fits in with how I see Iresa's childhood.

GM: Okay, just a bit more background. You're a good athlete, but you're having a lot of problems with your studies. One day you walk into the classroom where you're learning the old classic languages, and you see the teacher, Josek, and a couple of other students laughing. The teacher sees, you, puts his hand on his mouth and turns a little red. What are you doing?

Peggy: He did that because he saw me?

GM: It would seem so. Josek glances at you, whispers something to the other students and they break out laughing even harder. What are you doing?

Peggy: I guess I'll just ignore it and go sit down.

GM: The teacher hushes up the students, and things settle back down to normal. A couple of hours later the class is over. Are you doing anything unusual?

Peggy: No... Well, do I know any of the students who were laughing?

GM: Sure, they're all classmates. You know one of them well. Hmmm. Let's call her Vina.

Peggy: I'll go up to Vina and ask her about the joke.

GM: She seems embarrassed and she tells you it was nothing.

Peggy: Right.

GM: Do you want to press her on it?

Peggy: No, I'll let it go.

GM: Fine. A couple of days later in the dining room you see some older kids laughing and passing around a note. The one who has the note sees you and starts shoving the note into a book. What are you doing?

Peggy: Can I grab the note?

GM: If you want to step over, probably.

Peggy: I'll take the note.

GM: The guy doesn't want to give it up. He holds his hand out to block you. What are you doing?

Peggy: Can I twist his arm and make him give it to me?

GM: With your strength? In a flash you've got the note. What now?

Peggy: What's on the note?

GM: It's in Latinate. Not exactly your best subject. However, you do spot your name. It's used like a rhyme, at the end of every sentence.

Peggy: What does it mean?

GM: As far as you can tell, it's some kind of play on words. What are you going to do?

Peggy: I'll go find that slime, Josek.

GM: You see him on the other side of the dining room.

Peggy: I'm going to push this in his face and ask him if it's his work.

GM: He looks pretty scared when you walk over. He's kind of a wimp, and you're at that age where the girls are mostly bigger than the boys. He looks at the note and kind of gulps. What are you doing?

Peggy: (in Iresa's angry voice) If you ever, ever, write anything about me ever, ever again, I'm going to pound you into the ground, rip off your arms, and break your legs!

GM: Well, that had an impact! Everybody in the dining room goes deathly quiet. A couple of teachers are looking your way, what are you doing?

Peggy: I'll just leave.

GM: Fine. That's the last clear memory you have of Josek.

Tricks of the Story-Telling Trade.

Earlier in the book, near the beginning of the section on Campaign Building, there is a discussion on the basic elements of a story. Here are some tricks that a Game Master can use to enhance the story in progress.

Foreshadowing. Literally, this means seeing the shadow before seeing whatever is casting it. A classic screen device is to have a shadow cast from a streetlight, foreshadowing whoever is approaching from around the corner. In a serious drama, it is followed by some vicious thug, or by the comic relief of someone tiny and harmless.

Either way, foreshadowing is a useful trick in role-playing. It sets the players up for what is about to come, increases tension, and makes the outlandish more believable.

Trump Scrying. Built into *Amber* is the concept of reading the future with Trump. No one expects an exact prediction, so this is a great opportunity for the Game Master to throw in hints of things to come.

The Red Herring. Throughout every role-playing session there are countless opportunities to throw in "red herrings." There are hints or clues of things that may or may not become important later on in a campaign. The mention of a name, a peculiar event, a dead body, a bullet from somewhere unexpected, an item found or lost. Just about anything can be turned into a "hook" for either a current story, or as groundwork for something that can be run in the future.

Retrofiting the Story. Sometimes we have to change things in mid-stream. In woodshop you might start out building a lamp, and, after a few accidents, end up with a passable ashtray. The same is true with stories in role-playing. Things sometimes work out better if you change things behind the scenes, to make everything make more sense at the end.

Did Roger do it? Roger Zelazny wrote the *Chronicles* over several years (and well I remember my impatient waiting from one year to the next for the subsequent books).

What seemed reasonable in 1970, when *Nine Princes in Amber* was released, may have seemed absurd in 1977, during the writing of *The Courts of Chaos*. Nothing wrong with that. We expect writers to keep thinking about their stories, and to come up with more and better ideas as they write.

In the second book, *The Guns of Avalon* (1972), Roger introduces the character Ganelon, a lovable rogue. Ganelon becomes a major character and, so it seems, just about the only person the hero can really trust.

Ganelon, loyal and steadfast, gradually becomes more and more important throughout *Sign of the Unicorn* (1975). Ganelon asks all the right questions, and he tags along on many of the journeys of discoveries. In *The Hand of Oberon* (1976) Ganelon becomes even more important, and starts doing some pretty wierd things. He beats up Gerard, the strongest man in the universe. He runs across the blood-stained Primal Pattern. He gets his hands on a set of Trump, and starts using them freely. In this last book we also discover the power of Shape Shifting for the first time, when Dworkin mistakes Corwin for a disguised Oberon.

In the last line of *The Hand of Oberon* we find out that Oberon *is* Ganelon. He set the whole thing up, to the extent of creating the Shadow Lorraine as an elaborate trap designed specifically for Corwin so he could tag along in his disguise.

Did Roger know, back in the late sixties, that Ganelon was actually Oberon? Or did he decide on it after writing the first book? Or did he, sometime between 1972 and 1975, decide that Ganelon was just perfect for the role? Did Roger retrofit the story?

I don't know. And Roger ain't telling.

Unless you've got the genius of Roger Zelazny, you're going to have to accept that you may need to occasionally retrofit a campaign. It's not a trick you can use very often. It's just a trick that you should know you can use.

Since you don't know which casual by-standers are be accepted as important friends and followers, and which will be by-passed by your player characters. Some of your important characters will be ignored or destroyed. On the other hand, the most casual of encounters, based on characters you slap together at a moment's notice, can end up as important servants and confidants, lovers or rivals.

Building Player Character Suffering into a Campaign.

Hopefully it should be pretty clear, both to the Game Master, and to the players, that it's very hard to die in an *Amber* campaign. That's how it should be, in a role-playing system where each player invests so much time and effort in character creation and development.

However, the lack of death as a threat seems to give some players a feeling of invulnerability. It's as if, by removing the specter of the grim reaper, *Amber* has lost its sting.

Nothing could be more wrong.

More than once, I've heard words to the effect of, "Erick, this isn't much fun. I'm more involved with this character than any other I've ever had. But the horrible things that are happening to my character are almost more than I can stand."

Hmmm. I guess I'm doing things right.

In Amber, player characters almost never die.

Not quite *never*, because a player can always commit a series a errors, culminating in a final, fatal, error. So it is possible to die. Just difficult.

Not having to worry much about dying doesn't mean Amber is easy. Quite the opposite.

After all, consider the outcome of Corwin's first assault on Amber. For his troubles he gets the eyes burned from his head, numerous beatings, and a stay of several years in a filthy dungeon.

Think it's *easy* role-playing through something like that?

Of course not. It would be a lot easier if the character just died, and if the player could just get on with a new one.

Making a mistake in Amber won't usually result in a player character's death, but can often mean something worse than death.

Imprisonment & Torture. Corwin had his eyes burned from his head, and was left to rot in a nasty dungeon. Getting out took something like three years. Certainly, from the point of view of a player, a much worse fate than simply dying.

Destruction of Trust. At one point in the *Chronicles* Corwin asks Martin a really hard question, one that he is pretty sure that Martin will have to avoid. Sure enough, to cover a friend, Martin avoids the truth. Making Corwin feel incredibly guilty. You see, Random, Martin's father, a good enough poker player to notice the deception, was present at the time. As Corwin put it, "...I had just bought us a piece of instant insurance at the small price of putting a father on guard against his long-lost son."

Death of Comrades & Friends. Yes, give the player characters plenty of friends, family, servants and employees. Get them used to the attention, the companionship, and all the perks of being loved and adored. Perhaps you can then demonstrate why Amberites are so leery of getting involved emotionally.

Hatred, Fear & Loathing. Being an Amberite, with all the associated power, not to mention the perk of immortality, does rub some people the wrong way.

Guilt. Definitely a biggie. It's so easy, really. All the player characters have such great power. It's only reasonable that they'll want to use it. So what happens if they read the cards wrong? What happens if they finger someone innocent for the villain? Or just take offense at a belligerent or hostile encounter? It an easily come down to a moral decision. Do they kill, or not?

Anyone killed by a player character is an opportunity for delivering guilt. In a brightly ribbon, gaily wrapped package, the player character can be given a present of a box of evidence. Evidence proving that they killed the wrong person.

WRAP-UP: ULTIMATE AMBER ROLE-PLAYING

"The Amber RPG is an exercise in pure role-playing. It is not a game. The rules and numbers are not important. Only your character is important."

Don Woodward

Ultimately, I hope you can toss this book.

The best kind of role-playing is pure role-playing. No rules, no points, and no mechanics.

If there is such a thing as an "improved" version of *Amber*, it's something that goes straight for the story-telling.

Don't like something here? Toss it out!

Find a way something works better? Use it!

Here's a few guidelines to point you toward an even more radical form of role-playing:

1. *Dumping the Character Creation Process.*

Try running a campaign without any sort of character creation "rules" at all. Instead of engaging in an Attribute Auction, and having the players work with points and powers, just role-play the process.

Start by just asking the player what kind of character they'd like to play. Personality, looks, ambitions, whatever. Gradually build a picture of what kind of character the player wants. Work your way up to weaknesses, strengths, and foibles. Get to the point of figuring out what kind of upbringing the character might have had.

Then, flash back to the past.

Do a bit of one-on-one role-playing. Start with some seminal experience, something in their childhood. Not threatening, but challenging. We've all had to face rough situations as children. Draw on it, find some situation that has no particular "right" solution.

It could be a bully or a difficult playmate, somebody annoying, but potentially valuable. Or a bit of role-play involving a beloved pet ("he followed me home, can't I keep him?"), that the character's parents or guardians either don't want, or find untrained or uncontrolled. Moral decisions start in childhood, when there is no "right" decision about how friends might do dangerous or bad things.

Eventually, as the character progresses, the choices of careers, skills, powers, family connections, and all the other facets of the character's background can be built up.

2. *Dumping the Points.*

Fact is, the points in *Amber* are artificial. Points are a way of keeping players on track, of balancing role-playing elements so there is a sense of fairness. The system is flexible enough for a wide variety of things to be done with a hundred or so points.

Which is really a limit on creativity.

Why can't one character be a world-wise, ancient seer, with fingers in power, while another character is a mere lad, weak and uncertain, and a mere initiate in the mysteries of the Pattern? No reason, except for an old-fashioned approach to role-playing. It's really the same logic that mandates that both players in a chess game start with the same sixteen pieces on the board.

And sometimes it doesn't really work right.

Some characters are young, fresh and brash. Dynamic enough in their personalities so that the addition of extra power and facility add nothing to the fun of role-playing their personalities. In which case, why bother with adding points?

Other characters are intrinsically interesting as they explore the fringes and outer limits of fantastical powers. What seems most interesting is letting them stretch themselves, pushing far beyond what any reasonable reward system would provide in points. Again, what's wrong with this?

In both cases, there's nothing wrong.

Sure, it's "unfair" to the players, that one should get more points than the other. But if the campaign is interesting enough, if the story is engaging, if both players are fully engaged, the one in overcoming his limits, and the other in exploiting his potential, then the magic of the role-playing experience is heightened.

3. *Dumping the Magic System.*

All the rules for magic spells were created long after the rest of the system had been developed. In the meantime plenty of Game Masters ran *Amber* without any formal magic system, pretty much letting players invent their own spells. The only restrictions were the explanations of the players and the Game Master's view of what would and wouldn't work.

It worked that way for years, and, if you care to experiment, give it a try.

4. *Dumping the Rules.*

I guess dumping dice from role-playing got people thinking, and experimenting.

As you've seen, *Amber* has been played across the country since 1985. Frankly, most of the Game Masters were winging it, since details were sketchy, and rules were almost nonexistent. In effect, each Game Master invented their own version of *Amber* and the rules needed to role-play.

In almost every case, something interesting happened. The rules part started to disappear, and the *Amber* part got more important.

A good example is the whole set of rules for casting magical spells. What's printed here is just one of a dozen or more systems developed in the course of developing *Amber*. So why not trash it? Let players cast any spells they like!

5. *Dumping the Game Master.*

Another interesting development among some *Amber* role-players was the creation of a new style of play.

Players, impatient with the pace of a Game Master's creative efforts, unwilling to wait days or weeks between sessions, would take off on their own, role-playing their conversations and explorations, without the moderation of the Game Master. They discovered that there was tons of fun just role-playing with each other.

Since most Game Masters have their own beloved characters, they were often the instigators and innovators of these player-to-player role-playing experiments.

THE THRONE WAR

Looking for a way to experiment? A way to play *Amber* without committing a lot of time? Want to shake down the system, and run a practice session, without taking it too seriously?

Then a *Throne War* is just what you're looking for. It's fast and deadly, uses all of the regular *Amber* role-playing system mechanics, and it won't take forever.

In other words, *Throne Wars* are almost the exact opposite of the usual *Amber* role-playing experience.

What a *Throne War* Involves.

There is only one point to this exercise, only one goal. Each player is out to become King of Amber. Winning is everything.

What you Need to Play.

Make sure you've read through the rules, have copies of the character sheets and worksheets on hand.

Before you start up the Bidding War, you'll have to decide

what you want to auction off. In addition to the usual four Attributes, you've also got the choice of auctioning off some advantages, ranging from positions of power, to important artifacts, to the alliance of potent elder Amberites.

Next step, get the players together.

That's it. You're ready to play.

Throne War Auctions.

In a Throne War, the auction format is even more important than in regular campaign character creation. Aside from the conventional Attribute Auctions, it's also fun to have auctions, or "bidding wars," for any or all of the following.

Still, while all these auctions can be fun, running more than seven or eight bidding wars can be exhausting. It's best to limit your Throne War's auctions. It's also best to announce *exactly what auctions will be conducted, and in what order, before the game begins.*

Here's an example of how a Game Master might kick off a Throne War:

"Hi guys! Today we're going to have a happy little game. A Throne War. That's where you each do your best to put together a tough character, and then try to kill all the other characters. Whoever ends up on the Throne wins. Everyone else loses. And, chances are, most of you will die.

"Your characters will be put together in the usual way. One hundred points, as usual. Use your points for Attributes, Powers, Items, Shadows, and Allies. Good Stuff and Bad Stuff is also available, but in this game they're used mostly in combat."

"Let me warn you about a couple of things. First, you want to aim for a character that's powerful, but also with few weaknesses. We're going to be playing pretty fast, so anybody with lousy Attributes who gets into a fight, whether based on Strength, Psyche or Warfare, is going to go down fast. Second, watch out for sneaky attacks. If someone attacks you from the back, or with some power that you have no defenses against, well, so long Charlie!

"Along that same line, I'd advise you to be pretty secretive. Don't tell anyone, even your best buddy about your character's Attributes or Powers. And, if you're going to pull something underhanded and nasty, slip me a note, or talk to me about it in private. Don't expect people to play completely in character when they overhear something they shouldn't.

"Cabals, unholy alliances, teams, and partnerships are allowed. I don't care how you gang up on each other. Just don't come whining to me when you get stabbed in the back.

"Now a word on the auctions. Yes, we're going to auction off the Attributes. As usual, whatever you bid, you pay, and that will determine your rank in that Attribute. However, we're also going to have a couple of other auctions. In the other auctions only the winner will spend any points, so arrange your strategy accordingly. Let me tell you what they are, and in what order we're going to run things..."

At which point, the Game Master should list each Auction in turn, describing both the Attribute Auctions and the others. Possibilities are auctions for Attributes, Positions, Artifacts and Allies.

1. Attributes. The usual four of Psyche, Endurance, Strength, and Warfare. It's always interesting to mix these up in a Throne War, arranging to bid first for Strength, then for one of the items below, then Endurance, then perhaps the Jewel of Judgement, then Psyche, another odd item, and finally Warfare.

Or, for another strange alternate version, cut down the number of auctions. A single auction, where players bid on the Jewel of Judgement, and then let them pick their own points, is really interesting. Another good possibility is to bid on Warfare alone, again leaving the other point selections to "secret" bidding.

2. Positions. Each player can bid to control various offices within the defense system of Amber. Examples include:

Captain of the Castle Guard. The most common position, and the most valuable, is becoming the Captain of the Castle Guard. The character controls the Castle, and has been doing it for some years. All of the guardsmen are sworn to obey their Captain (though, if the Captain isn't around, they can be intimidated by any Amberite). The character who gets this position is given the responsibility of describing all the Castle's defenses, traps and guards.

Here's an example of what you might read before the auction begins:

"Next up is the bid for Captain of the Castle Guard. This is a winner-take-all auction. That means that only the winner actually pays any points, unlike the Attribute Auctions. Whoever bids the most points will pay that amount to assume the role of Captain of the Castle Guard.

"As Captain you'll have several important advantages. You'll have control over the Castle itself, it's defenses, guardians, keys, gates and dungeons. All the guards will obey you and will owe you first allegiance. You'll decide where guards are posted, the location of secret passages, and of traps.

In short, being Captain gives you the power to design the Castle itself. Not being Captain means not knowing what threats might be hidden in the walls."

General of Arden's Forest Patrol. Unless one travels by mystic means (Pattern, Trump, etc.), or by the harbor, the only way in or out of Amber is through the Forest of Arden. Taking the position traditionally held by Julian (though without Morgenstern, Hellhounds or Julian's Psychic Hawks), the character who gets this position commands the armed woodsmen will know all the various roads, trails and paths. The character also decides the distribution of the woodsmen, and designates any traps and barriers that are to be placed in the Forest. The General also issues commands concerning who of his siblings will be allowed to pass, who will arrested, and who will be attacked.

Another option is to auction off Julian, as an ally, along with his command of Arden.

Sheriff of the City of Amber. The Sheriff commands all the officials of the City, ranging from Deputies, charged with keeping the peace, and all the volunteer militia, who are trained to defend the City from rebels and invaders. There is also a harbor fleet under the Sheriff's command, consisting of those ships and rescue boats that work the dock and shipyard, assisting and guiding the Navy and Merchant craft. The Sheriff also controls the administration and bureaucracy of the City.

Admiral of the Amber Navy. The primary route to Amber is by sea, and that sea route is completely controlled by Amber's vast fleet. The character holding the Admiralty controls not only the ships and sailors, but also all the trade and commerce in and out of Amber. Characters can arrange the distribution of the fleet, and give commands to every ship captain.

Another option is to split the Fleet. In *Nine Princes in Amber* both Caine and Gerard commanded Fleets, Caine taking the northern armada, and Gerard the southern. Imitating that, you might want to conduct an auction for each of the two armadas.

3. Artifacts. It's usually just one artifact, the most important item in the Amber Universe, the Jewel of Judgement. At the start of the bidding for the Jewel of Judgement, the Game Master should announce:

> "We'll now be bidding for the Jewel of Judgement. The auction will proceed along the usual lines, but only the winner will actually spend any points. Whoever succeeds in this auction will gain the Jewel, and will also start the game as someone who has been attuned to the item. The winner will be briefed in some of the Jewel's powers and potentials. However, be warned that winning the auction is no guarantee that you'll keep the Jewel. If it is stolen, then it is stolen. It doesn't matter how many points you spend, if you don't protect it well, it can be lost. If someone else manages to walk the Pattern with the Jewel of Judgement, they will become attuned to it and may be able to use its powers."

NOTE: Although the player who gains the Jewel of Judgement will be attuned to it, and trained in its use, they will not actually start play holding on to it. The player will have to show up at the dying King's bedside to receive the Jewel.

Mystical weapons, like Corwin's Pattern sword, *Greyswandir*, are another possibility.

Trump decks can be auctioned off in a variety of ways. Either charge for decks, or have an Auction where only the top two or three vote getters have Trump decks.

4. Allies. An interesting twist is to allow the players to bid on the services and loyalty of one or more of the elder Amberites. It's best to limit this to just one or two, making sure to announce them early in the proceedings.

Benedict is the probably the most common character made available in a Throne War auction. Since his Warfare is so superior, he makes for a wonderful and highly sought after asset.

> "We'll now auction off the loyalty and allegiance of Amber's greatest warrior, Benedict. Winner of the bid will pay those points to gain Benedict's services, while losers will pay nothing. Of course, as we all know, Benedict is an incredible fighter, capable of defeating just about anyone in armed combat. Still, he is vulnerable to Psyche and other mystical attacks."

A good counterpoint to Benedict is *Corwin*. And *Dworkin* is another interesting possibility, as is *Fiona* in a similar role. In each case, one of these knowledgeable characters combined with possession of the Jewel of Judgement can be devastating.

Player Conferences

The Game Master should have two private conferences with each player.

1. Shake-Out Chat. This happens right after the player has put together the character. It's at this point that the Game Master can help the player put the character together, pointing out any problems or such. If the character looks okay, then the Game Master should give a briefing on how the player can set things up before play begins.

Before wrapping up the Shake-Out Chat, get the player (and writing up) the answers to the questions you'll be asking in the Pre-War Briefing.

2. Pre-War Briefing. Don't start any of the Pre-War Briefings until all the Shake-Out Chats are complete. That way you'll know exactly what Attributes, Powers, and so forth each player has.

During the Pre-War Briefing, just before the mayhem begins, the idea is for the player to brief the Game Master on any secret plans, codes, signals, and other deviousness. For example, if the player has set up a group of followers to attack any other prince or princess who, say, tries to get to the Pattern Room in the dungeons of the Castle. Just so the player can remind the Game Master, this might be called "Plan A," so that later the player can remind the Game Master of those forces. To avoid confusion later, it's a good idea for both the player and the Game Master to jot down notes on these devious plans and schemes.

Before asking the final question, inform the player that the King of Amber is very ill, perhaps near death. Then find out where they'd like their character to be at the start of play. Anywhere is possible, from the King's bedside in Castle Amber, to a remote location out in Shadow.

Starting the *Throne War*

Make the following announcement to the assembled players:

> "Although his ill-health has been well known, you receive a message that the king has but an hour left to live. He asks that you come to his side to receive his final blessing, and hear his final bequest. What are you each doing?"

Attendance isn't mandatory, so it's up to each player to make the choice of appearing or not.

Even the character who bought attunement to the Jewel of Judgement may choose not to show up. However, that means that "Dad" will hand the Jewel over to someone else, for "safe keeping." And, if the player doesn't regain the Jewel in time, there's the chance that another player character may have the opportunity to get attuned.

When everyone has had a chance to declare their intentions, being in attendance or not, shift the action to the King's bed chamber. Read the following:

"You enter the room, smelling the too-sweet smell of some drug or potion, and an acrid smell as well. Your Father, looking small and frail, is covered by heavy quilts. Only his grey-thatched head is visible, and his right hand, at his breast, clutching the throbbing red Jewel of Judgement. Where will you stand?"

Each player should declare exactly where they are standing in the room. The Game Master should have a clear picture of the location of each character before proceeding.

The King will either name the buyer of the Jewel of Judgement, or, if that character is not present, then another favorite (someone with a lot of Good Stuff), and will say, in a quavering and frail voice:

"Come, take this bauble. Use it well, in the service of the Kingdom."

Should someone press the old King to name a successor, read the following:

"The old King struggles to speak. He manages to croak out the words, `among you the heir must be...' and then he coughs bloodily and goes still. What are you doing?"

Wait a bit to see what happens. There should be an opportune moment, like when it's decided that he's dead, when the King can utter his last words, as follows:

"Suddenly, the old King bolts into a sitting position, and says, `the best! The best among you should be King!' What are you doing?"

At this point the King falls back, dead. The script is over, and it's up to the players to determine the action from this point on.

Play Ball!

Since anybody can end up anywhere, anytime, it's a good idea to put together a list of the players, so you can go around-the-table without leaving anyone out. Players who are out on their own are easy enough, you just deal with them when you get to them on the list. When two or more players are together, then you've got to remember to deal with them as a group, and skip over those in the group who come later in the list.

Speaking of groups, Cabals are a wonderful side-effect of a Throne War. Players spontaneously decide to work together, ganging up against others, and combining their advantages.

As with anything else in the *Throne War* there are no "rules" about alliances and cabals. Always keep open the option for any player to back-stab any other. This is particularly critical when the characters happen to be standing next to each other.

In this case, and in most others, the Game Master should encourage note passing. If you get players passing notes around constantly, players will come to accept it as normal.

Subsequent private talks. After the first two conferences, players should be content to either discuss their plans publicly (what about "Plan A?"), or resort to note passing. If a player **must** have another private consultation, the Game Master, in the interests of fairness, should announce that everyone should get the same number of private sessions.

Resolving the *Throne War*

If all but one player character has died (or if everyone has died, another strong possibility), then the resolution of the Throne War is pretty obvious.

However, players being unpredictable, you can pretty much count on running into endings that are a little more vague. Here are some of the possibilities.

Run and hide in infinite Shadow. Commonly a player character will find that the odds are just too terrible, and that discretion turns into the better part of valor. Since Shadow is infinite, escape is usually an option for any character on the losing end of a Throne War.

A Cabal takes the Throne. Sometimes people just are too good to be true. They'll crown their favorite, chase out the opposition, and declare the Throne War over.

It's always best, in situations like this, for the Game Master to consult with each of the surviving Cabal members privately. Give each player the clear opportunity to do dirt, secretly, to the others. After all, this is a Throne War!

Leaving Things Open

It's interesting that one of my (me, the ubiquitous Game Designer) favorite characters, Lloyd, was created for a Throne War. The Game Master, Scott Thoms, had billed the event as a Throne War, but, as sometimes happens, he threw in a few curves. He decided to have a threat from the Courts of Chaos. When we, the players, didn't immediately start killing each other, he threw in a couple of mysteries. That got the action rolling, but in the wrong direction.

We forgot about killing each other, and ended up cooperating. The Game Master rose to the challenge and gave us some great role-playing. I don't know how much Scott had worked out in advance, and how much was spontaneously created as a result of the way the players reacted. It all turned into a very satisfying piece of role-playing, and I had a ton of fun! Especially since I managed to end up on the throne by mutual consensus, without spilling a drop of Amber (or, in this case, Amethyst) blood.

Throne Wars are supposed to be about killing, playing with the Amber role-playing system, and exploring techniques of Game Mastering. Most of the time Throne Wars go as expected. It's just a good idea to keep your options open, and, if things work out well, let the players lead the story into unexpected directions.

It doesn't really matter how things work out, in open bloodshed, in ambiguous and uncertain conclusions, or in a direction that seems to lead to more role-playing. Remember that the whole point of a Throne War is to have fun.

BATTLEGROUND ON SHADOW EARTH

The Threat Summary: Faced with an oddly shrinking universe, and a confrontation between the elder Amberites and a very angry Courts of Chaos, it's up to the player characters to track down the real problem.

Getting Started.

Each player should be instructed to write up a description of where their character might be. They can start anywhere they like, in Amber, Shadow, or the Courts of Chaos.

If there is more than one of the player characters in any one location, then try to role-play the characters into a group. For example, if two or more players state they'll start in Castle Amber, then they might happen to be seated together at dinner when the fun begins.

Anyone on Shadow Earth. Any player characters on Shadow Earth will be witness to the opening of the battleground. Forces from the Courts of Chaos, leading large numbers of Chaos creatures, will open gateways to any Amberites and immediately attack. Counterforces, lead by Bleys, Julian, or other elder Amberites will counter-attack. Shadow Earth itself will not change in size.

- "There is a crash and a blinding flash of green light. Appearing right in the middle of a brick wall you see a strange tunnel that seems to lead all the way through to the Courts of Chaos. Strange creatures are spilling out of the tunnel, led by a couple of demonic forms you take to be Lords of Chaos. What are you doing?"

Anyone elsewhere in Shadow. All of Shadow will shrink in size by a fifth. Player characters with the Blood of Amber will not change, and may perceive that they have suddenly "grown." Player characters who have Logrus, but not Pattern, will shrink even more, by almost half.

- "Your perspective changes suddenly, as if you just stepped up onto a chair. Your clothing feels tight, and you realize that everyone around you is staring in disbelief, since you seem to have suddenly grown. What are you doing?"

Anyone in Amber. Amber itself will not change in size, but visiting characters will seem to shrink. Characters in Amber who lack Pattern and who are not born of Amber will lose a fifth of their size. Characters with Logrus and not Pattern will suddenly be halved. If none of the player characters have the requirements to "shrink," then have them be present when the visitors from other Shadows, or from the Courts of Chaos, suddenly seem to shrivel up.

- "There seems to be something wrong. Your fork just got a lot bigger, and you have a lot more leg room under the table. Everyone else just got a bit bigger. What are you doing?"

Amberites visiting the Courts of Chaos. Anyone with Pattern will suddenly seem to double in size, relative to the incredible shrinking of the Courts of Chaos.

- "Just when you think you've gotten used to all the strange changes constantly moving through the Courts of Chaos, you get hit with a stranger one. This time you seem to have turned into a giant twice your normal size. What are you doing?"

Background.

The characters start scattered throughout Shadow, living their day to day lives as they see fit. Suddenly, as if a great wave moved out from the Primal Pattern, everything in their environment shrinks. In every Shadow the people, things, and places are reduced by about a fifth (20%). So a character's five foot person is shrunk to four feet, the size ten shoes on the character's feet shrink down to a size eight, and the ceiling seems to drop from eight feet to an uncomfortable six-and-a-half.

This effect is even worse out in the Courts of Chaos, where everything is reduced by half (50%). Visiting Amberites will suddenly turn into giants.

Amber itself, and those with Pattern, will be unaffected by the strange shrinking. The Lords of Chaos will see Amber's resistance to the affect as proof that the whole thing is an attack directed at them from Amber.

The only other place unaffected by the shrinking is Shadow Earth. Obviously, from the point of view of both the Lords of Chaos and the elder Amberites, that must mean that Shadow Earth is somehow the source of the problem. Chaos will move to attack, focusing in anywhere they can locate characters with Pattern. Amberites will counter, moving their own forces to Shadow Earth to take up the battle.

What Happens Next.

Open war breaks out on Shadow Earth, with the forces of Chaos breaking in all over the place, and with various elder Amberites leading armies in from Shadow.

If the player characters fail to resolve things in an hour or so of role-playing, then assume that the effect ripples out again, making everything even worse. Shadow reduces by another quarter, and the Courts of Chaos by half again. Then, if things continue to go unresolved, the effect can start accelerating, to the point where Shadow and Chaos just start visibly shrinking.

Elder Amberites & Lords of Chaos.

Any version of the elder Amberites will do for this scenario, but it's a good idea to put some time into designing a couple of Lords of Chaos.

The elder Amberites will be puzzled, but all too ready to respond to aggression from their old enemies in the Courts of Chaos. They may even be a little bit gleeful about the whole thing, since the battleground is to be on Shadow Earth, where those with Pattern have far greater control over the stuff of Shadow. It's also likely that characters like Bleys will find miniature Lords of Chaos to be pretty funny.

Out in the Courts of Chaos the Lords are furious. They can see that they are somehow being attacked. It seems that the shrinking effect may be just the first step in an assault, and they don't intend to just sit around. Since Shadow Earth is the only place out in Shadow unaffected, they'll deduce it is somehow the source of their problem.

Each Lord of Chaos should be roughly a 100 to 200 point character. They will each have collected up a force of Chaos Creatures. Named & Numbered Chaos Creatures should be around thirty points, while Hordes should be based on five to fifteen point Chaos Creatures.

Possible Resolutions.

The root cause of all the problems is something that each Game Master should customize for each group of player characters. In each case the resolution should involve something happening on Shadow Earth, where many Amberites have visited at one time or another. Here are two of the many possibilities:

Corwin's Doodle.

In the early 1920s, obsessed with his lost memory, Corwin tried several different desperate measures. One of his experiments verged on the occult, when he travelled to a castle ruin in Scotland (it looks something like a small Castle Amber), and attempted to "draw his fear." The result was a strangely warped Pattern drawn on the floor of a hidden dungeon. Originally, Corwin's construct had no particular power, but now Corwin is off in his new Pattern, and his attempts at seizing control there are causing the doodle in Scotland to act as a conduit, draining away the substance of both Pattern and Chaos.

Special Conditions: Corwin must be off doing something in his new Pattern, and unreachable by normal means.

Sockets: Each player characters who has Corwin for a father will likely serve a special role, as the key to finding Corwin. However, to balance things out, Corwin's children might be identified and detected by the forces of Chaos as somehow being connected to their problem.

Solutions: There is nothing that can be done on Shadow

Earth. The solution is to track down Corwin wherever he is in his own Pattern.

Brand's Artifact.

Brand, after learning in Tir-na Nog'th that Corwin might be the agent of his destruction, had gone off to observe the still-amnesiac Corwin.

While on Shadow Earth, Brand also constructed an artifact based on the vision of the Pattern he had planned to create. Brand basically wanted to examine the Pattern he was building in his own mind. Having looked through it, he then left it in his room.

Although it could be called a gateway to another version of Pattern, the other version doesn't exist. What has happened is that Brand's "simulation" is good enough to generate an effect that will gobble up much of the real universe. There's really no limit to how long this will last or how small things can get.

Somewhere on Shadow Earth, wherever the Game Master would like, there will be the room that Brand used for his experiment. It can be a hotel room, an apartment, or the bedroom of a house. Here's what the player characters might see when they track it down:

"Inside you see nothing more remarkable than a rather messy bedroom. A four-cornered bed, covered with rumpled covers, open books, papers, and a tray with a half-empty bottle of wine and two upset glasses. Aside from the bed there are two small side tables, a free-standing closet, and a large sea chest."

The player characters will be able to enter the room, walk around, and even touch things, but they'll be unable to move or affect any of the objects in the room. They could even blow up the entire building, leaving the contents of Brand's room suspended, unaffected, in mid-air.

Inside the chest Brand has placed his artifact, the item he created to examine his hypothetical Pattern. Should a player character look into the chest, by means of Magic, Logrus sight, or Advanced Pattern, here is what they might see:

"You see a shimmering fragment of Pattern. It seems to be whirling around in a small red gem, something like a reduced version of the Jewel of Judgement, but with no fittings or chain. The Pattern fragment is moving, rotating inside the gem, so that only a small section, perhaps a twentieth of its form, is visible at any one time."

If anyone manages to look through the Pattern fragment (something that should be possible for those with Advanced Pattern), they'll be able to look into Brand's imaginary universe. What they see is not real, but only a fantasy based on Brand's vision of himself as absolute ruler. An interesting variation might be to have the "imaginary" Brand looking back out at the player characters.

Sockets: Someone with Brand's blood, one of his children, may be needed. Just as only the descendants of Oberon can do anything to Amber's Primal Pattern, so it might take a kid of Brand to stop or destroy Brand's artifact.

Solutions: One possibility is that a child of Brand could turn off the artifact. Another way to solve things might be to get one of the elder Amberites to study the problem. For example, the player characters could tract down Dworkin, haul him off to Shadow Earth, and get him to solve the problem.

OPENING THE ABYSS

The Threat Summary: Brand has seized the Power of the Abyss and kidnapped many of the elder Amberites. Caine, Bleys, Fiona, Julian and Dworkin are attempting to mobilize other forces against Brand and need the player characters to run interference.

Thanks to that first group, caught in the Tendrils of the Abyss back in 1986: Felicia L. Baker (KELSEY), Greg Bellinger (DAMIEN), Carol Dodd (BRONWYN), Rob Justice (ARGENTIS), Erik Kittlesen (DELIAN), Terry O'Brien (DAMARIAN), Steven Schafer (TEMAR), Kevin Shore (LANCE), Eric Snider (HARLAN), John Speck (GODFREY), Mike Sutton (KAYEN), and Peter Taylor (ALEXANDER).

Getting Started.

Gathering of player characters. Bleys, having

made himself regent in Amber, decides that it's time to collect up all those pesky younger Amberites from out in Shadow and put them to some kind of use. His plan is to stash them all in the dungeon, chained up in the same room, and see what happens.

Options. The idea is to get the player characters chained up in the dungeon. There are several ways to go about this. The fastest is to simply start the scenario with the characters already there. Other variations might have characters like Bleys, Fiona, and Julian actually going out and collecting (kidnapping) the player characters from wherever they'd like to start the campaign.

In Dungeons Amber. Once all the player

characters are together, here's how the situation might be described. Read each of the following in response to the player characters' questions:

"You are chained to a stone wall, with just enough slack so that your arms can rest in your lap when you sit down. The chains, and the manacles around your wrists, are of some strange black metal and seem quite sturdy. You are in a large area, occupied by other prisoners, and lit by a central fire. What are you doing?

"Looking around, you see you are in a square chamber that is perhaps ten paces across. Three of the walls are stone, and fitted with the same chains and manacles that you and your fellow prisoners are wearing. In the center are hanging chains, a fire pit, and some kind of metal chair. The other wall consists of both upright and horizontal metal rods, forming something like the bars of a cage. Beyond the bars you can see two guards idly chatting with each other."

"Chained to a metal chair in the center of the room you see a filthy madman. His clothing hang in tatters, he is smeared with grime and dirt, and he is violently rocking back and forth to the limit of his chains. Some low noises come from his frothing mouth, but they are obviously no more than moans."

"Yes, you realize that the man in the chair is probably Caine, but unshaven and bent-over and terribly gaunt.

What are you doing?"

The player characters will now have the opportunity to get to know each other. It's a good idea to have each player speak up and describe the appearance of their own character. Conversations between characters can occur freely from this point on. Remember that Bleys and/or Fiona are keeping a watchful eye on the proceedings.

Caine's "Demonstration." After a short time, long enough for the player characters to get a feel for their situation, and to take in an impression of their fellow prisoners, but before things get comfortable, read the following:

"You see the guards look up, startled. They brandish their spears and run off to your right. What are you doing?"

From this point on the action should be fairly fast. Players will have time to do relatively little. For example, they won't have the time to either bring the Logrus or the Pattern to mind, or, for Advanced Trump Artists, to make contact with a memorized image. Certainly things happen far too quickly for the casting of any magical spells. Each of the following descriptions should be alternated with an opportunity for the player characters to respond:

"Something comes flying back from off to the right of the bars. You see that it is one of the guards, all bloodied and mangled. Strange lights seem to flash here and there, and there are inhuman roars that seem to be getting closer."

"A monstrous creature, humanoid, but nine feet tall and covered with bulky muscles, strides into view. It stops just outside the bars and looks inward toward you. What are you doing?"

"A total of three of the creatures have appeared, and even more of the strange lights. Two of them grab onto a section of the bars and start to heave. The other is staring in toward you."

"With a shriek of protest, the metal bars are bent and pulled from their stone fittings. Immediately the third creature advances, reaching out toward the metal chair and the chained madman."

"A gruesome scene assails your eyes! The monster rips the madman apart. As it does so its voice changes, rising from the guttural to something of a higher pitch. You realize that the creature is laughing. What are you doing?"

"You see that the ogre seems to be shrinking as it laughs, changing its shape. As you watch it slips down to something more human in size and form. The other creatures, as well as the strange lights are leaving, heading back the way they came."

"His shape shifting completed, you see before you the exact duplicate of the man who was recently chained to the metal chair. It is Caine, dressed in a shiny black outfit and laughing like crazy. He looks around and seems to study each of you prisoners in turn."

At this point Caine will take a few minutes to chat with the player characters. If asked, he will say nothing about Brand, and will imply that Bleys is the real villain responsible for their being chained up. He'll point out that it would be in their own best interest to describe the "monster" and the killing of Caine, but to keep their mouths shut about the shape shifting and their conversation.

"Caine says, `I think it would be best if you were all to stay chained up. After all, it provides a perfect alibi. And I doubt that Bleys will keep you imprisoned after you have experienced such a terrible thing as the murder of dear sweet Caine. Besides, you can claim that he has proven himself unfit by so grossly failing to protect you!'"

If any player characters have managed to free themselves, Caine will chain them back up, and then:

"Caine walks out, following the path of the creatures, and disappears from view. What are you doing?"

A bit later help will appear, in the form of several more guards. Eventually Julian will appear on the scene, and, if urged on by the player characters, he'll release everyone, on the condition that they stay in the castle "until Bleys and Fiona have a chance to speak with you."

Background.

Here are the events leading up the player characters' involvement in the adventure. Brand and Dierdre experience the power of the Abyss, and are extended into a threat upon Amber. Caine, discovering in the Unicorn's Grove an omen of Brand's return, makes his own preparations. Meanwhile, the rest of the elder Amberites react to the threat in a variety of ways.

Brand & Dierdre Exploring the Potential of the Abyss. Brand should have died quickly from his wounds. Theoretically, a fall in the Abyss should be fatal to both Brand and Dierdre. This has turned out not to be the case.

Brand's Fall. Immediately after falling into the Abyss, Brand was able to make use of one of the special properties of the Jewel of Judgement, that of slowing time. He used the Jewel to slow the fabric of his body, an also that of Dierdre, so that they were effectively frozen in time. The arrows in his neck and chest, should have killed Brand, but he was preserved by the slowing of his body. Physically, he could not move, but mentally, using his powerful Psyche, he had full access to his formidable Powers. Had it not been for the slowing, the entropic forces of the Abyss would have killed both Brand and Dierdre.

Brand's eye-opener came just a few moments later. Somehow the Jewel of Judgement was taken from him by the Unicorn. Loss of that bauble made his work much more difficult, and gave him no way of reversing the "slowing" that had been done on to body.

With his contact to the Pattern lost, Brand's Psyche could no longer range out of the Abyss. Having no other choices, he was forced to explore the composition of the Abyss itself.

This turned out to be a major new source of power for Brand. After years of experimenting, he eventually achieved a type of attunement with the Abyss itself.

While his mind could not reach, unaided, beyond the Abyss, his new power allowed him to push holes outward, creating openings to the Abyss anywhere in Shadow.

Dierdre, Behind Mental Walls. Meanwhile, Dierdre, Brand's captive, remained a puppet to Brand's ambitions. Being inferior in the powers of the mind, she could not battle his mind directly, but she was strong enough to resist his attempts to control her. Gradually, as Brand's control over the Abyss expanded, his attention was occasionally diverted from battling Dierdre. She used these lulls to overhear his thoughts, and to learn something of his new Power.

Brand's Discovery of Chaos Creatures. When Brand happened across the crude lifeforms of the Abyss he was at first disappointed. Until, of course, he realized that he could shape the creatures into any form. After a bit of experimentation he settled on a form where the creatures would have a hard crusty shell, containing a superheated Abyss "plasma." Hundreds of the creatures are now under Brand's control, throughout the Abyss, and ready to be dispatched anywhere he extends an Abyss opening.

Dierdre Altered & Imprisoned. Shortly after Brand gained the Advanced Abyss Power, he realized that in altering a body with Abyss he could also alter minds. His first victim was Dierdre. She was powerless to resist, though her own understanding of the Abyss allowed her to keep a small portion of her brain.

Brand's New Plan. Brand has decided that the Unicorn is really the secret to the destruction of Amber and the Primal Pattern. He will use the power of the Abyss to blot out the Patterns, and to capture his enemies, but his main goal, and that which the player characters must prevent, is the destruction of the Unicorn of Amber.

Brand's Captives. Brand first captured Corwin using an image of Dierdre as "bait." When Random, Gerard and Benedict attempted to rescue Corwin they were likewise captured. Martin followed shortly after, looking for his father. Brand is looking to expand the collection...

Caine Exploring the Meaning of the Unicorn's Grove. Caine, a bundle of complexities, had many reasons for checking out the Unicorn's Grove. Should he be questioned about this, and should he have the inclination to give a civil answer, he might say:

"In all honesty, I thought if I figured out the Unicorn, it's tricks, or maybe just how it could be bought, I could make my own arrangements for the next time the Horn of the Unicorn picked a King of Amber."

Caine's "Madness." To cover his movements, Caine had already arranged to have a "mad" Shadow of himself wandering around in Castle Amber. This is the creature that Caine slays at the start of things.

Caine's Grove Knowledge. Having gained some understanding of the Unicorn, and the Unicorn's Grove, Caine has figured out that the Unicorn is not of the same sort of "reality" as the rest of the Amber universe. Fortunately Caine was present in the Unicorn's Grove when Brand first attempted to investigate there.

Caine now has control of various fairy forces, including fairies, Trolls, and others. He is suspicious of anyone, having become aware that Brand has altered Dierdre's mind. He'll not allow anyone to enter the Grove, nor will he leave. Still, Caine could be a source of important information.

What Happens Next.

Free Time. Once released by Julian, the player characters will be free to do whatever they like. Although they have been asked to remain at Castle Amber, there is nothing to stop them from leaving. Eventually they will each, one by one, be invited to attend a formal dinner to be held that night in Amber.

Brand's First Appearance. Should any characters attempt to contact any of the captured Amberites, they will find that the Trump will grow hot. If they choose to push the contact, they are likely to experience the following:

> "Suddenly a great black spot appears on the ground all around you. The floor seems to be losing its substance. What are you doing?"

Player characters should be given the opportunity to escape, or could be pursued by Brand's creatures. Effectively this encounter should serve as something of a foreshadowing of things to come.

Bleys at Dinner. Probably most, if not all, of the player characters will attend. It's up to the Game Master to set the scene. In addition to the player characters, this is also an opportunity to introduce any other non-player Amberites. Eventually, when things are settled down, read the following:

> "You hear a blare of trumpets, a tune usually reserved for the entrance of the king. Bleys enters the room, with Fiona on his arm, and with Julian serving as guard of honor. This should be the signal for you to take your place at the dinner table. What are you doing?"

Bleys will seat himself in the king's traditional position, with Fiona in the queen's seat. Julian will take the chair of honor to the right of Bleys. After everyone has been seated, and all servants and guards have been cleared from the room, Bleys will make the following announcement:

> "Unfortunately some dire fate seems to have befallen our beloved King Random. We are also mourning the loss of Corwin, Benedict and Gerard, who, as far as we know, were Random's companions in his ill-met attempt at rescuing Dierdre from the Abyss. I, Bleys, have decided to take responsibility for Amber in the King's absence. I hereby declare myself regent, and name Princes Fiona as my Minister of State, and Julian as Lord Protector.
>
> "Do I have your consent?"

The reactions are, of course, up to the player characters from this point on. If asked, inform the players that a pledge of loyalty is traditional at this time. Conversation and dinner will follow. Neither Bleys, Fiona or Julian will say a word about Brand, except in an effort to get the player characters talking. They will admit that Random and his companions were lost "by the Abyss," but will deny that they know anything more.

After Dinner. Once the dinner is over, the player characters are free to do whatever they like. This is when the campaign begins in earnest, and all the options are open. Contact with other Amberites and encounters with Brand or his creatures, are all up to the actions of the players and the responses of the Game Master.

Disappearance of Bleys and Fiona. Shortly after dinner, Bleys and Fiona will make their excuses and leave. That will likely be the last that anyone sees of them for some time. They will take off to join Dworkin at the Primal Pattern. From that vantage point they will keep a eye on events, but will mainly take a defensive stance, stopping anyone who attempts to travel to Primal Pattern or Dworkin.

Caught by Brand. Player characters who are caught by Brand may undergo a number of changes. In order to keep their players active there are a number of possible "outs" to permanent capture.

Escape of the Spirit. While their body may be trapped, characters might be able to escape with their Psyche. In this disembodied form they might take possession of another body, or share a body with a player character.

Trumping Out. Another possibility is that the character might slip out of Brand's clutches and into a Trump. Literally. This means that Brand will take control of their body, changing it into the fire form, and "programming" it to attack the other player characters, while the Psyche of the characters get trapped in their own Trump. Effectively this means the player will be nothing more than a spirit that can move from one Trump to another, that can talk and listen, but has no physical body.

Summary of the elder Amberites.

G/B	Character	PSY	STR	END	WAR
0/0	BLEYS	[35]	[35]	[35]	[95]
0/5	BRAND	[135]	A	A	A
0/0	CAINE	[43]	[10]	[25]	[40]
0/1	DIERDRE	[25]	[35]	[25]	[90]
0/5	DWORKIN	[95]	[10]	C	A
0/0	FIONA	[122]	A	A	A
0/0	JULIAN	[40]	[25]	[25]	[87]
0/1	MERLIN	[4]	C	C	C

BRAND (Mad Visionary of Amber Version)

Controlling his Abyss creatures, and using the Abyss to capture anyone who falls into his hands, Brand may soon come into possession of Castle Amber itself.

Trump. Brand will instantly be aware of any attempt to Trump him, or even to spy upon him using Trump.

Brand's Abyss Creatures [54 Points] - Psychically, the Abyss Creatures are very weak. If left away from the Abyss for more than a few hours, Abyss Creatures will start to erode away. First their skill will thin and peel, then they will collapse into Abyss plasma, a substance that will

burn itself out in an hour or so.
- Immense Vitality [4 Points]
- Tireless Stamina [4 Points]
- Combat Training [1 Point]
- Destructive Force Damage [8 Points] - This comes into play when the creatures strike with the Abyss plasma contained in their bodies. With enough of a punch, a creature's plasma can burst through the skin of the knuckles and rupture into a target.
- Able to Speak [1 Point]
- Horde [*3 Points]

DIERDRE (Corwin's Paragon Version)

In her current form Dierdre no longer has the Power of Pattern, and has lost her Power Words. Instead she has a limited form of Abyss Power, and is trapped in a fire form. She will obey Brand in a rather flat, automatic fashion.

Trump. Brand will keep Dierdre's Trump hot.

CAINE (Protector of Amber Version)

Caine walked in on the prison of the player characters to spread panic and confusion, and to arrange for their release. He figured that alarming the young Amberites might actually get them to do something to counter Brand. Now he intends to remain in the Unicorn's Grove, to stand between Brand and the Unicorn. Should Caine be removed, there will be nothing to prevent Brand from entering.

Trump. His Trump is "hot" by his own design.

BLEYS (Dworkin's Apprentice Version) & FIONA (Sorceress of Amber Version)

Operating as a team, and a deadly team at that. They, along with Dworkin, stand guard over the Primal Pattern itself. In that place they are out of reach of the Trumps and they will prevent anyone from coming to them.

Trump. After the dinner, both Bleys and Fiona will arrange for their Trump to be hot in order to avoid detection.

JULIAN (Arden's Champion Version)

Julian knows that Brand is doing something, but he is uncertain exactly what. Julian remains in contact with both Caine and Bleys, and will likely mobilize some of his forces to protect every possible Shadow entry to the Grove of the Unicorn. His relations with the player characters will be open, but cold. He simply doesn't trust anyone right now, but he'll listen to any new information.

Trump. Unless he is taken by Brand, Julian will remain a free agent and will take all Trump calls that come his way.

DWORKIN (Dealer of Trump Version)

Dworkin will be primarily concerned about protecting the Primal Pattern, Brand's earlier target. He'll treat the younger, player character, Amberites as well-intended children. Basically harmless, and with nothing much to add to the conversation. Dworkin will work with Bleys and Fiona, and will only leave the Primal Pattern if they can convince him that the Unicorn is truly in danger.

Trump. Fiona will have arranged for Dworkin's Trump to be hot even before the campaign begins.

RANDOM (Friendly Younger Brother Version),

BENEDICT (Ideal Warrior Version), CORWIN (Champion of Amber Version), MARTIN (Heir to Amber Version) & GERARD (Strongman of Amber Version)

All are caught by Brand. They are each in separate flaming pentacles, protected only by focusing on their own Trump. If freed from their traps, and rescued from the Abyss, they will be grateful and will assist the players in whatever way they can.

Trump. Will be hot unless Brand chooses to intercede.

FLORA (Faithful Servant of the Crown Version), LLEWELLA (Reluctant Princess Version) & MERLIN (The Ultimate Neutral Version)

Llewella will remain in Rebma, and will be very reluctant to leave. If asked she will explain that she is needed to protect the Pattern in Rebma. An observer of the various phenomena, she is trying to analyze the hot Trump.

Florimel is on Shadow Earth, and she sees no good reason to leave. From her point of view things are a little too "hot" in Amber right now. Besides, she is scared to death of Brand. If pushed, she could tag along and be something of a help to the player characters.

Merlin starts the campaign involved in his own problems related to some conflicts with others in the Courts of Chaos. He'll be willing to help the player characters, but will also be subject to occasional Trump calls from elsewhere.

Trump. All three will be initially willing to take Trump calls. However, once any of them have personal contact with the Abyss or with Brand they'll shut themselves off.

Player Character "Sockets."

"Plugs" and "Sockets" are descriptions of how player characters can fit into a scenario. Characters are plugs, defined according to their parents, their Attributes, their Powers, their Good and Bad Stuff, or by any of the other things that go into describing a character. Sockets are the player character's roles in the scenario. For example, in "Abyss" any player character who is an offspring of Dierdre is considered a plug that connects to the socket of the dreams that Dierdre has been influencing.

Offspring of Bleys or Fiona. Witnessing the blatant power grab of their parents, the children of Bleys or Fiona will likely think the worst. After all, being raised by one of these power-hungry characters you could hardly not suspect them of doing away with Random in an attempt to take the throne.

Offspring of Dierdre. Her servitude to Brand has made her subject to his will. However, there are times when she will be able to partially shake off Brand's control. In particular, should she be put in a situation where she is forced to attack her own child, she might be able to temporarily shake off Brand's control.

Offspring of Brand. Brand has big plans, and so he'd like to have admirers. Perfect in his ideal cast is a son or daughter for him to groom as his replacement. Brand may even capture one of his offspring, then treat them very well, and even take them into his confidence.

Possible Resolutions

So, where do we go from here? How can the player characters resolve things, foil Brand, free Dierdre, and generally put things right? It's up to you.

This setting can be the opening to an entire long-term campaign, as it was the first time it ran. Their solution involved two of the player characters learning various aspects of the Abyss Power (one with Advanced Pattern, the other with Shape Shifting), and using the knowledge to counter Brand, and to free the captive elder Amberites. At the last minute, when Brand was about to slay the Unicorn, another player (Nick Olah), playing Brand's son Moradin, stalled Brand long enough for the other characters to come to the rescue.

However, the aftermath of the defeat of Brand had one character infected with Primal Chaos and another trapped in a Trump. Another character ended up with a body that had been shape shifted into a more muscular, "warlike" form, purged of Pattern and with only the Power of the Abyss available.

There is no one "solution" to the problem of Brand and the Abyss. Instead, each group of player characters will find their own way, and each Game Master will offer a variety of alternatives. The story may end, or it may continue indefinitely. It's all up to you.

THE POWER OF THE ABYSS

There are three forms of Abyss Power. The basic form lets characters control the forces and creatures and is roughly equivalent to a fifty point Power. Combining Abyss with Shape Shifting is nothing more than expanding the Shape Shifting to the tune of a fifteen point advancement. The advanced version of Abyss Power is worth something like seventy or eighty points.

Basic Abyss Power. Once an initiate of the Abyss, a character gains a new power source. Abyss Power is in conflict with the powers of Pattern, Logrus, or Trump. It is, however, somehow related to Shape Shifting. Here are some of the powers of the Abyss.

Create Abyss Opening. A hole is punched through Shadow, across the boundaries of the Abyss, and opens up into the void. Whatever material falls into the hole is shunted to the Abyss, and can then be controlled by one wielding Abyss power. An initiate of the Abyss can also use Abyss Holes as a conduit, opening one up from outside to tap the Abyss power.

Protect Contents from the Abyss. Once things, or creatures, or characters, fall into the realm of the Abyss, they are gradually destroyed by the entropic forces that dwell there. However, it is possible for an Initiate of the Abyss to stave off these effects.

The first method, using a brute application of Abyss Power, is where the initiate simply prevents the Abyss from contacting the foreign matter. This can last for as long as the Initiate concentrates on the subject.

The second method of protecting things from the Abyss is to attune to them, changing them slightly so they resonate with the Abyss, and are therefore unaffected by it.

Summon and Control Abyss Creatures. In their natural state, the creatures of the Abyss are unintelligent, sessile, and inoffensive creatures, whose main purpose in life was to survive and reproduce in the Abyss environment. Under the influence of an Abyss master they can be molded into Artifacts and Creatures, and given the powers of the Abyss.

Combining Abyss Power with Shape Shifting.

Initiates of the Abyss, if they also have Shape Shifting, are able to adjust their bodies to heightened states, where they can some of the powers of the Abyss.

The Fire Form (Fast Time). Using the Abyss, one changes temporal form, accelerating the molecules of their body, turning into a super-heated plasma. To an observer it seems that the character has burst into flame. In this form the character is not truly solid. It becomes possible to punch through physical objects, inflicting great damage, or to move directly through walls and doors, superheating them, but without leaving a gap. Physical attacks directed against someone in this form often pass through without doing damage to the plasma character, but hurting the attacker.

With a touch, the character can heat objects, to just about any temperature. With a bit of practice, the character can learn to touch or grab objects without heating them. With more experience in Fire Form, a character can change other things into Fire Form, allowing weapons, clothing, and other possessions to be used.

Concentrating on the energy, the character can radiate heat all around, though outside of the Abyss this is very draining. This concentration can also be used, in the Abyss, as a means of flying. The character generates a "bubble" of air, where each molecule is vibrating at hyper-speed. Using this technique, a character in Fire Form can fly, zooming around at great speed, hover, or float.

The Ice Form (Slow Time). A reversal of the fast time effect, this effect slows down the molecules of the body, making it solid and exceptionally resistant to physical or energy attacks. It's also possible to convey extreme cold to other objects by touch.

Advanced Abyss Power. This represents Brand's degree of control over the Abyss.

Corrupt/Overwhelm Trump. The subject of a Trump, once in the grip of the Abyss, still has a connection with their card. Brand is able to alter things so that the Trump will react in bizarre new ways.

Brand has set things up so that, automatically, any attempt at contacting one of his captives will cause the Trump to become hot. Pushing the contact merely increases the Trump energy, which generates even more heat. Eventually, if pushed, or if the character doing the contacting is powerful Psychically, then the card will be hot enough to burn flesh and ignite fabric. Pushed even farther, the Trump will eventually consume itself altogether, in a small blast of combustion.

If Brand has the time, he can focus his will on one of these Trump contacts. Then, instead of the heat effect, he can manipulate the contact in various ways.

Brand can, for example, make it seem as if a Trump is wroking okay, but that the subject of the Trump is somehow distracted or not paying attention to the caller. If the character making the call yells, or uses Psychic force, there will be no reaction from the subject, except the sense of the contact will become clearer. Then, if the character using the Trump attempts to pull out the card's subject, Brand can either come himself, send out one of his servants, or send out something else. If a character attempting to step through the Trump, then Brand has sufficient control to either block them, bring them to his own presence, or shunt them someplace else in the Abyss.

Time Manipulation. So long as the Abyss Initiate is in contact with the Abyss, it's possible to slow down, or speed up, the temporal flow of the molecules of their body. The effect can be centered on the initiate, covering any size area, or can be directed at a particular object or creature.

Inflict Abyss Powers. Taking possession of a character, within the realm of the Abyss, it's possible to alter them, switching their normal Attributes and Powers around, and giving them forms more appropriate to the Abyss. This also can be used to change their personalities.

Countering the Power of the Abyss. Here is how other Powers work when characters try using them against the Brand's Abyss Power.

Pattern. Within the Abyss, or at the edge of one of the Abyss Openings, Pattern is relatively weak. However, any Abyss entity outside of the influence of the Abyss can be manipulated by Pattern. Characters with Advanced Pattern Imprint will be able to force back the "stain" of the Abyss on the Pattern, or shove away Abyss openings, on the basis of Psyche battles. In other words, the characters driving the force of the Pattern will have to match Psyche with whoever is driving the Abyss.

Logrus. Masters of Logrus will be powerless against the Power of the Abyss and can do nothing to dispel Abyss Openings or Abyss Creatures. However, while in the Abyss, Logrus Masters can continue to use their power, reaching out with Logrus tendrils and using the ability of the Logrus lens.

Trump. While Trump is an absolute defense against the power of the Abyss, Brand's great power over Trump tends to even things out.

Shape Shifting. Shape Shifters have the potential to imitate the Abyss forms. However, when in Abyss form they become vulnerable to manipulation by Brand or anyone with Advanced Abyss Power.

Magic is generally powerless against the Abyss itself, and raw Abyss power. Spells cast in the region of the Abyss tend to fizzle out unless the caster has some form of Abyss power. Spells can be used normally against Abyss creatures or items.

Hungry for more?

Each issue of *Amberzine* is packed with at least 160 pages of great *Amber* character diaries, campaign logs, Trump art, and helpful write-ups on diceless role-playing! *Amberzine* is available only by mail, and is limited to 1,000 copies per issue. Send $10 (U.S. funds) for one, or $40 for a five-issue subscription, to:

Phage Press
P.O. Box 519
Detroit MI 48231-0519

INDEX

Character Name

Player Name: Phone:

PHYSICAL CHARACTERISTICS

HEIGHT: WEIGHT: BUILD:

HAIR: EYES:

DESCRIPTION:

CLOTHING COLORS/STYLE:

PERSONAL SYMBOL (OPTIONAL):

ATTRIBUTES

PSYCHE: ❏ Ranked Rank:___ [___ Points] | ❏ Amber | ❏ Chaos | ❏ Human
STRENGTH: ❏ Ranked Rank:___ [___ Points] | ❏ Amber | ❏ Chaos | ❏ Human
ENDURANCE:❏ Ranked Rank:___ [___ Points] | ❏ Amber | ❏ Chaos | ❏ Human
WARFARE: ❏ Ranked Rank:___ [___ Points] | ❏ Amber | ❏ Chaos | ❏ Human

POWERS

PATTERN IMPRINT ❏ 50 Points   ADVANCED PATTERN IMPRINT ❏ 75 Points
LOGRUS MASTERY ❏ 45 Points   ADVANCED LOGRUS MASTERY ❏ 70 Points
TRUMP ARTISTRY ❏ 40 Points   ADVANCED TRUMP ARTISTRY ❏ 60 Points
SHAPE SHIFTING ❏ 35 Points   ADVANCED SHAPE SHIFTING ❏ 65 Points
POWER WORDS ❏ 10 Points   SORCERY ❏ 15 Points   CONJURATION ❏ 20 Points

ARTIFACTS & CREATURES

Total Points, from Artifact & Creature Worksheet: ____ Points

PERSONAL SHADOWS

Total Points, based on Shadow Construction Guidelines (page 73-74): ____ Points

ALLIES

❏ Ally in Amber. 1 point each.   ❏ Court Friend. 2 points each.
❏ Chaos Devotee. 4 points each.   ❏ Amber Devotee. 6 points each.

PLAYER CONTRIBUTION

❏ 10 Point Diary.   ❏ 10 Point Trump.   ❏ 10 Point Game Log.   ❏ Other [___ Points]

❏ **GOOD STUFF:** [___ Points]   ❏ **ZERO STUFF**   ❏ **BAD STUFF:** [+___ Points]

AMBER
Artifact & Creature Worksheet

Character Name

Item Name & Description

QUALITIES

Vitality
- ❏ 1 Point - Animal Vitality
- ❏ 2 Points - Chaos Vitality
- ❏ 4 Points - Amber Vitality

Movement
- ❏ 1 Point - Confers Mobility
- ❏ 2 Points - Double Speed
- ❏ 4 Points - Engine Speed

Stamina
- ❏ 1 Point - Double Stamina
- ❏ 2 Points - Amber Stamina
- ❏ 4 Points - Endless Stamina

Aggression
- ❏ 1 Point - Combat Training
- ❏ 2 Points - Combat Reflexes
- ❏ 4 Points - Combat Mastery

Armor
- ❏ 1 Point - Versus Weapons
- ❏ 2 Points - Versus Guns
- ❏ 4 Points - Invulnerable

Damage
- ❏ 1 Point - Extra Hard
- ❏ 2 Points - Doubling Damage
- ❏ 4 Points - Deadly Damage

Intelligence/Communication
- ❏ 1 Point - Able to Speak
- ❏ 2 Points - Speak & Sing
- ❏ 4 Points - Tongues

Psychic Sensitivity
- ❏ 1 Point - Sensitivity
- ❏ 2 Points - Danger Sense
- ❏ 4 Points - Extraordinary

Psychic Resistance
- ❏ 1 Point - Chaos Resistance
- ❏ 2 Points - Psychic Neutral
- ❏ 4 Points - Psychic Barrier

Total Point Cost

POWERS

Shadow Movement
- ❏ 1 Point - Follow Shadow Trail
- ❏ 2 Points - Follow Shadow Path
- ❏ 4 Points - Seek in Shadow

Shadow Manipulation
- ❏ 1 Point - "Mold" Shadow Stuff
- ❏ 2 Points - "Mold" Shadow Folk
- ❏ 4 Points - "Mold" Reality

Item Healing
- ❏ 1 Point - Self-Healing
- ❏ 2 Points - Rapid Healing
- ❏ 4 Points - Regeneration

Item Shape Shifting
- ❏ 1 Point - Alternate Form
- ❏ 2 Points - Named & Numbered
- ❏ 4 Points - Limited Shape Shift

TRANSFERAL
- ❏ 5 Points per Quality
- ❏ 10 Points per Power

QUANTITY MULTIPLIERS
- ❏ Unique (*1)
- ❏ Named & Numbered (*2)
- ❏ Horde (*3)
- ❏ Shadow Wide (*4)
- ❏ Environmental (*5)
- ❏ Ubiquitous (*6)

TRUMP POWERS
(Requires Trump Artistry)
- ❏ 1 Point - Contains Trump Image
- ❏ 2 Points - Trump Deck
- ❏ 4 Points - Trump Powered

MAGIC POWER
(Requires both Sorcery & Conjuration)
- ❏ 1 Point - Rack a Spell
- ❏ 2 Points - Named & Numbered
- ❏ 4 Points - Rack & Use Spells